Birkhäuser

Trends in Mathematics

Trends in Mathematics is a series devoted to the publication of volumes arising from conferences and lecture series focusing on a particular topic from any area of mathematics. Its aim is to make current developments available to the community as rapidly as possible without compromise to quality and to archive these for reference.

Proposals for volumes can be submitted using the Online Book Project Submission Form at our website www.birkhauser-science.com.

Material submitted for publication must be screened and prepared as follows:

All contributions should undergo a reviewing process similar to that carried out by journals and be checked for correct use of language which, as a rule, is English. Articles without proofs, or which do not contain any significantly new results, should be rejected. High quality survey papers, however, are welcome.

We expect the organizers to deliver manuscripts in a form that is essentially ready for direct reproduction. Any version of $\TeX$ is acceptable, but the entire collection of files must be in one particular dialect of $\TeX$ and unified according to simple instructions available from Birkhäuser.

Furthermore, in order to guarantee the timely appearance of the proceedings it is essential that the final version of the entire material be submitted no later than one year after the conference.

More information about this series at http://www.springer.com/series/4961

Eder Kikianty • Mokhwetha Mabula •
Miek Messerschmidt • Jan Harm van der Walt •
Marten Wortel
Editors

Positivity and its Applications

Positivity X, 8-12 July 2019, Pretoria, South Africa

Birkhäuser

Editors
Eder Kikianty
Department of Mathematics
University of Pretoria
Hatfield, South Africa

Mokhwetha Mabula
Department of Mathematics
University of Pretoria
Hatfield, South Africa

Miek Messerschmidt
Department of Mathematics
University of Pretoria
Hatfield, South Africa

Jan Harm van der Walt
Department of Mathematics
University of Pretoria
Hatfield, South Africa

Marten Wortel
Department of Mathematics
University of Pretoria
Hatfield, South Africa

ISSN 2297-0215 ISSN 2297-024X (electronic)
Trends in Mathematics
ISBN 978-3-030-70973-0 ISBN 978-3-030-70974-7 (eBook)
https://doi.org/10.1007/978-3-030-70974-7

This book is published under the imprint Birkhäuser, www.birkhauser-science.com, by the registered company Springer Nature Switzerland AG.
The registered company address is: Gewerbestrasse 11, 6330 Cham, Switzerland

Dedicated to the memory of
Coenraad Christoffel Andries (Coen)
Labuschagne (1958 to 2018).

Preface

The tenth iteration of the Positivity series of conferences (Positivity X) was held from 8 to 12 July 2019 at the Haftield campus of the University of Pretoria in Pretoria, South Africa. The conference was co-organised by the University of Pretoria and the University of Johannesburg. The organizing committee consisted of Ronalda Benjamin (University of Stellenbosch), Retha Heymann (University of Stellenbosch), Eder Kikianty (University of Pretoria), Coenraad Labuschagne (University of Johannesburg), Miek Messerschmidt (University of Pretoria), Mokhwetha Mabula (University of Pretoria), Andrew Pinchuck (Rhodes University), Jan Harm van der Walt (University of Pretoria) and Marten Wortel (University of Pretoria).

At the conference, thirteen plenary talks and forty-five contributed talks were delivered, covering a range of topics related to positivity and its applications. The plenary speakers were Youssef Azouzi (Tunis, Tunisia), Jacek Banasiak (Pretoria, South Africa), David Blecher (Houston, USA), Marcel de Jeu (Leiden, the Netherlands), Jochen Glück (Passau, Germany), Koos Grobler (Potchefstroom, South Africa), Anke Kalauch (Dresden, Germany), Ali Khan (Baltimore, USA), Bas Lemmens (Canterbury, the UK), Sonja Mouton (Stellenbosch, South Africa), Vladimir Troitsky (Edmonton, Canada), Bruce Watson (Johannesburg, South Africa), and Foivos Xanthos (Toronto, Canada). The topics covered include harmonic analysis, operator algebras, semi-group theory, economics, stochastic processes, the geometry of cones, partial differential equations, and applications of positivity to mathematical modeling and the life sciences. PDF files of many of the talks, both plenary and contributed, can be found at the conference website, http://positivitymathematics.org/slides. The papers in this volume represent a sample of the topics discussed at the conference. All contributions have been peer reviewed for quality and correctness, according to the usual standards of the mathematical community.

Positivity X was supported financially by the DST-NRF Center of Excellence in Mathematical & Statistical Sciences (CoE-MaSS), South Africa, the Absa Chair in Actuarial Science at the University of Pretoria, DST-NRF SARChI Chair in Mathematical Models and Methods in Bioengineering and Biosciences, the Gottfried Wilhelm Leibniz Basic Research Institute, the University of Johannesburg and the University of Pretoria. We gratefully acknowledge the support of these institutions, without which the hosting of the conference would not have been possible.

The tenth Positivity conference is a cause for celebration. Also to be celebrated is the vitality of the field, which is demonstrated by the wide variety of topics covered at the conference, all with positivity as a central theme. The above notwithstanding, the conference was tinged with great sadness. On 10 July 2018, our friend and colleague Coenraad (Coen) Labuschagne passed away following a brief illness. Coen was not only a member of the organizing committee, but in fact the driving force behind the hosting of the conference in South Africa. As the senior member of the team it would have been natural for him to chair the organizing committee. But to Coen it was important to put the limelight on his younger colleagues, and so he declined the chairmanship. For his contributions to this conference, to the Positivity community at large, and for the impact he had on our lives, we dedicate this volume to his memory.

Hatfield, Pretoria, South Africa — Eder Kikianty
Hatfield, Pretoria, South Africa — Mokhwetha Mabula
Hatfield, Pretoria, South Africa — Miek Messerschmidt
Hatfield, Pretoria, South Africa — Jan Harm van der Walt
Hatfield, Pretoria, South Africa — Marten Wortel
December 2020

Contents

Coenraad Christoffel Andries Labuschagne

J. J. Grobler

Coenraad Christoffel Andries (Coen) Labuschagne was born on May 16, 1958. He was married to Laetitia Lourens, an anaesthesiologist. They were a high school couple who met at the Roodepoort High School. After their wedding they settled in Florida on the West Rand in South Africa. Coen and Laetitia had a daughter Minette, and they experienced the trauma of losing her unexpectedly while she was still a student. On July 10, 2018 Laetitia had to cope with a similar tragedy when Coen, after a short illness, passed away.

His undergraduate studies, up to the Honours level, were completed at the Rand Afrikaans University. He then moved to Potchefstroom where he completed the M.Sc. (Mathematics) in 1981. Not being able to postpone his compulsory military training any longer he was, however, fortunate to do it teaching Mathematics at the Saldanha Bay Military Academy in 1982–1983. During these 2 years it became clear to him that he wanted to pursue an academic career and he commenced his doctoral studies in 1984 at Potchefstroom. He was awarded in May 1986 the D.Sc. (Mathematics) degree at the Potchefstroom University for CHE with a thesis entitled: *The Riesz tensor product of Archimedean Riesz spaces* under the supervision of the author.

Coen started his academic career as a lecturer at the University of the Witwatersrand in January 1986. He was promoted to full professor in 2011 in the Department of Computational and Applied Mathematics. He had the unique distinction that he was promoted to Associated Professor twice: for the first time in 2006 in the Department of Mathematics and, after choosing in 2008 to be transferred to the Department of Computational and Applied Mathematics as a senior lecturer, again in 2009. The reason for changing departments was his interest in Mathematics of Finance that was taught in the latter department. There he was responsible for an Honours course in Mathematics of Finance. His alma mater, which changed its

J. J. Grobler (✉)
North-West University, Potchefstroom, South Africa

E. Kikianty et al. (eds.), *Positivity and its Applications*, Trends in Mathematics,
https://doi.org/10.1007/978-3-030-70974-7_1

name to the University of Johannesburg, offered him a position in their Department of Finance and Investment Management; a position he occupied until 2018.

Coenraad supervised at least 22 M.Sc. and 8 Ph.D. students. Two of these were registered at the El-Manar University at Tunis. He acted as examiner for Ph.D. theses at almost all South African universities and also for two Ph.D. theses at the Chiang Mai University in Thailand, a country he loved to visit. He was a research visitor at various locations: Australia, Palermo in Italy (5 visits), Thailand, the USA, China and Germany (2 visits). His international collaborators included Valeria Marraffa (Palermo, Italy), Hung Nguyen (New Mexico and Chiang Mai), Karim Boulabiar and Habib Ouerdiane (Tunisia), and Olivier Pamen (Liverpool). Coenraad was also invited as a key-note speaker at the International Conference on Intelligent Technologies, Sydney, 2007 to speak on his research on set-valued stochastic processes. The organizers of Positivity VII asked him to deliver a key-note address in Tunis, May 2011. The conference was canceled due to political unrest, but he did visit Karim Boulabiar where he and Bruce Watson gave a series of four talks each at a workshop in Tunis that replaced the conference. In July 2013 he delivered the address at the Positivity VII Conference (Zaanen Centennial Conference) in Leiden, the Netherlands. He was invited by Vladimir Troitsky to a workshop at Oaxaca, Mexico where he, his student Wen-Chi Kuo, and the author were asked to present their research on Stochastic Processes in Riesz spaces in a 1 day session. This workshop took place in April 2018. Coen had to withdraw at the last minute due to illness, an illness that proved to be fatal, for he passed away 3 months later.

Coen was interested in three main areas of research. The first was tensor products, an interest stemming from his doctoral thesis. One of his main feats here was his definition of Riesz reasonable cross norms in terms of reasonable cross norms and the Fremlin cone induced on the tensor product of two Banach lattices. In collaboration with Paul Allenby and Theresa Offwood, he applied his theory of tensor products to a number of interesting problems.

The second area of research in which Coen made a major contribution was in the area of generalized Stochastic Processes. In the Riesz space case he, his student Wen-Chi Kuo and her husband Bruce Watson proved many results for discrete processes and with the author he studied the case of continuous processes and Brownian motion. They studied the quadratic variation of these processes (with Valeria Maraffa), Itô's integral, Itô's rule and Lévy's theorem, and the Girsanov theorem. All these results were cast in the setting of Riesz spaces. In the setting of Banach spaces and lattices, he developed with Stuart Cullender a theory of discrete stochastic processes. Valeria Marraffa also collaborated in this area with him. With Andrew Pinchuck, Clint Van Alten and Valeria Marraffa, he studied set-valued stochastic processes. They derived a lattice version of Rädström's embedding theorem.

The third area of research to which Coen made a contribution was to the field of Mathematics of Finance. With Theresa Offwood he gave an elementary proof for the fact that there is a connection between the Escher-Girsanov transform and the Wang transform. Collaborators in this field included A. Kotzé, M.L. Nair and

N. Padahyachi. They also contributed to pricing of convertible bonds and of exotic options via the Wang transform.

Finally, besides these main areas, Coen was always open to collaborate on interesting problems. He wrote papers with colleagues Mareli Korostenski and Clint Van Alten on various subjects. Coen was also interested in Markov operators on Banach lattices, with special focus on the family of copulas. His student Peter Hawke wrote a M.Sc. dissertation on this topic, and with Eder Kikianty, Coen wrote a survey paper on the structure of copulas and quasi-copulas from the point of view of Banach spaces and Banach lattices. This paper was presented as a plenary lecture by Coen at the Symposium on Mathematics for Uncertainty and Fuzziness at the prestigious Research Institute for Mathematical Sciences in Kyoto, Japan, in 2014.

Coen was very enthusiastic about mathematical research. He was extremely organized, hardworking and known for his good lectures. He had a special way of inspiring students and colleagues to collaborate with him in research projects. He had a lasting impact on many people, in particular his students, for whom he cared greatly. A conversation with any of his former students reveals the impact he had on their lives, not only mathematically but also on a personal level. In this respect, I quote some of his former students below.

> Coenraad was a caring and supportive supervisor. Even after I left academia, whenever I needed his help, he was there. Our relationship as supervisor and postgraduate students didn't end when we graduated. He treated us like his extended family. – Wen-Chi Kuo
>
> Coenraad helped me and provided me with a lot of support, especially during my first year of postdoc in South Africa. Realising that I didn't have a family in the country, he decided to become my family. He was incredibly generous with his time, advice, and support. I always sought for his wisdom, and he was always there for me. – Eder Kikianty
>
> I have wonderful memories of my time as a Ph.D. student in Mathematics of Finance under Coenraad's guidance. Not only did I complete my Ph.D. in record time, Coenraad taught me all I know about the best wines in South Africa and took me along to many conferences all around the world from Canada to Australia. He is greatly missed! – Theresa Offwood
>
> I will always remember Coenraad as an excellent mathematician, teacher and much more. He was always supportive and encouraging over the time that I knew him and he seemed to have a positive influence on everybody he encountered. Coenraad's memory will live on through the many people whose lives were impacted by him. – Andrew Pinchuck
>
> Coenraad was not only my supervisor in academia, but also my mentor. He was one of the most genuine, kindest and most supportive people. He had a great sense of humor, and could be serious as well as light hearted. His brilliance and unrivaled intellectual capacity was evidenced on numerous occasions - he had a great talent of simplifying complex problems, which in itself is an art, and this made him a brilliant teacher. He will sorely be missed. There is so much more to Coenraad than what I can put in words. I feel like I'm not doing him justice with what I wrote. – Thorsten von Boetticher

Having been the driving force behind the hosting of the conference "Positivity X" in South Africa, it is befitting that this Proceedings is dedicated to him.

Inverse Monotonicity of Elliptic Operators in Variational Form

Roumen Anguelov and Nduduzo Majozi

Abstract The inverse monotonicity of elliptic operators in classical formulation is a well known consequence of the maximum principle. This result is often formulated as a comparison theorem for solutions of linear elliptic PDEs as it derives order in the solution space from the order in the space of data. Variational formulation and the associated concept of weak solution is widely used in the theory, applications and numerical analysis of elliptic PDEs.

The space of weak solutions as well as space of data are Sobolev spaces, which are wider than the respective spaces of solutions and data in the classical formulation. This paper proves inverse monotonicity, or equivalently comparison theorems, for this much more general formulation of the operators and respective equations. Since the maximum principle does not apply to weak solutions, the presented here theory is a useful set of tools that can be used in its place to derive order in the space of weak solutions from the order in the generalized data space. We specifically discuss the case of a single equation and the case of weakly coupled system of equations.

Keywords Elliptic PDEs · Comparison theorem · Variational formulation · Weak solutions · Inverse Monotone operators

Mathematics Subject Classification (2010) Primary 35J50; Secondary 47H07

The authors acknowledge the support of the DST-NRF SARChI Chair in Mathematical Models and Methods in Biosciences and Bioengineering at the University of Pretoria and the DSI-NRF Centre of Excellence in Mathematical and Statistical Sciences, South Africa. Opinions expressed and conclusions arrived at are those of the authors and are not necessarily to be attributed to the acknowledged here institutions.

R. Anguelov (✉) · N. Majozi
Department of Mathematics and Applied Mathematics, University of Pretoria, Hatfield, Pretoria, South Africa
e-mail: roumen.anguelov@up.ac.za

E. Kikianty et al. (eds.), *Positivity and its Applications*, Trends in Mathematics,
https://doi.org/10.1007/978-3-030-70974-7_2

1 Introduction

Let $P : Z \to Y$, where Z and Y are partially ordered spaces. The operator P is called *inverse monotone* if for any $u, v \in Z$

$$Pu \leq Pv \Longrightarrow u \leq v. \quad (1.1)$$

The name inverse monotone is derived from the fact that if the operator P is invertible, then (1.1) is equivalent to P^{-1} being monotone. Mathematical models of real life phenomena are commonly formulated as Partial Differential Equations (PDEs) or systems of PDEs. These types of models can be written in the form

$$Pu = g, \quad (1.2)$$

where the operator P involves both a differential operator on a function space and a boundary condition on u. In such setting, statements of the form (1.1) are referred to as comparison theorems. The name reflect the fact that (1.1) provides means of proving order relation between functions in the solution space of equations of the form (1.2). This type of theorems is useful from both theoretical, e.g. proving uniqueness, and practical, e.g. constructing lower/upper bounds, applications, points of view.

Comparison theorems are typically derived from maximum principles associated with the operator P in classical formulation, that is the domain of P comprises sufficiently smooth functions so that all derivatives involved in P exist in the classical sense, [7]. In practice, very often the operator P is extended to a larger domain in order to accommodate a wider class of physically meaningful solutions. One of the very popular approaches is to formulate the operator P and, respectively, Eq. (1.2), in variational form. The solutions of Eq. (1.2) in variational from are called weak solutions. Since the maximum principle does not apply to weak solutions, there is no theory providing comparison theorems in the more general setting of variational formulation. The results presented in the sequel fill to some extend this gap. More precisely, the paper provides comparison theorems for the operators of elliptic PDEs in variational form, thus providing order relations in the wider space of weak solutions.

In order to illustrate more clearly the ideas, in the next section we recall the comparison theorem derived from the maximum principle for elliptic operators. Sections 3 and 4 deal with one dimensional and multidimensional operators in variational form related to a single elliptic PDE or a weakly coupled system of elliptic PDEs, respectively. Some concluding remarks are given in the final section.

2 Operators Associated with Elliptic PDEs: The Classical Case

Let Ω be an open bounded subset of $\mathbb{R}^n$. We consider the following Dirichlet boundary value problem:

$$\begin{cases} L[u] = f & \text{in } \Omega, \\ u = g & \text{on } \partial\Omega, \end{cases} \tag{2.1}$$

where L is a second-order differential operator given by

$$L[u] \equiv -\sum_{i,j=1}^{n} \frac{\partial}{\partial x_j}\left(a_{ij}\frac{\partial u}{\partial x_i}\right) + \sum_{i=1}^{n} b_i \frac{\partial u}{\partial x_i} + cu, \tag{2.2}$$

where a_{ij}, b_i and c are functions of $x \in \Omega$. We assume the symmetry condition that $a_{ij} = a_{ji}$ for $i, j = 1, 2, \ldots, n$. Using matrix and vector notations

$$A(x) = \begin{pmatrix} a_{11}(x) & a_{12}(x) & \ldots & a_{1n}(x) \\ a_{21}(x) & a_{22}(x) & \ldots & a_{2n}(x) \\ \vdots & \vdots & \ddots & \vdots \\ a_{n1}(x) & a_{n2}(x) & \ldots & a_{nn}(x) \end{pmatrix}, \quad b(x) = \begin{pmatrix} b_1(x) \\ b_2(x) \\ \vdots \\ b_n(x) \end{pmatrix}, \tag{2.3}$$

the operator L can be written compactly as

$$L[u] = -\nabla \cdot (A\nabla u) + b \cdot \nabla u + cu. \tag{2.4}$$

A classical solution of (2.1) is a function $u \in C^2(\Omega) \bigcap C(\overline{\Omega})$. Hence, we associate with (2.1) the operator $P : C^2(\Omega) \cap C(\overline{\Omega}) \to C(\Omega) \times C(\partial\Omega)$ given by

$$P(u) = \begin{pmatrix} L[u] \\ u|_{\partial\Omega} \end{pmatrix}. \tag{2.5}$$

Then problem (2.1) can be written as the equation

$$Pu = \begin{pmatrix} f \\ g \end{pmatrix}. \tag{2.6}$$

The main property of the operator L of interest here, and which we assume in the sequel, is that it is uniformly elliptic. Let us recall

Definition 2.1 We say that L is *uniformly elliptic* if there exists a constant $\mu_0 > 0$ such that

$$\xi \cdot A(x)\xi \geq \mu_0|\xi|^2$$

for all $x \in \Omega$ and all $\xi \in \mathbb{R}^n$.

For a uniformly elliptic operator L and open and bounded Ω we have the maximum principle as follows.

Theorem 2.2 (Maximum Principle [5, Theorem 1, Chapter 6.4.1]) *Let $c \geq 0$ and $u \in C^2(\Omega) \cap C(\bar{\Omega})$ satisfy on Ω the differential inequality*

$$L[u] \equiv -\sum_{i,j=1}^{n} \frac{\partial}{\partial x_j}\left(a_{ij}\frac{\partial u}{\partial x_i}\right) + \sum_{i=1}^{n} b_i \frac{\partial u}{\partial x_i} + cu \leq 0.$$

Then

$$\max_{x \in \bar{\Omega}} u(x) = \max_{x \in \partial\Omega} u(x).$$

A direct consequence of the maximum principle is the Comparison Theorem stated here below:

Theorem 2.3 (Comparison Theorem) *If $c \geq 0$, for every $u, v \in C^2(\Omega) \cap C(\bar{\Omega})$ we have*

$$L[u] \leq L[v],\, u|_{\partial\Omega} \leq v|_{\partial\Omega} \implies u \leq v \text{ on } \Omega.$$

Using the notation (2.5), the comparison theorem can be restated as

$$P(u) \leq P(v) \implies u \leq v,$$

for every $u, v \in C^2(\Omega) \cap C(\bar{\Omega})$.

Theorem 2.4 *If $c \geq 0$, under the stated assumptions the operator P is inverse monotone on $C^2(\Omega) \cap C(\bar{\Omega})$.*

Let us note that Theorems 2.3 and 2.4 refer to functions in the space $C^2(\Omega) \cap C(\bar{\Omega})$, which contains the classical solution of (2.1). However, they do not deal with any specific solutions of (2.1). They apply to any two functions in this space. Hence, they have a possibly wider scope of application than the theory related to the PDE (2.1).

As it is well known, physically meaningful solutions of the problem (2.1) are often not classical. One way to assimilate nonclassical solutions is through a variational formulation of the problem. These are referred to as weak solutions. We note that Theorems 2.3 and 2.4 are based on the maximum principle and apply only to the space of classical solutions $C^2(\Omega) \cap C(\bar{\Omega})$. In the next section we show that these important results can be extended to the function space of weak solutions of (2.1) using a different method of proof.

3 Operators Associated with Elliptic PDEs in Variational Formulation

The weak solutions are defined in Sobolev spaces, [1]. For completeness of the exposition we introduce here the main relevant concepts.

Definition 3.1 (Weak Derivative) Let $u, v \in L^1_{loc}(\Omega)$, and α is a multi-index. We say that v is the αth-weak partial derivative of u and write $v = D^\alpha u$, provided

$$\int_\Omega u D^\alpha \varphi dx = (-1)^{|\alpha|} \int_\Omega v\varphi dx$$

for all functions $\varphi \in C_c^\infty(\Omega)$.

In the sequel all derivatives are considered in the weak sense. If a derivative exists in the classical sense then it equals the weak derivative. Hence, we use for weak derivatives the same notation as for classical ones.

The space $C_c^\infty(\Omega)$ of all infinitely continuously differentiable functions with compact support in Ω is usually called the space of test functions. We denote by $L^2(\Omega)$ the Hilbert space of square integrable functions. The scalar product in $L^2(\Omega)$ is denoted by $(\cdot, \cdot)$ and the associated norm by $||\cdot||$.

The Sobolev space of particular relevance here is

$$H^1(\Omega) = \left\{ u \in L^2(\Omega) \,\middle|\, \frac{\partial u}{\partial x_i} \in L^2(\Omega), i = 1, 2, \ldots, n \right\}. \tag{3.1}$$

The Sobolev space $H^1(\Omega)$ is a Hilbert space with respect to the scalar product

$$(u, v)_{H^1} = \int_\Omega \left(u(x)v(x) + \sum_{i=1}^n \frac{\partial u(x)}{\partial x_i} \frac{\partial v(x)}{x_i} \right) dx$$

and therefore a Banach space with norm $||u||_{H^1} = \sqrt{(u, u)_{H^1}}$.

Let us recall the concept of trace as given in the following theorem.

Theorem 3.2 (Trace Theorem [8, Theorem 1.3.1]) *Let Ω be a bounded open subset of $\mathbb{R}^n$ with Lipschitz continuous boundary $\partial\Omega$.*

(a) *There exists a unique linear continuous map $T : H^1(\Omega) \mapsto H^{1/2}(\partial\Omega)$ such that $Tu = u|_{\partial\Omega}$ for each $u \in H^1(\Omega) \cap C(\bar{\Omega})$.*

(b) *There exists a linear continuous map* $T^{-1} : H^{1/2}(\partial\Omega) \mapsto H^1(\Omega)$ *such that* $TT^{-1}(\varphi) = \varphi$ *for each* $\varphi \in H^{1/2}(\partial\Omega)$.

Using the concept of trace given in (a) we define

$$H_0^1(\Omega) = \{u \in H^1(\Omega) : Tu = 0\}.$$

The space $H_0^1(\Omega)$ can equivalently be defined as the closure of $C_c^\infty(\Omega)$ in $H^1(\Omega)$. The dual space of $H_0^1(\Omega)$, that is the space of all bounded linear functionals on $H_0^1(\Omega)$, is denoted by $H^{-1}(\Omega)$. The dual pairing between $H^{-1}(\Omega)$ and $H_0^1(\Omega)$ is denoted by $\langle f, \varphi\rangle$, where $f \in H^{-1}(\Omega)$ and $\varphi \in H_0^1(\Omega)$. To make this definition more tangible for non-experts in Sobolev spaces, let us recall that $L^2(\Omega) \subseteq H^{-1}(\Omega)$ and if $f \in L^2(\Omega)$ the dual pairing is just the scalar product in $L^2(\Omega)$, that is $\langle\cdot,\cdot\rangle = (\cdot,\cdot)$.

We assume in the sequel that all coefficients in L are measurable and uniformly bounded on Ω.

Let $f \in L^2(\Omega)$. Multiplying the first equation in (2.1) by a test function $\varphi \in C_c^\infty(\Omega)$, integrating over Ω and applying Green's formula we obtain that every classical solution of this equations satisfies

$$B(u,\varphi) := \int_\Omega \sum_{i,j=1}^n a_{ij}\frac{\partial u}{\partial x_i}\frac{\partial \varphi}{\partial x_j} + \int_\Omega \sum_{i=1}^n b_i \frac{\partial u}{\partial x_i}\cdot\varphi + \int_\Omega cu\varphi = (f,\varphi) \tag{3.2}$$

Since the space of test functions is dense in $H_0^1(\Omega)$, Eq. (3.2) holds for all $\varphi \in H_0^1(\Omega)$. Then the problem (2.1) with homogeneous boundary conditions, that is when $g = 0$, can be generalized to the problem:

$$\text{Find } u \in H_0^1(\Omega) \text{ such that (3.2) holds for all } \varphi \in H_0^1(\Omega). \tag{3.3}$$

Let us note that the bilinear form B in is defined on a much wider domain than in (3.2). More precisely, it is defined for all $u, \varphi \in H^1(\Omega)$.

The existence and uniqueness theory is derived for an even more general problem than (3.3) as follows:

$$\begin{aligned}&\text{Given } f \in H^{-1}(\Omega), \text{ find } u \in H_0^1(\Omega) \text{ such that}\\ &B(u,\varphi) = \langle f,\varphi\rangle \text{ for all } \varphi \in H_0^1(\Omega).\end{aligned} \tag{3.4}$$

The theory is given in detail in many books, e.g. [5]. The procedure is basically as follows: Under the assumptions made for L and Ω, the bilinear form B is bounded on $H^1(\Omega) \times H^1(\Omega)$, that is there exists $\alpha > 0$ such that

$$B(u,v) \le \alpha||u||_{H^1(\Omega)}||v||_{H^1(\Omega)}, \quad u, v \in H^1(\Omega). \tag{3.5}$$

Under the additional assumption

$$\operatorname{essinf}(c - \nabla \cdot b) \geq 0, \tag{3.6}$$

the bilinear form B is coercive on $H_0^1(\Omega)$, that is there exists $\mu_1 > 0$ such that

$$B(u,u) \geq \mu_1 ||u||^2_{H^1(\Omega)}, \;\; u \in H_0^1(\Omega). \tag{3.7}$$

Then by the Lax-Milgram lemma [5, Theorem 1, Section 6.2.1], for every $f \in H^{-1}(\Omega)$ the problem (3.4) has a unique solution.

If the problem (2.1) is non-homogeneous, using Theorem 3.2(c), we define a function $\tilde{g} \in H_1(\Omega)$ such that $T\tilde{g} = g$. Then the non-homogeneous problem (2.1) can be transformed to a homogeneous one for $\tilde{u} = u - \tilde{g}$. Using this result the non-homogeneous problem (2.1) can be generalized to the following variational formulation

$$\begin{aligned} &\text{Given } f \in H^{-1}(\Omega) \text{ and } g \in H^{\frac{1}{2}}(\Omega), \text{ find } u \in H^1(\Omega) \text{ such that} \\ &B(u,v) = \langle f, v\rangle \text{ for all } v \in H_0^1(\Omega) \text{ and} \\ &Tu = g, \end{aligned} \tag{3.8}$$

where the existence and uniqueness of a solution follows from the respective theory for homogeneous problems. If $f \in L^2(\Omega)$, this solution of (3.8) is called a weak solution of (2.1).

Considering problem (3.8), the operator P in (2.5) can be extended to $P : H^1(\Omega) \to H^{-1}(\Omega) \times H^{\frac{1}{2}}(\partial\Omega)$ as

$$Pu = \begin{pmatrix} B(u,\cdot) \\ Tu \end{pmatrix}. \tag{3.9}$$

Let us note that for any $u \in H^1(\Omega)$, $B(u,\cdot)$ is a linear functional on $H_0^1(\Omega)$ defined via

$$v \to B(u,v), \;\; v \in H_0^1(\Omega).$$

It follows from (3.5) that this linear functional is bounded. Therefore, $B(u,\cdot) \in H^{-1}(\Omega)$.

Using this notation, problem (3.8) can be written in the form (2.6), where both the data and the solution belong to much larger spaces than in the classical case, specifically $f \in H^{-1}(\Omega)$, $g \in H^{\frac{1}{2}}(\Omega)$ and $u \in H^1(\Omega)$. The existence and uniqueness of a solution of Eq. (3.9) implies that P is actually a bijection. Our primary interest is in the preservation of order. The goal of this section is to prove that P is inverse monotone.

The proofs of the order related properties use the fact that the space of $H^1(\Omega)$ is a lattice, [3, Chapter 4.7]. Consider a function $w \in H^1(\Omega)$. Then

$$w^+ = \sup\{w, 0\} \in H^1(\Omega), \; w^- = \sup\{-w, 0\} \in H^1(\Omega), \tag{3.10}$$

and $w = w^+ - w^-$. Clearly, $(w^+, w^-) = 0$. Furthermore, it is also shown in the proof of [3, Chapter 4.7, Propostion 6] that for any $j = 1, 2, \ldots n$

$$\frac{\partial}{\partial x_j} w^+(x) = \begin{cases} \frac{\partial}{\partial x_j} w(x) & \text{if } w(x) > 0 \\ 0 & \text{if } w(x) \leq 0 \end{cases} \tag{3.11}$$

and

$$\frac{\partial}{\partial x_j} w^-(x) = \begin{cases} 0 & \text{if } w(x) \geq 0 \\ -\frac{\partial}{\partial x_j} w(x) & \text{if } w(x) < 0. \end{cases} \tag{3.12}$$

Theorem 3.3 (Positivity Theorem) *Let the bilinear form B be coercive on $H_0^1(\Omega)$. Then, if for some $u \in H^1(\Omega)$ we have*

(a) *$B(u, \varphi) \geq 0$ for all $\varphi \in H_0^1(\Omega)$ with $\varphi \geq 0$ and*
(b) *$Tu \geq 0$*

then $u \geq 0$ a.e. in Ω.

Proof Let $u \in H^1(\Omega)$ satisfy (a) and (b). Let $u = u^+ - u^-$, where u^+ and u^- are given in (3.10). Using (3.10), (3.11) and (3.12) we obtain

$$B(u^+, u^-) = \int_\Omega \sum_{i=1}^n a_{ij} \frac{\partial u^+}{\partial x_i} \frac{\partial u^-}{\partial x_j} + \int_\Omega \sum_{i=1}^n b_i \frac{\partial u^+}{\partial x_i} \cdot u^- + \int_\Omega c u^+ u^- = 0$$

since in all products, the factors have disjoint support.

Using that $Tu \geq 0$ we obtain that $Tu = Tu^+$ and $Tu^- = 0$. Therefore $u^- \in H_0^1(\Omega)$. Then, taking $\varphi = u^-$ in (a) we have

$$B(u^+ - u^-, u^-) \geq 0,$$
$$B(u^+, u^-) - B(u^-, u^-) \geq 0,$$
$$B(u^-, u^-) \leq 0.$$

Using the coercivity of B given in (3.7), we have

$$\mu_1 ||u||_{H^1(\Omega)} \leq B(u^-, u^-) \leq 0.$$

Therefore $u^- = 0$ a.e. in Ω, which implies that $u = u^+ \geq 0$ a.e. in Ω. □

Theorem 3.4 (Comparison Theorem) *Let the bilinear form B be coercive on $H_0^1(\Omega)$. Then, if for some $u, v \in H^1(\Omega)$ we have*

(a) *$B(u, \varphi) \leq B(v, \varphi)$ for all $\varphi \in H_0^1(\Omega)$ with $\varphi \geq 0$ and*
(b) *$Tu \leq Tv$*

then $u \leq v$ a.e. in Ω.

The proof follows by applying Theorem 3.3 to $z = v - u$.

Theorem 3.5 *Let the bilinear form B be coercive on $H_0^1(\Omega)$. Then the operator P defined in (3.9) is inverse monotone.*

Proof The inverse monotonicity means that the order in the space $H^{-1}(\Omega) \times H^{\frac{1}{2}}(\partial\Omega)$ is preserved in the sense of (1.1). Let us recall that the positive cone of the space $H^{-1}(\Omega)$ is defined through the positive cone of $H_0^1(\Omega)$. More precisely, let K be the positive cone of $H_0^1(\Omega)$, that is

$$K = \{\varphi \in H_0^1(\Omega) : \varphi(x) \geq 0, x \in \Omega\}.$$

Then the positive cone in the dual space $H^{-1}(\Omega)$ is defined by

$$K^* = \{f \in H^{-1}(\Omega) : \langle f, \varphi \rangle \geq 0 \text{ for all } \varphi \in K\}.$$

Equivalently, this means that for any $f_1, f_2 \in H^{-1}(\Omega)$

$$f_1 \leq f_2 \iff \langle f_1, \varphi \rangle \leq \langle f_2, \varphi \rangle \text{ for all } \varphi \in K.$$

Specifically with reference to the operator P in (3.9), we have

$$B(u, \cdot) \leq B(v, \cdot) \iff B(u, \varphi) \leq B(v, \varphi) \text{ for all } \varphi \in H_0^1(\Omega) \text{ with } \varphi \geq 0.$$

Then the proof follows directly from Theorem 3.4. □

Remark 3.6 The assumption (3.6) looks a bit strange within what appears to be a quite elegant theory, particularly when compared to the simple condition $c \geq 0$ in the classical case. When b is a constant vector, there is not a problem as $\nabla \cdot b = 0$. To simplify the condition (3.6) when b is a function of $x \in \Omega$ we can re-write the operator as follows

$$\begin{aligned} L[u] &= \nabla \cdot (A\nabla u) + b \cdot \nabla u + cu \\ &= \nabla \cdot (A\nabla u) + \frac{1}{2} b \cdot \nabla u + \frac{1}{2} \nabla \cdot (bu) + \left(c - \frac{1}{2} \nabla \cdot b\right) u. \end{aligned}$$

Then we have

$$L[u] = \nabla \cdot (A\nabla u) + \tilde{b} \cdot \nabla u + \nabla(\tilde{b}u) + \tilde{c}u, \tag{3.13}$$

where $\tilde{b} = \frac{1}{2}b$ and $\tilde{c} = c - \frac{1}{2}\nabla \cdot b$. The bilinear form associated with (3.13) is

$$\tilde{B}(u, \varphi) = \int_\Omega \sum_{i,j=1}^{n} a_{ij} \frac{\partial u}{\partial x_i}\frac{\partial \varphi}{\partial x_j} + \int_\Omega \sum_{i=1}^{n} \tilde{b}_i \left(\frac{\partial u}{\partial x_i}\varphi - u\frac{\partial \varphi}{\partial x_i}\right) + \int_\Omega \tilde{c}u\varphi. \quad (3.14)$$

Then

$$\tilde{B}(u, u) = \int_\Omega \sum_{i,j=1}^{n} a_{ij} \frac{\partial u}{\partial x_i}\frac{\partial u}{\partial x_j} + \int_\Omega \tilde{c}u^2$$

does not depend on b at all and it is easy to see that coercivity follows from the uniform ellipticity of L and $\tilde{c} \geq 0$.

In the next section, which is more technical, we will use the operator L in the form (3.13).

4 Operators Associated with Weakly Coupled System of PDEs in Variational Form

We consider in this section operators associated with weakly coupled systems of elliptic PDEs. In the classical form such a system is formulated for $u \in \left(C^2(\Omega) \bigcap C(\overline{\Omega})\right)^m$ as

$$L_k[u] = f_k \text{ in } \Omega, \quad k = 1, \ldots, m, \quad (4.1)$$

$$u_k|_{\partial\Omega} = g_k, \quad k = 1, \ldots, m, \quad (4.2)$$

where

$$L_k[u] = -\sum_{i,j=1}^{n} \frac{\partial}{\partial x_j}\left(a_{ij}^{(k)} \frac{\partial u_k}{\partial x_i}\right) + \sum_{i=1}^{n} b_i^{(k)} \frac{\partial u_k}{\partial x_i} + \sum_{i=1}^{n} \frac{\partial}{\partial x_i}\left(b_i^{(k)} u_k\right) + \sum_{\ell=1}^{m} c_{k\ell} u_\ell, \quad (4.3)$$

for $k = 1, \ldots, m$. The weak coupling refers to the fact that the equations are not coupled in the differential part of the operators L_k. If we denote $A^{(k)} = (a_{ij}^k)_{i,j=1}^n$ and $b^{(k)} = (b_1^{(k)}, \ldots, b_n^{(k)})'$, then (4.3) can be written in following more compact form

$$L_k(u) = -\nabla \cdot (A^{(k)} \nabla u_k) + b^k \cdot \nabla u + \nabla \cdot (b^{(k)} u) + \sum_{\ell=1}^{m} c_{k\ell} u_\ell. \quad (4.4)$$

We assume that the uniform ellipticity condition is satisfied, namely there exists a $\mu_0 > 0$ such that for every $k = 1, \ldots, m$ we have

$$\xi \cdot A^{(k)}(x)\xi \geq \mu_0|\xi|^2, \quad x \in \Omega, \ \xi \in \mathbb{R}^n.$$

Using vector notation

$$u = \begin{pmatrix} u_1 \\ \vdots \\ u_m \end{pmatrix}, \quad L[u] = \begin{pmatrix} L_1[u] \\ \vdots \\ L_m[u] \end{pmatrix}, \quad f = \begin{pmatrix} f_1 \\ \vdots \\ f_m \end{pmatrix}, \quad g = \begin{pmatrix} g_1 \\ \vdots \\ g_m \end{pmatrix},$$

the system (4.1)–(4.2) can be presented as a single vector equation and boundary condition as

$$L[u] = f, \tag{4.5}$$

$$u|_{\partial\Omega} = g, \tag{4.6}$$

Further, the operator admits the following convenient vector representation

$$L[u] = -\sum_{i,j=1}^{n} \frac{\partial}{\partial x_j}\left(A_{ij}\frac{\partial u}{\partial x_i}\right) + \sum_{i=1}^{n} B_i \frac{\partial u}{\partial x_i} + \sum_{i=1}^{n} \frac{\partial}{\partial x_i}(B_i u) + Cu, \tag{4.7}$$

where the partial derivatives are implemented coordinate-wise, A_{ij}, $i, j = 1, \ldots, n$, and B_i, $i = 1, \ldots, n$, are diagonal $m \times m$ matrices given by $A_{ij} = diag(a_{ij}^{(1)}, \ldots, a_{ij}^{(m)})$, $B_i = diag(b_i^{(1)}, \ldots, b_i^{(m)})$, and finally $C = (c_{k\ell})_{k,\ell=1}^{m}$.

In order to derive a variational formulation we assume that all coefficients in L are measurable and uniformly bounded.

Let $f \in (L^2(\Omega))^m$. Multiplying Eq. (4.7) by any $\varphi \in (C_c^\infty(\Omega))^m$, integrating over Ω and applying Green's formula we obtain that any solution of (4.5) satisfies

$$\begin{aligned} B(u, \varphi) := &\sum_{i,j=1}^{n} \int_\Omega (A_{ij}\frac{\partial u}{\partial x_j}) \cdot \frac{\partial \varphi}{\partial x_i} \\ &+ \sum_{i=1}^{n} \int_\Omega \left(\frac{\partial u}{\partial x_i} \cdot (B_i\varphi) - (B_i u) \cdot \frac{\partial \varphi}{\partial x_i}\right) + \int_\Omega (Cu) \cdot \varphi \\ = &\int_\Omega f \cdot \varphi. \end{aligned}$$

Similar to the one dimensional case we generalize (4.1)–(4.2) to the problem

$$\begin{aligned}
&\text{Given } f \in (H^{-1}(\Omega))^m,\, g \in (H^{\frac{1}{2}}(\partial\Omega))^m, \text{ find } u \in (H^1(\Omega))^m \text{ such that}\\
&B(u,\varphi) = \langle f, \varphi\rangle \text{ for all } \varphi \in (H_0^1(\Omega))^m \text{ and} \qquad (4.8)\\
&Tu = g.
\end{aligned}$$

Then we can associate with the problem (4.8) the operator $P : (H^1(\Omega))^m \to (H^{-1}(\Omega))^m \times (H^{\frac{1}{2}}(\partial\Omega))^m$ defined through

$$Pu = \begin{pmatrix} B(u,\cdot) \\ Tu \end{pmatrix}. \qquad (4.9)$$

Our goal is to prove that the operator P is inverse monotone. To obtain this result we need to make the following assumptions about the matrix C:

$$(i) \quad c_{k\ell} \le 0 \text{ for } k \ne \ell, \qquad (4.10)$$

$$(ii) \quad \text{the matrix } C + C' \text{ is positive semi-definite for any } x \in \Omega. \qquad (4.11)$$

The assumption (4.10) is essential for deriving the intended monotonicity. A similar assumption is made in [9] for obtaining a comparison theorem in the space of classical solutions. The assumption (4.11) generalizes the condition $c \ge 0$ in the one dimensional case. From this assumption we obtain that for any $x \in \Omega$ and $\eta \in \mathbb{R}^m$ we have

$$(C(x)\eta)\cdot\eta = \frac{1}{2}((C(x) + C'(x))\eta)\cdot\eta \ge 0. \qquad (4.12)$$

With this inequality we can obtain the coercivity of the bilinear form on $(H_0^1(\Omega))^m$. Indeed, for any $u \in (H_0^1(\Omega))^m$ we have

$$\begin{aligned}
B(u,u) &= \sum_{k=1}^{m} \int_\Omega \left((A^{(k)}\nabla u_k)\cdot\nabla u_k\right) + \int_\Omega (Cu)\cdot u \\
&\ge \mu_1 \left(\sum_{k=1}^{m} ||\nabla u_k||^2\right).
\end{aligned}$$

Then, following the standard approach of using the Poincaré inequality we obtain that there exists a constant μ_2 such that

$$B(u,u) \ge \mu_2 \left(\sum_{k=1}^{m} ||u_k||^2_{H^1(\Omega)}\right) = \mu_2 ||u||^2_{(H^1(\Omega))^m}. \qquad (4.13)$$

The boundedness of $B(u, v)$ is obtained similar to the one dimensional case. Then, the existence and uniqueness of solution of (4.8) follows from the Lax-Milgram Lemma—first for the homogeneous problem and then extended to the non-homogeneous one. More detailed existence and uniqueness theory of variational formulation of general systems of elliptic PDEs is given in [6].

As mentioned, our main interest is the order properties. All product spaces in this section are considered with the associated coordinate-wise partial order.

Theorem 4.1 (Positivity Theorem) *If for some $u \in (H^1(\Omega))^m$ we have*

(a) $B(u, \varphi) \geq 0$ *for all* $\varphi \in (H_0^1(\Omega))^m$ *with* $\varphi \geq 0$ *and*
(b) $Tu \geq 0$

then $u \geq 0$ *a.e. in* Ω.

Proof Let $u = u^+ - u^-$, where u_k^+ and u_k^-, $k = 1, \ldots, m$, are as given in (3.10). Using (3.10), (3.11), (3.12) and (4.10), we obtain

$$B(u^+, u^-) = \sum_{k=1}^{m} \left(\sum_{i,j=1}^{n} \int_\Omega (a_{ij}^{(k)} \frac{\partial u_k^+}{\partial x_j}) \cdot \frac{\partial u_k^-}{\partial x_i} + \sum_{i=1}^{n} \int_\Omega b_i^{(k)} \left(\frac{\partial u_k^+}{\partial x_i} u_k^- - u_k^+ \frac{\partial u_k^-}{\partial x_i} \right) \right)$$
$$+ \sum_{k=1}^{m} \int_\Omega c_{kk} u_k^+ u_k^- + \sum_{k \neq \ell} \int_\Omega c_{k\ell} u_k^+ u_\ell^-$$
$$= \sum_{k \neq \ell} \int_\Omega c_{k\ell} u_k^+ u_\ell^- \leq 0.$$

Let us note that $Tu \geq 0$ implies that $Tu^- = 0$, so that $u^- \in (H_0^1(\Omega))^m$. Using (a) with $\varphi = u^-$ and the coercivity (4.13), we obtain

$$0 \leq B(u^+ - u^-, u^-) = B(u^+, u^-) - B(u^-, u^-) \leq -B(u^-, u^-) \leq -\mu_2 ||u^-||^2_{(H^1(\Omega))^m}.$$

Hence, $||u^-||^2_{(H^1(\Omega))^m} = 0$, or equivalently $u^- = 0$ a.e. on Ω. Then, $u = u^+ \geq 0$ a.e. on Ω. □

Theorem 4.2 (Comparison Theorem) *Let C satisfy* (4.10)–(4.11). *Then, if for some $u, v \in (H^1(\Omega))^m$ we have*

(a) $B(u, \varphi) \leq B(v, \varphi)$ *for all* $\varphi \in H_0^1(\Omega)$ *with* $\varphi \geq 0$ *and*
(b) $Tu \leq Tv$

then $u \leq v$ *a.e. in* Ω.

The proof follows by applying the Theorem 4.1 to $z = v - u \in (H^1(\Omega))^m$.

Theorem 4.3 *Let C satisfy* (4.10)–(4.11). *The the operator P defined in (4.9) is inverse monotone.*

Proof Let K be the positive cone of $(H_0^1(\Omega))^m$, that is

$$K = \{\varphi \in (H_0^1(\Omega)^m) : \varphi_k(x) \geq 0, x \in \Omega, k = 1, \ldots, m\}.$$

Then the positive cone in $(H^{-1}(\Omega))^m$ is defined by

$$K^* = \{f \in (H^{-1}(\Omega))^m : \langle f, \varphi \rangle \geq 0 \text{ for all } \varphi \in K\}.$$

Equivalently, this means that for any $f_1, f_2 \in (H^{-1}(\Omega))^m$

$$f_1 \leq f_2 \iff \langle f_1, \varphi \rangle \leq \langle f_2, \varphi \rangle \text{ for all } \varphi \in K.$$

Specifically with reference to the operator P in (4.9), we have

$$B(u, \cdot) \leq B(v, \cdot) \iff B(u, \varphi) \leq B(v, \varphi) \text{ for all } \varphi \in (H_0^1(\Omega))^m \text{ with } \varphi \geq 0.$$

Then the proof follows directly from Theorem 4.2. □

5 Conclusion

This paper presents results related to order properties of the operators associated with elliptic PDEs and some systems of elliptic PDEs in variational form. Specifically, weakly coupled systems have been considered, since the maximum principle has been extended to such systems in the classical case, [4, 10]. We note that the conditions on the matrix C in [4] and [10] are very similar to the conditions (4.10)–(4.11) assumed in this paper. The maximum principle does not have a natural extension to strongly coupled systems. The existing results provide order for norms—property more associated with norm-boundedness than with the order in the solution space, [2].

The results in this paper extend the classical order properties of the operators in the mentioned elliptic problems, namely single PDEs or weakly coupled systems, to the wider space of weak solutions. Alternatively, these properties can be seen as a "pull back" of the order in the target space (space of data) to the operator's domain (the solution space) as given by the concept by inverse monotonicity. The theory has a natural extension to parabolic PDEs and systems of parabolic PDEs, which will be the subject of future work.

References

1. H. Brezis, *Functional Analysis, Sobolev Spaces and Partial Differential Equations* (Springer, Berlin, 2011)

2. P. Cannabsa, On a maximum principle for elliptic systems with constant coefficients. Rendiconti del Seminario Matematico della Università di Padova **64**, 77–84 (1981)
3. R. Dautray, J.L. Lions, *Mathematical Analysis and Numerical Methods for Science and Technology: Functional and Variational Methods*, vol. 2 (Springer, Berlin, 2000)
4. D.G. De Figueiredo, E. Mitidieri, Maximum principles for linear elliptic systems. Rendiconti dell'Istituto di Matematica dell'Universita di Trieste, March 1990. https://doi.org/10.1007/978-3-319-02856-9_21
5. L.C. Evans, *Partial Differential Equations* (American Mathematical Society, Providence, 1998)
6. B.M. O'Connor, Elliptic systems of partial differential equations and the finite element method. J. Math. Anal. Appl. **94**, 283–295 (1983)
7. M.H. Protter, H.F. Weinberger, *Maximum Principles in Differential Equations* (Springer, Berlin, 1984)
8. A. Quarteroni, A. Valli, *Numerical Approximation of Partial Differential Equations* (Spinger, Berlin, 1991)
9. W. Walter, *Differential and Integral Inequalities* (Spinger, Berlin, 1970)
10. C. Zhou, Maximum principles for elliptic systems. J. Math. Anal. Appl. **159**, 418–439 (1991)

On Compact Operators Between Lattice Normed Spaces

Youssef Azouzi and Mohamed Amine Ben Amor

Abstract In this paper we continue the study of compact-like operators in lattice normed spaces started recently by Aydin, Emelyanov, Erkurşun Özcan and Marabeh. We show among others, that every p-compact operator between lattice normed spaces is p-bounded. The paper contains answers of almost all questions asked by these authors.

Keywords Compact operator · Order topology · Uo-convergence

Mathematics Subject Classification (2010) 47B07, 47S30

1 Introduction

Today the theory of operators on lattice-normed spaces is an active area of Functional Analysis [1, 2, 4–6].

In [6], the authors introduced a new notion of compact operators in Lattice-normed spaces and studied some of their properties. These operators act on spaces equipped with vector valued norms taking their values in some vector lattices. Recall that an operator from a normed space X to a normed space Y is said to be compact if the image of every norm bounded sequence (x_n) in X has a norm convergent subsequence. This notion has been generalized in the setting of lattice normed spaces giving rise to two new notions: sequentially p-compactness and p-compactness (p referred to the vector valued norm). Notice that these notions coincide in the classical case of Banach spaces. In general setting with vector lattice valued norms boundedness and convergence are considered with respect to these 'norms'. Also as notions of relatively uniform convergence and almost order

Y. Azouzi (✉) · M. A. Ben Amor
Research Laboratory of Algebra, Topology, Arithmetic, and Order, GOSAEF, Faculty of Mathematical, Physical and Natural Sciences of Tunis, Tunis-El Manar University, El Manar, Tunisia
e-mail: youssef.azouzi@ipest.rnu.tn; mohamedamine.benamor@ipest.rnu.tn

E. Kikianty et al. (eds.), *Positivity and its Applications*, Trends in Mathematics,
https://doi.org/10.1007/978-3-030-70974-7_3

boundedness have been generalized, new properties for the operator are considered like semicompactness. Recall that if (E, p, V) and (F, q, W) are Lattice-normed spaces and T is a linear operator from E to F, then T is said to be *p-compact* (respectively, *rp-compact*) if for every p-bounded net (x_α) in E, there is a subnet $\left(Tx_{\varphi(\beta)}\right)$ that p-converges (respectively, rp-converges) to some $y \in F$. The operator is said to be *sequentially p-compact* if nets and subnets are replaced by sequences and subequences above. In this paper we prove some new results in this direction. Namely we show that every p-compact operator is p-bounded. As a consequence we get that every rp-compact is p-bounded. Also we give an example of sequentially p-compact operator which fails to be p-bounded. As a consequence we deduce that a sequentially p-compact need not be p-compact. In fact these two notions are totally independent. Example of p-compact operators that fail to be sequentially p-compact is given. As mentioned above the study of p-compact operators between lattice normed spaces was started in [6]. That paper contains several new results but also some open questions. Almost all these questions will be answered in our paper. The structure of the paper is as follows. Section 2 contains some preliminaries and definitions. In Sect. 3 we prove that p-compact operators between lattice normed spaces are p-bounded. We give an example showing that this may not happen for sequentially p-compact operators. As a consequence we derive that sequentially p-compactness does not imply p-compactness. We show also the p-compactness does not imply sequentially p-compactness. Section 4 is devoted to compare p-compactness with p-semicompactness. We show that these two notions are completely independent. In the last section we investigate rp-compact operators. As an immediate consequence of our main theorem, we deduce that every rp-compact operator is p-bounded. We add our stone to the building, by studying the connection between (sequentially) p-compactness and (sequentially) rp-compactness.

2 Preliminaries

The goal of this section is to introduce some basic definitions and facts. For general informations on vector lattices, Banach spaces and lattice-normed spaces, the reader is referred to the classical monographs [3] and [10].

Consider a vector space E and a real Archimedean vector lattice V. A map $p : E \to V$ is called a *vector norm* if it satisfies the following axioms:

1. $p(x) \geq 0$; $p(x) = 0 \Leftrightarrow x = 0$; $(x \in E)$.
2. $p(x_1 + x_2) \leq p(x_1) + p(x_2)$; $(x_1, x_2 \in E)$.
3. $p(\lambda x) = |\lambda| p(x)$; $(\lambda \in \mathbb{R},\ x \in E)$.

A triple (E, p, V) is a *lattice-normed space* if $p(.)$ is a V-valued vector norm in the vector space E. When the space E is itself a vector lattice the triple (E, p, V) is called a *lattice-normed vector lattice.* A set $M \subset E$ is called *p-bounded* if $p(M) \subset [-e, e]$ for some $e \in V_+$. A subset M of a *lattice-normed vector lattice*

(E, p, V) is called *p-almost order bounded* if, for any $w \in V_+$, there is $x_w \in E_+$ such that $p((|x| - x_w)^+) = p(|x| - x_w \wedge |x|) \leq w$ for any $x \in M$.

Let $(x_\alpha)_{\alpha\in\Delta}$ be a net in a *lattice-normed space* (E, p, V). We say that $(x_\alpha)_{\alpha\in\Delta}$ is *p-convergent* to an element $x \in E$ and write $x_\alpha \xrightarrow{p} x$, if there exists a decreasing net $(e_\gamma)_{\gamma\in\Gamma}$ in V such that $\inf_{\gamma\in\Gamma}(e_\gamma) = 0$ and for every $\gamma \in \Gamma$ there is an index $\alpha(\gamma) \in \Delta$ such that $p(x - x_\alpha) \leq e_\gamma$ for all $\alpha \geq \alpha(\gamma)$. Notice that if V is Dedekind complete, the dominating net $\left(e_\gamma\right)$ may be chosen over the same index set as the original net. We say that (x_α) is *p-unbounded convergent* to x (or for short, *up-convergent* to x) if $p(x_\alpha - x) \wedge u \xrightarrow{p} 0$ for all $u \in V_+$. It is said to be *relatively uniformly p-convergent* to $x \in X$ (written as, $x_\alpha \xrightarrow{rp} x$) if there is $e \in E_+$ such that for any $\varepsilon > 0$, there is α_ε satisfying $p(x_\alpha - x) \leq \varepsilon e$ for all $\alpha \geq \alpha_\varepsilon$.

When $E = V$ and p is the absolute value in E, the p-convergence is the order convergence, the up-convergence is the unbounded order convergence, and the rp-convergence is the relatively uniformly convergence. We refer to [9] and [8] for the basic facts about nets in topological spaces and vector lattices respectively. We will use [10, 12] as unique source for unexplained terminology in Lattice-Normed Spaces. Since the most part of this paper is devoted to answer several open questions in [6], the reader must have that paper handy, from which we recall some definitions.

Definition 1 Let (E, p, V) and (F, q, W) be two lattice-normed spaces and $T \in L(E, F)$. Then

1. T is called *p-compact* if, for any p-bounded net (x_α) in E, there is a subnet x_{α_β} such that $Tx_{\alpha_\beta} \xrightarrow{q} y$ in F for some $y \in F$.
2. T is called *sequentially p-compact* if, for any p-bounded sequence x_n in E, there is a subsequence $\left(x_{n_k}\right)$ such that $Tx_{n_k} \xrightarrow{q} y$ in F for some $y \in F$.
3. T is called p-semicompact if, for any p-bounded set A in E, the set $T(A)$ is p-almost order bounded in F.

3 *p*-Compact Operators Are *p*-Bounded

It is well known that compact operators between Banach spaces are bounded. This result remains valid for general situation of p-compact operators as it will be shown in our first result, which answers positively Question 2 in [6].

Theorem 2 *Every p-compact operator between two Lattice-normed spaces is p-bounded.*

Proof Assume, by contradiction, that there exists a p-compact operator $T : (E, p, V) \longrightarrow (F, q, W)$ which is not p-bounded. Then there exists a p-bounded subset A of E such that $T(A)$ is not q-bounded. So, for every $u \in W^+$ there exists some $x_u \in A$ satisfying $q(T(x_u)) \nleq u$. Since the net $(x_u)_{u\in W^+}$ is p-bounded there is a subnet $\left(y_v = x_{\varphi(v)}\right)_{v\in\Gamma}$ and an element $f \in F$ such that $(Ty_v) \xrightarrow{q} f$. It follows that the net (Ty_v) has a q-bounded tail, which means that for some v_0 in Γ and some $w \in W^+$ we have,

$$q\left(Tx_{\varphi(v)}\right) \leq w, \qquad \text{for } v \geq v_0. \tag{1}$$

Pick v_1 in Γ such that $\varphi(v) \geq w$ for all $v \geq v_1$. It follows that for $v \geq v_0 \vee v_1$, we have $q(Tx_{\varphi(v)}) \nleq \varphi(v)$ and so

$$q(Tx_{\varphi(v)}) \nleq w,$$

which is a contradiction with (1). and the proof comes to its end. □

The following lemma, which connects unbounded order convergence with pointwise convergence, is a known fact, although a quick proof is included for the sake of completeness.

Lemma 3 *Let $E = \mathbb{R}^X$ be the Riesz space of all real-valued functions defined on a nonempty set X. The following statements are equivalent:*

(i) *The net $(f_\alpha)_{\alpha\in A}$ is uo-convergent in E.*
(ii) *for every $x \in X$, the net $(f_\alpha(x))_{\alpha\in A}$ is convergent in $\mathbb{R}$.*

Proof

(i) $\Longrightarrow$ (ii) Assume that $f_\alpha \xrightarrow{uo} f$ in the Dedekind complete Riesz space E. Then there is a net $(g_\alpha)_{\alpha\in A}$ which decreases to 0 and for some α_0 we have

$$|f_\alpha - f| \wedge 1 \leq g_\alpha \text{ for all } \alpha \geq \alpha_0. \tag{2}$$

Since $(g_\alpha(x))$ decreases to 0 for every $x \in X$, it follows easily from (2) that $f_\alpha(x) - f(x)$ converges to 0, as desired.

(ii) $\Longrightarrow$ (i) Assume now that f_α converges simply to some $f \in E$ and let $h \in E^+$. Define a net (g_α) by putting

$$g_\alpha(x) = \sup_{\beta\geq\alpha}\left(\left|f_\beta - f\right| \wedge h\right)(x), \quad x \in X.$$

it is clear that g_α decreases to 0 and $|f_\alpha - f| \wedge h \leq g_\alpha$. This shows that $f_\alpha \xrightarrow{uo} f$ and we are done.

□

Consider the Riesz space F of all bounded real valued functions defined on the real line with countable support and denote by E the direct sum $\mathbb{R}\mathbf{1} \oplus F$, where $\mathbf{1}$ denotes the constant function taking the value 1. This example will be of great interest for us. The following lemma establishes some of its properties. Recall that a vector sublattice Y of a vector lattice X is said to be *regular* if every subset in Y having a supremum in Y has also a supremum in X and these suprema coincide. For more information about this notion and nice characterizations of it via unbounded order convergence the reader is referred to [8].

Lemma 4 *The space E introduced above has the following properties.*

(i) *E is a regular vector sublattice of $\mathbb{R}^{\mathbb{R}}$.*
(ii) *E is Dedekind σ-complete but not Dedekind complete.*

Proof

(i) It is clear that E is a vector sublattice of $\mathbb{R}^{\mathbb{R}}$. To show that it is regular assume that $(g_\alpha)_{\alpha\in A}$ is a net in E satisfying $g_\alpha \downarrow 0$ in E. Let $g = \inf_\alpha g_\alpha$ in $\mathbb{R}^{\mathbb{R}}$ and $x \in \mathbb{R}$. Then $h = g(x)\mathbf{1}_{\{x\}} \in E$ and $0 \le h \le g_\alpha$ for all α. This implies that $h = 0$ and then $g(x) = 0$. Hence $g = 0$ and the regularity is proved.
(ii) Let (g_n) be an order bounded sequence in E and write $g_n = \lambda_n + f_n$, with $\lambda_n \in \mathbb{R}$ and $f_n \in F$. Let Ω be the union of the supports of f_n, then Ω is countable. Let g be the supremum of (g_n) in $\mathbb{R}^{\mathbb{R}}$, that is,

$$g(a) = \sup g_n(a), \text{ for all } a \in \mathbb{R}.$$

It will be sufficient to show that $g \in E$. To this end observe that $g(x) = \alpha := \sup \alpha_n$ for all $x \in \mathbb{R}\backslash\Omega$. Now put $f = (g - \alpha)\mathbf{1}_\Omega$. Then $f \in F$ and $g = \alpha + f \in E$ as required. Next we show that E is not Dedekind complete. Consider the net $(g_x)_{x\in[0,1]}$ in E defined by $g_x = x\mathbf{1}_{\{x\}}$. It is a bounded net in E and its supremum in $\mathbb{R}^{\mathbb{R}}$ does not belong to E. As E is regular in $\mathbb{R}^{\mathbb{R}}$ this net can not have a supremum in E.

□

Remark 5 Consider the following operator:

$$T : L_1[0,1] \longrightarrow c_0; \qquad f \longmapsto Tf = \left(\int_0^1 f(t)\sin nt\,dt\right)_{n\ge 1}.$$

It is mentioned in [3], that T is not order bounded; it is perhaps more convenient to consider the same operator defined on $L_1[0, 2\pi]$. In this case if we define u_n by $u(t) = \sin nt$ for $t \in [0, 2\pi]$, then $|u_n| \le 1$, however $(Tu_n) = (e_n)$ is not order bounded in c_0, where (e_n) denotes the standard basis of c_0. This statement implies also that T is not sequentially order compact. Because (e_n) has no order bounded subsequence, it follows that (Tu_n) can not admit an order convergent subsequence. So the statement made in [6] that T is p-bounded is not correct.

The above example is presented in [6] to show that sequentially p-compact operators need not be p-bounded. Although the operator given in that example fails to be sequentially p-compact, the assertion that sequentially p-compact operators need not be p-bounded is true. This will be shown in our next example.

Example 6 Consider the Riesz spaces E and F defined just before Lemma 4 and let T be the projection defined on E with range F and kernel $\mathbb{R}\mathbf{1}$. We claim that T is sequentially order compact, but not order bounded. Let (f_n) be an order bounded sequence in E. Then $|f_n| \leq \lambda$ for some real $\lambda > 0$ and for all n. Write $f_n = g_n + \lambda_n$ with λ_n real and $g_n \in F$ and observe that $|g_n| \leq 2\lambda$ for all n. We have also $|g_n| \leq 2\lambda\mathbf{1}_A \in F$ where A is the union of the supports of g_n, $n = 1, 2, \ldots$ A standard diagonal process yields a subsequence $\left(g_{k_n}\right)$ of (g_n) which converges pointwise on A and then on $\mathbb{R}$ since all functions g_{n_k} vanish on $\mathbb{R}\backslash A$. Hence $\left(g_{k_n}\right)$ is uo-convergent in $\mathbb{R}^{\mathbb{R}}$. As $\left(g_{k_n}\right)$ is order bounded this implies that $\left(g_{k_n}\right)$ is order convergent in $\mathbb{R}^{\mathbb{R}}$. Observe moreover that $\sup\limits_{p\geq n} g_{k_n}$ belongs to F, which shows that $\left(g_{k_n}\right)$ is order convergent in F. The fact that T is not order bounded is more obvious: it is clear that the image of the net $\left(\mathbf{1}_{\{x\}}\right)_{x\in[0,1]}$ by T is not order bounded in F.

As an immediate consequence of Theorem 2 and Example 6 we deduce that sequentially p-compactness does not imply p-compactness. At this stage one might expect that the converse is true. Does p-compactness imply sequentially p-compactness? This is an open question left in [6]. Unfortunately the answer is again negative.

Example 7 Let X be the set of all strictly increasing maps from $\mathbb{N}$ to $\mathbb{N}$ and $E = \mathbb{R}^X$ be the space of all real-valued functions defined on X, equipped with the product topology.

1. First we will prove that the identity map, $\mathfrak{I}$, is a p-compact operator on the lattice-normed space $(E, |\ |, E)$. To this aim, pick a p-bounded net $(f_\alpha)_{\alpha\in A}$ in E, that is, $|f_\alpha| \leq f$ for some $f \in E^+$ and for every $\alpha \in A$. It follows that

$$f_\alpha \in \prod_{x\in X}[-f(x), f(x)].$$

   Notice that the space $\prod_{x\in X}[-f(x), f(x)]$, equipped with the product topology, is compact by Tychonoff's Theorem. Thus (f_α) has a convergent subnet $\left(g_\beta\right)_{\beta\in B}$ in $\prod_{x\in X}[-f(x), f(x)]$ to some g. This means that

$$g_\beta(x) \longrightarrow g(x) \text{ for all } x \in X.$$

   According to Lemma 3, g_β is *uo*-convergent to g in E. Since bounded uo-convergent nets are order convergent, we have that $g_\beta \xrightarrow{o} g$. This proves that $\mathfrak{I}$ is a p-compact operator.
2. We prove now that $\mathfrak{I}$ is not sequentially p-compact. Let (φ_n) be a sequence in $\{-1,\ 1\}^X$ which has no convergent subsequence (see Example 3.3.22 in [11]). This sequence is order bounded in E and every subsequence (ψ_n) of (φ_n) does not converges in $\{-1,\ 1\}^X$, that is, for some $x \in X$, $\psi_n(x)$ diverges. According to Lemma 3, (ψ_n) is not uo-convergent in E. Since (ψ_n) is order bounded it does not converge in order. This finishes the proof.

In classical theory of Banach spaces the identity map is compact if and only if the space is finite-dimensional. In contrast of this the situation is not clear in general case. We already have seen an example of infinite-dimensional space on which the identity map is p-compact. This question has been investigated in [6] where the authors showed that $I_{L_1[0,1]}$ fails to be compact however, I_{ℓ_1} is p-compact. In the next example we show that $I_{L_\infty[0,1]}$ is not p-compact, answering a question asked in [6].

Example 8 The identity operator I on the lattice normed space $(L_\infty[0,1], |\ .\ |, L_\infty[0,1])$ is neither p-compact nor sequentially p-compact. To this end, consider the sequence of Rademacher function given by :

$$\begin{aligned} r_n : [0,1] &\longrightarrow \mathbb{R} \\ t &\longmapsto sgn\,(\sin(2^n \pi t)) \end{aligned}$$

for all $n \in \mathbb{N}$, which is order bounded since $|r_n| = 1$. Suppose now that (r_n) has an order convergent subnet $\left(r_{n_\alpha}\right)_{\alpha\in\Gamma}$. Then $r_{n_\alpha} \xrightarrow{o} r$ for some $r \in L_\infty[0,1]$. Let $\alpha \in A$. For every $\beta > \alpha$, $\int_0^1 r_{n_\alpha} r_{n_\beta} d\mu = 0$. On the other hand $\left(r_{n_\alpha} r_{n_\beta}\right)_\beta$ converges in order to $r_{n_\alpha} r$ in $L_\infty[0,1]$ and then in $L_1[0,1]$. Since the integral is order continuous, we deduce that

$$\int_0^1 r_{n_\alpha} r\, d\mu = 0.$$

This equality holds for every $\alpha \in A$, and a similar argument leads to

$$\int_0^1 r^2 d\mu = 0,$$

which is a contradiction since $|r| = 1$, and the claim is now proved.

4 Semicompact Operators

The notion of semicompact operators has been introduced by Zaanen in [13] and extended in the framework of lattice normed spaces in [6].

Let (X, p, E) be a lattice normed space and (Y, q, F) be an lattice normed vector lattice. A linear operator $T : X \to Y$ is called *p-semicompact* if it maps p-bounded sets in X to p-almost order bounded sets in Y. We recall that a subset B of Y is said to be p-almost order bounded if for any $w \in F_+$, there is $y_w \in Y$ such that

$$q((|y| - y_w)^+) = q(|y| - y_w \wedge |y|) \leq w \text{ for all } y \in B.$$

Semicompact operators from Banach spaces to Banach lattices fail, in general, to be compact (see [3]). This yields trivially that p-semicompactness does not imply p-compactness. However, the converse is true in the classical case as has been shown in Theorem 5.71 in [3]. And one can expect to extend this result in general situation. This is already the subject of Question 4 in [6]. Unfortunately the answer is again negative. Before stating our counterexample let us recall that every order bounded operator from a vector lattice E to a Dedekind complete vector lattice F has a modulus [3].

Example 9 Let E be a Dedekind complete Banach lattice with order continuous norm and T be a norm-compact operator in $\mathcal{L}(E)$ such that T has no modulus, and therefore T can not be order bounded. For the existence of such operator we refer the reader to the Krengel's example in [3, p 277.]. Consider now the following lattice-normed vector spaces $(E, \|.\|, \mathbb{R})$ and $\left(E, p, \mathbb{R}^2\right)$, where $p(x) = \begin{pmatrix} \|x\| \\ 0 \end{pmatrix}$ for all $x \in E$. It is straightforward to prove that T is again p-compact operator and we claim that T is not p-semicompact. To this end we will argue by contradiction and we assume that T is p-semicompact. Fix an element $u \in E^+$ and let $w = \begin{pmatrix} 0 \\ 1 \end{pmatrix}$, then there exists z_w such that $p\left((T(x) - z_w)^+\right) \leq w$ for all $x \in [-u, u]$, which means that $(T(x) - z_w)^+ = 0$. Noting that this occurs for x and $-x$ we see that

$$|T(x)| \leq z_w \text{ for all } x \in [-u, u].$$

This shows that T is order bounded, a contradiction. and our proof comes to an end.

A slight modification of the proof of Example 9 leads to a more general result. The proof of it will be left for the reader.

Proposition 10 *Let (E, p, V) be a lattice normed space and (F, q, W) a lattice normed vector lattice. We assume that $q\ (F)^d$ is not trivial. Then every semicompact operator $T : (E, p, V) \longrightarrow (F, q, W)$ is p-bounded as an operator from (E, p, V) to $(F, |.|, F)$.*

5 *rp*-Compact Operators

As every rp-compact operator between lattice-normed spaces is p-compact, the following result is an immediate consequence of Theorem 2.

Theorem 11 *Let (E, p, V) and (F, q, W) be lattice-normed spaces and T be in $\mathfrak{L}(E, F)$. If T is rp-compact then T is p-bounded.*

For the sequentially case, we know that rp-sequentially compactness implies sequentially p-compactness but the converse is false (see Remark 4 in [6]).

As sequentially p-compact operators need not be p-bounded (Example 6), it follows from Theorem 11 that sequentially p-compactness does not imply rp-compactness. In the following example we will prove that even p-bounded sequentially p-compact operators fail to be rp-compact.

Example 12 Let E be the Riesz space defined above. We claim that the identity operator $I : E \to E$ is sequentially p-compact but fails to be rp-compact. Let (x_n) be a bounded sequence in E, that is, $|x_n| \leq x$ for some $x \in E^+$. Write $x = \alpha + f$, and $x_n = \alpha_n + f_n$ where $\alpha \in \mathbb{R}^+$ and $f \in F$ and $\alpha_n \in \mathbb{R}$, $f_n \in F$ for $n = 1, 2, \ldots$ It is easily seen that $|\alpha_n| \leq \alpha$, $|f_n| \leq x + \alpha$. By a standard diagonal argument there exists a subsequence such that $f_{\varphi(n)}(a)$ converges for every $a \in \mathbb{R}$ and $\alpha_{\varphi(n)}$ converges in $\mathbb{R}$. This shows that x_n converges pointwise on $\mathbb{R}$ and its limit is clearly in E. By Lemma 3, $x_n \xrightarrow{uo} x$ in $\mathbb{R}^{\mathbb{R}}$ and then $x_n \xrightarrow{o} x$ in $\mathbb{R}^{\mathbb{R}}$ as it is an order bounded sequence. Now using Lemma 27 in [7] and Lemma 4 we deduce that $x \in E$. On the other hand, let $\mathcal{F}$ be the collection of finite subsets of $\mathbb{R}_+$ ordered by inclusion and consider the net $(g_A)_{A \in \mathcal{F}}$ where $g_\alpha = \mathbf{1}_\alpha$. Then (g_α) is order bounded in E but has no convergent subnet. Since $g_\alpha \uparrow \mathbf{1}_{\mathbb{R}_+}$ in $\mathbb{R}^{\mathbb{R}}$ and E is regular in $\mathbb{R}^{\mathbb{R}}$, it follows that (g_α) is not order convergent in E and so are all its subnets.

References

1. N. Abasov, M. Pliev et al., Dominated orthogonally additive operators in lattice-normed spaces. Adv. Oper. Theory **4**(1), 251–264 (2019)
2. N. Abasov, A. Megahed, M. Pliev et al., Dominated operators from lattice-normed spaces to sequence Banach lattices. Ann. Funct. Anal. **7**(4), 646–655 (2016)
3. C.D. Aliprantis, O. Burkinshaw, *Positive Operators* (Springer, Berlin, 2006)
4. A. Aydı, E. Emelyanov, N. erkurşun-özcan et al., Unbounded p-convergence in lattice-normed vector lattices. Sib. Adv. Math. **29**(3), 164–182 (2019)
5. A. Aydı, S. Gorokhova, H. Gül, Nonstandard hulls of lattice-normed ordered vector spaces. Turk. J. Math. **42**(1), 155–163 (2018)
6. A. Aydı, E.Yu. Emelyanov, N. Erkurşun Özcan, M.A.A. Marabeh, Compact-like operators in lattice-normed spaces. Indag. Math. (N.S.) **29**, 633–656 (2018)
7. Y. Azouzi, Completeness for vector lattices. J. Math. Anal. Appl. **472**, 216–230 (2019)
8. N. Gao, V.G. Troitsky, F. Xanthos, Uo-convergence and its applications to Cesáro means in Banach lattices. Isr. J. Math. **220**(2), 649–689 (2017)
9. J. L. Kelley, *General Topology* (Courier Dover Publications, 2017)
10. A.G. Kusraev, *Dominated Operators* (Kluwer Academic Publishers, Dordrecht, 2000)
11. V. Runde, *A Taste of Topology* (Springer, Berlin, 2005)
12. B.Z. Vulikh, *Introduction to the Theory of Partially Ordered Spaces* (Wolters-Noordhoff Scientific Publications, Groningen, 1967)
13. A.C. Zaanen, *Riesz Spaces II* (North-Holland, Amsterdam, 1983)

How to Be Positive in Natural Sciences?

Jacek Banasiak

Abstract In many fields of science there is the chicken or the egg dispute—whether applications drive theory, or the theory makes applications possible. Actually, in mathematics, there is another option, when certain concepts existed both in applications and in pure theory, happily oblivious of each other. An example of such concepts are order and positivity which, together with compactness, created an important bridge between the finite and infinite dimensional spaces, allowing for a number of concepts from the undergraduate calculus, like the Bolzano–Weierstrass theorem, or the Lyapunov stability theorems, to be applied in probability theory and partial differential equations, before finding their place in the abstract Banach space theory.

In this paper we will illustrate how positivity methods can create such a bridge between finite and infinite dimensional population models, and what are potential pitfalls, within the framework of the theory of semigroups of operators. This paper is based on the lecture given at the conference Positivity X, Pretoria, 8–12 July 2019.

Keywords Mathematical modelling · Semigroup of operators · Population models · Markov processes · Substochastic semigroups · Positive semigroups · Honest semigroups · Non-uniqueness · Birth-and-death processes

1 Introduction

Differential, difference and integral equations, often combined, have been used to describe real world phenomena, from the evolution of the Universe to the fundamental particles' interactions, for well over two centuries. Since in most cases of interest the equations cannot be solved in explicit form, mathematicians have developed various theoretical tools for analyzing qualitative properties of solutions

J. Banasiak (✉)
University of Pretoria, Hatfield, Pretoria, South Africa

Technical University of Łódź, Łódź, Poland
e-mail: jacek.banasiak@up.ac.za

E. Kikianty et al. (eds.), *Positivity and its Applications*, Trends in Mathematics,
https://doi.org/10.1007/978-3-030-70974-7_4

to validate approximate and computational methods used to provide numerical and graphical results, required by the end users in the applied sciences. In this paper we shall confine ourselves to time-dependent models and their functional analytic treatment; that is, to the description of the process by a family of operators parametrized by time. These operators represent subsequent states of the process. Such a family of operators is called a (semi) dynamical system or, in the linear setting, a semigroup of operators.

The problem with such an approach (and, for that matter, with any other) is that the process of making the object of interest manageable by particular mathematical techniques, called the mathematical modelling, produces a model that is significantly different from the original object. It is often forgotten that the heat equation is not the same as the heat transfer, or that the Navier–Stokes system is not the flow of water. Both are approximate descriptions of the respective processes and we cannot be a priori certain that a given solution of the equation has any physical realization. To somehow address this problem, we extend the classical Hadamard's definition of the well-posedness of a problem by adding one more requirement. Thus, let us ask ourselves,

What Do We Want from a Mathematical Model?

1. *The existence of solutions.* This is the requirement that we haven't used any mutually exclusive postulates while building the model.
2. *The uniqueness of solutions.* This reflects the requirement that we have full information about the process and there is causality in the process.
3. *Continuous dependence on the input data.* Since our information is imperfect, we want small errors in our data to yield only small deviations of the output.
4. *Honesty.* We want solutions of equations of the model to faithfully reproduce the principles used to build the model.

To better explain the last point, we observe that many processes involve only nonnegative quantities such as density, energy, absolute temperature, pressure. Thus the corresponding dynamical system should give only nonnegative solutions for physically correct inputs. Such dynamical systems will be called positive. Also, many equations express conservation laws such as the conservation of mass or energy. It is thus natural to expect that the solutions of such equations should satisfy the same laws. However, as we shall see, many mathematical models, even linear, fail to have this property (that is, they are *dishonest*). The main aim of this survey is to describe how the interplay of the positivity of models and classical functional analysis leads to a comprehensive theory of honesty for linear infinite dimensional dynamical systems (semigroups) in population dynamics. While the theory we present is fairly general, we shall illustrate it on examples from the class of birth-and-death problems described by infinite systems of ordinary differential equations. A detailed analysis of such problems in a more probabilistic context can be found in [12]. Similar problems in the theory of fragmentation–coagulation equations are discussed in [8].

2 Population Balance Equations

By a population we understand a collection of objects interacting with each other and with the environment, structured by a set of attributes that can change in the interactions. For instance, moving and colliding particles of a gas are characterized by their position and velocity that can change upon collisions, [13], reacting polymers are characterized by their length, [8], animals in a population can be characterized by their age and geographical location, [5, 27], cells can be characterized by their maturity and the number of copies of a particular gene, [24, 30].

Population balance equations characterize the population using only the number (density) of objects with given attributes and are mathematical expressions of conservation laws. In fact, in any field of science the modelled processes must obey laws of physics and, in particular, the conservation laws. More precisely, if Q is a quantity of interest (e.g., the number of animals, the total amount of a pollutant, the amount of heat energy, the number of infected individuals) with the attributes in a fixed domain Ω, then, over any fixed time interval, in Ω

$$\begin{aligned}\text{the change of } Q &= \text{the inflow of Q} - \text{the outflow of Q}\\ &\quad + \text{the creation of Q} - \text{the destruction of Q}. \qquad (1)\end{aligned}$$

As mentioned before, we characterize populations using the density of the objects with respect to the attributes. The density, say $u(x)$, is either the number (often normalized) of elements with an attribute x (if the number of possible attributes is finite or countable), or gives the number of elements with attributes in a set A according to the formula

$$\int_A u(x)\mathrm{d}\mu, \qquad (2)$$

if x is a continuous variable (and the space of all attributes Ω can be equipped with some measure structure). Balancing, for a given set of attributes A,

- the loss of individuals from A due to the change of their attributes caused by internal or external interactions (that could include death),
- the gain in A due to the changes of individuals' attributes from outside A to the ones from A (that could include birth), results in the so-called Master Equation

$$\partial_t u(x,t) = (\mathcal{K}u)(x,t) := (\mathcal{A}u)x,t) + (\mathcal{B}u)(x,t), \qquad (3)$$

where $\mathcal{A}$ is referred to as the loss operator and $\mathcal{B}$ as the gain operator. The name comes from the theory of Markov processes, where it describes the evolution of probability of the system being in a particular state and the conservation law refers to the fact that the total probability, that is, the probability that a system is in one of the possible states, must be 1 at any give time.

The density u, by its physical meaning, should be non-negative. There are, however, population models having solutions that become eventually negative. It is then interpreted as the crash of the population, see e.g. [3].

3 Finite-Dimensional Population Equation

The simplest equations of this type occur when, at any given time, a system is in one of finitely many states and the switching between the states is determined by a matrix of migration rates. In the context of population dynamics, we consider a population divided into N classes, described by a vector $u(t) = (u_1(t), u_2(t), \ldots, u_N(t))$, where u_i is the number of individuals with attribute i, $i = 1, 2, \ldots, N$, at time t. The attribute may refer to a geographical location, financial status, number of genes of a particular type, etc. Over a short period of time Δt, the individuals can move from subpopulation i to subpopulation j with (approximate) probability $p_{ji}\Delta t$ but cannot die, emerge or leave the system, hence total population is constant.

Thus, (3) for the subpopulation with the attribute i takes the form

$$u_i'(t) = -u_i(t) \sum_{j=1, j\neq i}^{N} p_{ji} + \sum_{j=1, j\neq i}^{N} p_{ij} u_j(t), \quad 1 \le i \le N. \tag{4}$$

The equation expresses the principle of conservation of the total population. The left hand side gives the rate of change of the number of individuals with attribute i and the right hand side gives the explanation of this change: the positive terms give the total rate of the immigration to i from all other classes and the negative terms give the total emigration rate from i to all other classes. Thus, denoting $\mathcal{K} = -\mathrm{diag}\left(\sum_{j=1, j\neq i}^{N} p_{ji}\right)_{1\le i\le N} + (p_{ij})_{1\le i,j\le N, i\neq j} =: \mathcal{A} + \mathcal{B}$, we write

$$u' = \mathcal{A}u + \mathcal{B}u = \mathcal{K}u. \tag{5}$$

The unique solution, for an initial distribution $\mathring{u}$, is given by

$$u(t) = e^{t\mathcal{K}}\mathring{u},$$

so that the first three requirements of well-posedness are satisfied by the standard theory of systems of linear ordinary differential equations. As far as the honesty is concerned, first we recall that the model is to describe a population, so we should have $u_i(t) \ge 0$ provided $\mathring{u}_i \ge 0$ for all $0 \le i \le N$, that is, $u(t) \ge 0$ provided $\mathring{u} \ge 0$.

Indeed, using e.g. [17, Proposition VI.1.2] or [7, Proposition 2.1.4], we see that this true as $\mathcal{K}$ is positive off-diagonal.

By the construction of the model, the total size of the population at time t, given by

$$u(t) = u_1(t) + \ldots + u_N(t), \tag{6}$$

should be constant in time. In fact, adding the equations in (4) we see that the system correctly reflects the conservation principle

$$u'(t) = 0.$$

If we solve (4) and evaluate (6), then we also obtain $u(t) = u(0), t \geq 0$, confirming the above. Consider, for instance, the system

$$\begin{aligned} u_1' &= -u_1 + u_2, \quad u_1(0) = \mathring{u}_1, \\ u_2' &= u_1 - u_2, \quad u_2(0) = \mathring{u}_2. \end{aligned}$$

Clearly,

$$\begin{aligned} u'(t) &= u_1'(t) + u_2'(t) = (-u_1(t) + u_2(t)) + (u_1(t) - u_2(t)) \\ &= (-u_1(t) + u_1(t)) + (u_2(t) - u_2(t)) = 0, \end{aligned} \tag{7}$$

hence $u(t) = u(0) = \mathring{u}_1 + \mathring{u}_2$, so the system is conservative. Also, the solution

$$\begin{aligned} u_1(t) &= \frac{1}{2}e^{-2t}\left(1 + e^{2t}\right)\mathring{u}_1 + \frac{1}{2}e^{-2t}\left(-1 + e^{2t}\right)\mathring{u}_2, \\ u_2(t) &= \frac{1}{2}e^{-2t}\left(-1 + e^{2t}\right)\mathring{u}_1 + \frac{1}{2}e^{-2t}\left(1 + e^{2t}\right)\mathring{u}_2, \end{aligned}$$

satisfies

$$u_1(t) + u_2(t) = \mathring{u}_1 + \mathring{u}_2,$$

so it is conservative.

While the above calculations, such as (7), seem trivial, one should keep in mind that they tacitly depend on a number of properties such as the ability to differentiate a sum term by term, or the commutativity and associativity of summation. Such properties are often taken for granted but they are not as obvious in more complex situations, as we shall see below.

4 Making a Step to Infinity

As we already noted, the dynamics of (5) is fully determined by the matrix of rates (in the probabilistic context we say the matrix of intensities, see e.g. [12, Chapter 2]). In this section we shall provide several examples showing that this is no longer true in an infinite dimensional context.

Let us assume now that the number of possible attributes is arbitrary and, to simplify the exposition, assume that an attribute can change only to the neighbouring one, that is, the attribute i can change to either $i-1$, or to $i+1$. This results in the so-called birth-and-death system of equations

$$\begin{aligned} u_1'(t) &= -b_1u_1(t) + d_2u_2(t), \\ u_n'(t) &= -(d_n+b_n)u_n(t) + b_{n-1}u_{n-1}(t) + d_{n+1}u_{n+1}(t), \quad n \geq 2, \\ u_n(0) &= \mathring{u}_n, \end{aligned} \tag{8}$$

where $u = (u_n)_{n\geq 1}$ and $(d_n)_{n\geq 1}$ (with $d_1 = 0$) and $(b_n)_{n\geq 1}$ are sequences of nonnegative and, in general, unbounded coefficients. To make further calculations more compact, we let $b_0 = 0$ and $u_0 = 0$ whenever necessary. Extending the finite-dimensional ideas, we are interested in controlling the total population and thus we will take the space

$$X = l_1 = \left\{u;\ \sum_{n=1}^{\infty} |u_n| < \infty\right\}$$

as the state space. We observe that the right hand side of (8) has the same structure as (5), that is, as in (7),

$$\begin{aligned} \left(\sum_{n=1}^{\infty} u_n\right)' &= \sum_{n=1}^{\infty} u_n' = \sum_{n=1}^{\infty}(-(d_n+b_n)u_n + d_{n+1}u_{n+1} + b_{n-1}u_{n-1}) \\ &= (-b_1u_1 + d_2u_2) + (-(d_2+b_2)u_2 + b_1u_1 + d_3u_3) + \ldots \\ &= (-b_1u_1 + b_1u_1) + (d_2u_2 - d_2u_2) + \ldots = 0. \end{aligned} \tag{9}$$

Here, however, we should be more cautious, since to obtain the above result we differentiate term by term an infinite series and rearrange the order of summation in an infinite sum. This is allowed only under specific conditions on u which, a priori, is not known. To illustrate this point, we discuss two simpler special cases of (8) describing, respectively, only death and only birth process. An analysis of the full equation, based on the fundamental paper [23], can be found in [4, Chapter 7], see also Sect. 7.3 of the present paper, while detailed results for (8) with the coefficients considered below can be found in [12, Sections 2.4.10–16 & 3.4.2–6].

4.1 A Death Equation: Multiple Solutions

Consider, for $t \geq 0$,

$$\begin{aligned} u_1'(t) &= 3^2 u_2(t), \\ u_n'(t) &= -3^n u_n(t) + 3^{n+1} u_{n+1}(t), \quad n \geq 2, \\ u_n(0) &= \mathring{u}_n. \end{aligned} \tag{10}$$

If we are interested only in coordinate-wise solvability of (10), we can use the integrating factors to re-write it as

$$\begin{aligned} u_1(t) &= \mathring{u}_1 + 3^2 \int_0^t u_2(s)ds, \\ u_n(t) &= \mathring{u}_n e^{-3^n t} + 3^{n+1} \int_0^t e^{-3^n(t-s)} u_{n+1}(s)ds, \quad n \geq 2. \end{aligned} \tag{11}$$

To find a solution, following [29] we observe that if $0 \leq \mathring{u}^N = (\mathring{u}_1, \ldots, \mathring{u}_N, 0, \ldots)$, then the solution u to (11) is given by the solution u^N to the finite dimensional system

$$\begin{aligned} u_1^N(t) &= \mathring{u}_1 + 3^2 \int_0^t u_2^N(s)ds, \\ u_n^N(t) &= e^{-3^n t}\mathring{u}_n + 3^{n+1} \int_0^t e^{-3^n(t-s)} u_{n+1}^N(s)ds, \quad 2 \leq n \leq N-1, \\ u_N^N(t) &= e^{-3^N t}\mathring{u}_N, \end{aligned} \tag{12}$$

where we agree to identify elements of $\mathbb{R}^N$ with their extensions by 0 in l_1. It is easy to see that for any $N \geq 1$, u^N is nonnegative and $\|u^N(t)\|_{l_1} \leq \|\mathring{u}\|_{l_1}$ for $t \geq 0$ (note that (12) no longer is conservative). Further, if we consider u^{N+1}, then

$$u_N^{N+1}(t) = e^{-3^N t}\mathring{u}_N + 3^{N+1} \int_0^t e^{-3^N(t-s)} u_{N+1}^{N+1}(s)ds$$

from where $u_N^{N+1}(t) \geq u_N^N(t)$ on account of $u^{N+1} \geq 0$. Thus, going down with n from N to 1, we obtain $u^N(t) \leq u^{N+1}(t)$ for $t \geq 0$ and the sequence $(\|u^N(t)\|)_{N\geq 1}$ is convergent as nondecreasing and bounded. Using the properties of the l_1 norm, for $M \geq N$ we have

$$\|u^M(t) - u^N(t)\|_{l_1} = |\|u^M(t)\|_{l_1} - \|u^N(t)\|_{l_1}|$$

and thus $(u^N(t))_{N\geq 1}$ is a Cauchy sequence in l_1 for any t. Thus,

$$\lim_{N\to\infty} u^N(t) = u(t) \tag{13}$$

in l_1 for some coordinate-wise measurable $t \mapsto u(t) \geq 0$. Now, let us fix $n \geq 2$ and take $N > n$. Then, since

$$|u_{n+1}^N(s)| \leq \|u^N(s)\|_{l_1} \leq \|\mathring{u}\|_{l_1},$$

the Lebesgue dominated convergence theorem allows us to pass to the limit

$$\begin{aligned} u_n(t) = \lim_{N\to\infty} u_n^N(t) &= e^{-3^n t}\mathring{u}_n + 3^{n+1} \lim_{N\to\infty} \int\limits_0^t e^{-3^n(t-s)} u_{n+1}^N(s)ds \\ &= e^{-3^n t}\mathring{u}_n + 3^{n+1} \int\limits_0^t e^{-3^n(t-s)} u_{n+1}(s)ds, \end{aligned}$$

which shows that u is a continuous coordinate-wise solution to (11), hence it is differentiable coordinate-wise and thus solves (10). Summarizing, for any $\mathring{u} \in l_1$ there is $\mathbb{R}_+ \ni t \mapsto u(t) \in l_1$ that satisfies $u(t) \geq 0$, $\|u(t)\|_{l_1} \leq \|\mathring{u}\|_{l_1}$ and such that for each $n \geq 1$, $u_n \in C^1(\mathbb{R}_+)$ and (10) is satisfied.

Remark 4.1 In fact, u has stronger properties that follow from the general theory presented in Sect. 6 but for our present goal the above are sufficient.

On the other hand, for $\lambda > 0$ consider the system

$$\begin{aligned} \lambda v_1^\lambda &= 3^2 v_2^\lambda, \\ \lambda v_n^\lambda &= -3^n v_n^\lambda + 3^{n+1} v_{n+1}^\lambda, \quad n \geq 2. \end{aligned} \tag{14}$$

It is easy to see that, for a given v_1^λ,

$$v_n^\lambda = \frac{\lambda v_1^\lambda}{3^n} \prod_{i=1}^{n-2} \left(1 + \frac{\lambda}{3^i}\right), \quad n \geq 2,$$

with the convention that $\prod_{i=1}^{0} := 1$. We see that if $v_1^\lambda > 0$, then $v_n^\lambda > 0$ for all $n \geq 1$, $v^\lambda = (v_n^\lambda)_{n\geq 1}$ is monotonically increasing and, since $\sum_{i=1}^{\infty} \frac{1}{3^i} < \infty$,

$$\lim_{n\to\infty} \prod_{i=1}^{n-2} \left(1 + \frac{\lambda}{3^i}\right) =: P > 0.$$

Thus

$$\sum_{n=1}^{\infty} v_n^\lambda = v_1^\lambda \left(1 + \lambda \sum_{n=2}^{\infty} \frac{1}{3^n} \prod_{i=1}^{n-2} \left(1 + \frac{\lambda}{3^i}\right)\right) \leq v_1^\lambda \left(1 + \frac{\lambda P}{6}\right)$$

and $v^\lambda \in l_1$ for any $\lambda > 0$. But then it is clear that

$$v(t) := e^{\lambda t} v^\lambda \tag{15}$$

is a coordinate-wise (and also l_1) differentiable solution to (10) with $v(0) = v^\lambda$ and final total size (the l_1 norm) for any t. Similarly to the above, we also have

$$\sum_{n=1}^{\infty} v_n^\lambda \geq v_1^\lambda \left(1 + \lambda \sum_{n=2}^{\infty} \frac{1}{3^n}\right) = v_1^\lambda \left(1 + \frac{\lambda}{6}\right),$$

thus

$$\|v(t)\|_{l_1} \geq e^{\lambda t} v_1^\lambda \left(1 + \frac{\lambda}{6}\right)$$

and hence $v(t)$ cannot be the bounded solution constructed in the first part of the section for the same initial condition v^λ.

4.2 A Birth Process: Breach of the Conservation Law

Example 4.2 Consider, for $t \geq 0$,

$$\begin{aligned} u_1'(t) &= -3u_1(t), \\ u_n'(t) &= -3^n u_n(t) + 3^{n-1} u_{n-1}(t), \quad n \geq 2, \\ u_n(0) &= \mathring{u}_n. \end{aligned} \tag{16}$$

In this case it is easy to construct recursively a unique nonnegative (for $\mathring{u} \geq 0$) coordinate-wise solution

$$u_1(t) = e^{-3t}\mathring{u}_1,$$
$$u_n(t) = e^{-3^n t}\mathring{u}_n + 3^{n-1}\int\limits_0^t e^{-3^n(t-s)}u_{n-1}(s)ds, \quad n \geq 2.$$

Again, column sums of the coefficient matrix in (16) are zero, so the expectation is that the solution satisfies

$$\sum_{n=1}^{\infty} u_n(t) = \|\mathring{u}\|_{l_1}. \tag{17}$$

We estimate the solution to (16) for $\mathring{u} = (1, 0, 0, \ldots)$. We obviously have

$$u_1(t) = e^{-3t},$$
$$u_2(t) = 3e^{-3^2 t}\int\limits_0^t e^{3^2 s - 3s}ds = \frac{3}{3^2-3}\left(e^{-3t} - e^{-3^2 t}\right) \leq \frac{3}{3^2-3}e^{-3t}$$

and, by induction,

$$u_n(t) \leq e^{-3t}\prod_{i=2}^{n}\frac{3^{i-1}}{3^i - 3} = \frac{e^{-3t}}{3^{n-1}}\prod_{i=2}^{n}\left(1 + \frac{1}{3^{i-1}-1}\right). \tag{18}$$

We see that

$$\lim_{n\to\infty}\prod_{i=2}^{n}\left(1 + \frac{1}{3^{i-1}-1}\right) =: P < \infty$$

in a monotonic way and thus

$$\|u(t)\|_{l_1} \leq e^{-3t}\left(1 + P\sum_{n=2}^{\infty}\frac{1}{3^{n-1}}\right) = e^{-3t}\left(1 + \frac{P}{2}\right).$$

This shows that, on the one hand, $u \in l_1$ but, on the other, (17) is not satisfied for large t.

Example 4.3 As another example we consider, for $t \geq 0$,

$$\begin{aligned} u_1'(t) &= -u_1(t), \\ u_n'(t) &= -nu_n(t) + (n-1)u_{n-1}(t), \quad n \geq 2, \\ u_n(0) &= \mathring{u}_n, \end{aligned} \tag{19}$$

where, again, $\mathring{u} = (1, 0, 0, \ldots)$. By direct calculation, we find $u_1(t) = e^{-t}$, $u_2(t) = e^{-t}(1 - e^{-t})$, $u_3(t) = e^{-t} = e^{-t} - 2e^{-2t} + e^{-3t} = e^{-t}(1 - e^{-t})^2$ and thus we make the inductive assumption

$$u_n(t) = e^{-t}(1 - e^{-t})^{n-1}. \tag{20}$$

Then

$$\begin{aligned} u_{n+1}(t) &= ne^{-(n+1)t} \int_0^t e^{ns}(1 - e^{-s})^{n-1} ds = ne^{-(n+1)t} \int_0^t e^s(e^s - 1)^{n-1} ds \\ &= ne^{-(n+1)t} \int_0^{e^t - 1} z^{n-1} dz = e^{-(n+1)t}(e^t - 1)^n = e^{-t}(1 - e^{-t})^n \end{aligned}$$

and hence formula (20) has been proved. Thus we see that

$$\sum_{n=1}^{\infty} u_n = e^{-t} \sum_{n=1}^{\infty} (1 - e^{-t})^{n-1} = 1 \tag{21}$$

and the solution is norm conserving. At the same time, estimating as in (18), we obtain

$$u_n(t) \leq e^{-t}, \quad n \geq 1,$$

and we see that the solution converges coordinate-wise to 0 (even uniformly in n). Compared with (21), this example once again emphasizes the fact that in infinite dimensional systems the coordinate-wise description of the evolution does not provide the full picture of the dynamics.

5 Between Model and Its Analysis

The presented examples indicate that in the infinite dimensional scenario it is not sufficient to consider a model verbatim, as derived in applied sciences, since such a simplistic approach often yields pathological outputs and renders the model ill-posed. Instead, we should carefully re-interpret the model in an adequate mathematical setting keeping, however, constantly in mind that by doing so we could lose some important features of the original formulation. Thus the developed theory should be related the original problem and be able to explain those pathologies, even if it does not cover them.

Thus, in what follows we describe such a mathematical formalization, following the exposition in [8, Section 4.1]. We should, however, remember that mathematical modelling is not mathematics (one cannot prove that a model is correct) and thus, until the model is fully formalized in a mathematical setting, the modelling consists of judicious steps rather than proofs.

As we have seen above, equations derived in applied sciences typically are formulated point-wise, that is, all operations, such as differentiation and integration, are understood in the classical calculus sense and the equation should be satisfied for all values of the independent variables:

$$\partial_t u(t,x) = [\mathcal{K}u(t,\cdot)](x), \quad u(0,x) = \mathring{u}(x), \quad x \in \Omega, \tag{22}$$

where $\mathcal{K}$ is a differential, integral, or functional expression operating on functions defined on some set Ω (with obvious modifications if Ω is denumerable). We have presented several examples, and there is plenty of others, see [8, Section 4.1], showing that with such an approach, (22) may become ill-posed even if in the modelling process we took into account all information characterizing the process. This can bee seen by noticing that the same Eq. (22) behaves well for some ranges of coefficients, e.g. (19), and displays pathological features for others, such as (16). Thus, to analyze (22) we have to reformulate it in a mathematically rigourous way.

Our choice is to describe the evolution of a system by a family of operators $(\mathcal{G}(t))_{t\geq 0}$ parameterised by time, called the dynamical system or, in the linear setting, a semigroup. The operators act in some state space X, mapping an initial state $\mathring{u}$ of the system to all subsequent states in the evolution, that is, solutions are represented as

$$u(t) = \mathcal{G}(t)\mathring{u}. \tag{23}$$

Here we note that we shall identify functions of two variables $(t,x) \mapsto u(t,x)$ with functions $t \to u(t)$ taking values in the space X of functions of the variable x; in the context of the applications discussed here, this is possible by [8, Section 3.1.2] and should not lead to any misunderstanding.

To be able to talk about an evolution of the system, that is, about a motion in X, the latter must be equipped with a notion of distance and, if we are interested in linear models, the distance should be consistent with the algebraic structure of X. Thus, our state space should be a Banach space which we choose partly due to its relevance in the problem and partly for its mathematical convenience. This choice is not unique, it is a mathematical intervention into the model.

In our examples we were interested in the total size of the population and thus we have chosen the space l_1 as the state space; in general L_1 spaces are used in similar models if x is a continuous variable. If, however, we were interested in controlling the maximal concentration of particles, a better choice would be some space with the supremum norm. On the other hand, in investigations of long time behaviour of population models with growth the most useful space is the L_1 space weighted with the eigenvector of the adjoint problem, see e.g. [15], while l_1 and L_1 spaces in which the higher order moments are finite allowed for proving much stronger results for fragmentation–coagulation problems, see [8, Chapters 5 & 8].

The choice of the space is, of course, not sufficient — all our 'pathological' examples live in the original state space.

Once we select the state space X, the right-hand side of (22) can be interpreted as an operator $\mathsf{K} : D(\mathsf{K}) \to X$ defined on some subset $D(\mathsf{K})$ of X such that $x \to [\mathcal{K}u](x) \in X$ for $u \in D(\mathsf{K})$.

With this, (22) can be written as the Cauchy problem for an ordinary differential equation in X: find $u \in D(\mathsf{K})$ such that $t \mapsto u(t)$ is differentiable in X for $t > 0$ and

$$\begin{aligned} \partial_t u &= \mathsf{K}u, \quad t > 0, \\ \lim_{t\to 0^+} u(0) &= \mathring{u} \in X. \end{aligned} \tag{24}$$

Unfortunately, the domain $D(\mathsf{K})$ is also not uniquely defined by the model. Discussing this problem and, in general, operator realizations of $\mathcal{K}$, we focus on the Master equation (3).

For (8), the matrix $\mathcal{A}$ is the diagonal, defined as $\mathcal{A}u = -((b_n + d_n)u_n)_{n\geq 1}$, while $\mathcal{B}u = (b_{n-1}u_{n-1} + d_{n+1}u_{n+1})_{n\geq 1}$ (remember the convention $b_0 = u_0 = 0$), both defined for u belonging to the space of all sequences l_0.

Possibly the operator which is the closest to the formal expression $\mathcal{A} + \mathcal{B}$ is the *maximal* operator $K_{\max}$ which is $\mathcal{A} + \mathcal{B}$ defined on

$$D(K_{\max}) = \{u \in l_1;\ \mathcal{A}u + \mathcal{B}u \in l_1\}.$$

Here, it is possible that neither $\mathcal{A}u$ nor $\mathcal{B}u$ belongs to l_1.

Another natural choice is to consider $\mathcal{A} + \mathcal{B}$ on a domain which ensures that both $\mathcal{A}u$ and $\mathcal{B}u$ are in l_1. Thus, we define A as $\mathcal{A}|_{D(A)}$ on

$$D(A) = \left\{u \in l_1;\ \sum_{n=1}^{\infty}(b_n + d_n)|u_n| < \infty\right\}. \tag{25}$$

Then for $0 \leq u \in D(A)$, similarly to (9),

$$\|\mathcal{B}u\|_{l_1} = \sum_{n=1}^{\infty}(d_{n+1}u_{n+1} + b_{n-1}u_{n-1}) = \sum_{n=1}^{\infty}(d_n u_n + b_n u_n) = \|Au\|_{l_1}, \tag{26}$$

where this time the rearrangement of the summation is justified by the absolute summability of the right-hand-side. Thus, extending by linearity, we see that $B = \mathcal{B}|_{D(A)}$ is well-defined and we introduce

$$K_{\min} = (\mathcal{A} + \mathcal{B})|_{D(A)} = A + B,$$

where both terms on the right act in l_1.

Example 5.4 Let us have a preliminary look at what the domain of the generator has to do with the well-posedness of (24). If a solution u to (24) satisfies $0 \leq u \in D(K_{\min}) = D(A)$, then, by the l_1 differentiability required in the definition and (25),

$$\partial_t \|u(t)\|_{l_1} = \sum_{n=1}^{\infty}(Au + Bu)_n(t) = \sum_{n=1}^{\infty}(Au)_n(t) + \sum_{n=1}^{\infty}(Bu)_n(t) = 0 \tag{27}$$

and (9) holds, that is, the solution is conservative and the model is honest. Note that the right hand side holds also if u is not positive but then the left hand side does not have the interpretation of the derivative of the norm.

The conservativeness may be extended to the case, when the positive solutions stay in $D(\overline{K_{\min}})$, where $\overline{K_{\min}}$ is defined as $\overline{K_{\min}}u = \lim_{n\to\infty}(Au_n + Bu_n)$ with $D(K_{\min}) \ni u_n \to u \in D(\overline{K_{\min}})$, whenever both limits exist. Then it is easy to see that if $u(t) \in D(\overline{K_{\min}})$ for any $t \geq 0$, then

$$\sum_{n=1}^{\infty}\overline{(A + Bu)}_n(t) = 0 \tag{28}$$

and (9) also holds.

But can we be sure that $\mathring{u} \in D(\overline{K_{\min}})$ yields $u(t) \in D(\overline{K_{\min}})$ for all $t > 0$? Or, at least, can we find a $D(\mathsf{K})$ such that any u emanating from $\mathring{u} \in D(\mathsf{K})$ stays in $D(\mathsf{K})$ for all $t > 0$ and, for (24) to make sense, is differentiable in X? This leads us to the concept of the generator which, as we shall see below, is forced upon us by semigroup theory. To explain this, we have to formalize the above discussion.

Definition 5.5 A family $(G(t))_{t\geq 0}$ of bounded linear operators on a Banach space X is called a C_0-semigroup, or a strongly continuous semigroup, if

- (i) $G(0) = I$;
- (ii) $G(t + s) = G(t)G(s)$ for all $t, s \geq 0$;
- (iii) $\lim_{t\to 0^+} G(t)u = u$ for any $u \in X$.

A linear operator K is called the (infinitesimal) generator of $(G(t))_{t\geq 0}$ if

$$Ku = \lim_{h\to 0^+} \frac{G(h)u - u}{h}, \tag{29}$$

with $D(K)$ defined as the set of all $x \in X$ for which this limit exists.

By *(iii)*, we see that for $u \in D(K)$ and $t \geq 0$, the right hand side derivative of $t \mapsto G(t)u$ satisfies

$$\begin{aligned}\partial_t G(t)u &= \lim_{h\to 0^+} \frac{G(t+h)u - G(t)u}{h} = G(t) \lim_{h\to 0^+} \frac{G(h)u - u}{h} = G(t)Ku \\ &= \lim_{h\to 0^+} \frac{G(h)G(t)u - G(t)u}{h} = KG(t)u.\end{aligned} \tag{30}$$

With a similar calculation for the left hand derivative and $t > 0$, see e.g. the proof of [28, Theorem 1.2.4], we see that $G(t)u \in D(K)$ for any $t > 0$, $t \mapsto G(t)u$ is differentiable in X and satisfies (24). We observe that if $u \in X \setminus D(K)$, then in general $t \mapsto G(t)u$ is only continuous and thus it does not solve (24). It solves, however, the integrated version of (24) and thus it is called an integral, or mild, solution.

Finding the generator K usually is a challenge; some methods for doing this will be presented later in the paper. Here we mention that for a large class of problems (see Remark 6.14) associated with (3), the generator K always satisfies $K_{\min} \subset K \subset K_{\max}$, see e.g. [4, Theorem 6.20] (though this is not always the case, as we shall see in Theorem 7.18). The place of K on the scale between $K_{\min}$ and $K_{\max}$ determines the well-posedness of the problem (22). It turns out that all the following situations are possible

1. $K_{\min} = K = K_{\max}$,
2. $K_{\min} \subsetneq K = \overline{K_{\min}} = K_{\max}$,
3. $K_{\min} = K \subsetneq K_{\max}$,
4. $K_{\min} \subsetneq K = \overline{K_{\min}} \subsetneq K_{\max}$,
5. $\overline{K_{\min}} \subsetneq K \subsetneq K_{\max}$.

Each of these cases has its own specific interpretation in the model. If $K \subsetneq K_{\max}$, we don't have uniqueness, that is, there are $C^1([0,\infty), X)$ solutions to (22) emanating from zero and therefore they are not described by the semigroup, such as the one constructed in Sect. 4.1:

there is more to life, than meets the semigroup.

If $\overline{K_{\min}} \subsetneq K$, then despite the fact that the transition operators is formally conservative, the solutions are not: the modelled quantity leaks out from the system and the mechanism of this leakage is not present in the model.

In the remaining part of the paper we shall present a theory explaining these phenomena.

6 Substochastic Semigroup Theory

6.1 Generation Theorems

According to (30), if we have a semigroup $(G(t))_{t\geq 0}$, we can uniquely identify its generator and the equation the semigroup solves. In practice, however, we are faced with an expression $\mathcal{K}$ in (22) and we have to identify its realization in some Banach space that generates a semigroup. Though there is no general way for doing this, at least, given X and a realization K of $\mathcal{K}$, the Hille–Yosida theorem allows us to determine whether it is a generator or not.

Let $R(\lambda, K) := (\lambda I - K)^{-1}$ denote the resolvent of K defined for $\lambda \in \rho(K)$, the resolvent set of K.

Theorem 6.6 ([28, Theorem 3.1]) *K is the generator of a semigroup $(G_K(t))_{t\geq 0}$ if and only if K is closed and densely defined and there exist $M > 0, \omega \in \mathbb{R}$ such that $(\omega, \infty) \subset \rho(K)$ and for all $n \geq 1, \lambda > \omega$,*

$$\|R(\lambda, K)^n\| \leq \frac{M}{(\lambda - \omega)^n}. \tag{31}$$

Despite its theoretical importance, practical applications of the Hille–Yosida theorem are very limited due to that fact that it requires solving infinitely many equations of increasing complexity and estimating their solutions. Fortunately, for the class of contractive semigroups that are important in applications the conditions can be somewhat simplified. In fact, the proof of the full Hille–Yosida theorem is reduced to the contractive case, albeit not in a very constructive manner.

Let us recall that the duality set of $u \in X$ is defined as

$$\mathcal{J}(u) = \{u^* \in X^*;\ \langle u^*, u\rangle = \|u\|^2 = \|u^*\|^2\}, \tag{32}$$

where X^* is the dual to X. Then we say that an operator $(K, D(K))$ is *dissipative* if for every $u \in D(K)$ there is $u^* \in \mathcal{J}(u)$ such that

$$\Re\langle u^*, Ku\rangle \leq 0. \tag{33}$$

Theorem 6.7 ([16, Theorem II.3.15]) *For a densely defined dissipative operator $(K, D(K))$ on a Banach space X, the following statements are equivalent.*

- *(a) The closure $\overline{K}$ generates a semigroup of contractions.*
- *(b) $\overline{\mathrm{Ran}(\lambda I - K)} = X$ for some (and hence all) $\lambda > 0$.*

If either condition is satisfied, then K satisfies (33) for any $u^* \in \mathcal{J}(u)$.

Thus, once we know that (33) is satisfied, then instead of finding the continuous inverses to $(\lambda I - K)^n$ for all sufficiently large λ and $n \geq 1$, it suffices to find a solution u to

$$\lambda u - Ku = f \tag{34}$$

for any f from a dense subset of X and some $\lambda > 0$.

6.2 How Do Multiple Solutions Fit into the Theory of Semigroups?

While semigroup theory ensures the uniqueness of solutions, this applies only to a particular realization of (24) in which on the right hand side we have the generator K of the semigroup. However, as we know, K is one of many versions of $\mathcal{K}$ and this explains the existence of some nul-solutions, that is, solutions emanating from zero, [22, Section 27.3], and hence explains multiple solutions such as (15). Indeed, let $(K, D(K))$ be the generator of a C_0-semigroup $(G(t))_{t\geq 0}$ on a Banach space X. To simplify notation we assume that $(G(t))_{t\geq 0}$ is a semigroup of contractions, hence $\{\lambda : \Re\lambda > 0\} \subset \rho(K)$. Let us further assume that there exists an extension $\mathcal{K}$ of K defined on the domain $D(\mathcal{K})$. We have the following basic result.

Lemma 6.8 ([2]) *If $\mathcal{K}$ is closed, then for any λ with $\Re\lambda > 0$,*

$$D(\mathcal{K}) = D(K) \oplus \operatorname{Ker}(\lambda I - \mathcal{K}). \tag{35}$$

If we equip $D(\mathcal{K})$ with the graph norm, then $D(K)$ is a closed subspace of $D(\mathcal{K})$.

Furthermore, if $\operatorname{Ker}(\lambda I - \mathcal{K})$ is finite-dimensional for some λ with $\Re\lambda > 0$, then $\mathcal{K}$ is closed.

To explain the meaning of the lemma, consider the Cauchy problem

$$\partial_t u = \mathcal{K}u, \qquad \lim_{t\to 0^+} u(t) = \mathring{u}. \tag{36}$$

Then $u^\lambda(t) = e^{\lambda t}v^\lambda$, where $v^\lambda \in \operatorname{Ker}(\lambda I - \mathcal{K})$, is a $C^1([0,\infty), X)$ solution to (36) with $\mathring{u} = v^\lambda$. However, as $v^\lambda \notin D(K)$, $t \to G(t)v^\lambda$ in general is not differentiable, and thus neither is $t \mapsto v(t) = G(t)v^\lambda - u^\lambda(t)$. Hence, v is only a mild nul-solution of (36). A classical solution u to (36) can be obtained, [22, Theorem 23.7.2], [4, Theorem 3.48], by taking $y(\lambda) \in \operatorname{Ker}(\lambda I - \mathcal{K})$, multiplied by a suitable scalar function of λ to insure its appropriate integrability along $\gamma \pm i\infty$ for any $\gamma > 0$, and taking its inverse Laplace transform,

$$u(t) = \frac{1}{2\pi i}\int_{\gamma - i\infty}^{\gamma + i\infty} e^{\lambda t} y(\lambda) d\lambda, \qquad \gamma > 0. \tag{37}$$

6.3 Perturbation Theory for Positive Semigroups

Even if verifying assumptions of Theorem 6.7 is easier than of Theorem 6.6, solving (34) can be a formidable task. Thus, in practice, we try other methods among which perturbation techniques play an important role. In this approach we try to write (24) as

$$\partial_t u = Au + Bu, \quad u(0) = \mathring{u}, \tag{38}$$

where A is an 'easy' operator for which the generation result is easy to prove and find conditions on B such that $A + B$ (or its extension) is also a generator. It is easy to see that if A generates a semigroup and B is bounded, then $A + B$ is also the generator of a semigroup but obviously this class of perturbations is to restrictive for most applications.

So far our discussion has not involved positivity aspects, apart from the observation that the solutions to the equations discussed in Sect. 4 should be coordinate-wise nonnegative if such is the initial condition. It turns out that, whenever a semigroup has this property, employing positivity can simplify many results.

Let X be a Banach lattice with partial order denoted by $\geq$. For any $Y \subset X$, $Y_+ := \{u \in Y;\ u \geq 0\}$. It is easy to see that l_1 with the coordinate-wise order is a Banach lattice. An operator $O : X \to X$ is called positive if $u \geq 0$ implies $Ou \geq 0$. A semigroup $(G(t))_{t\geq 0}$ on X is called positive if $G(t)u \geq 0$ for all $t \geq 0$ and $u \geq 0$. The Laplace transform representation of the resolvent on the one hand and the Hille formula, [28, Theorems 3.1 & 8.3], on the other, show that $(G(t))_{t\geq 0}$ is positive if and only if the resolvent of its generator is positive for all large λ.

In what follows we shall develop a theory suitable for examples discussed in Sect. 4, that is, we consider problems of the form (38) in a Banach lattice X, where $(A, D(A))$ is the generator of a substochastic (positive and contractive) semigroup $(G_A(t))_{t\geq 0}$ and $(B, D(A))$ is a positive operator, though some results pertain to a more general situation.

As follows from Theorems 6.6 and 6.7, the first step in proving a generation result for a contractive semigroup is finding solutions to the resolvent equation

$$(\lambda I - (A + B))u = f, \quad f \in X, \lambda > 0. \tag{39}$$

Knowing that $R(\lambda, A)$ exists for $\lambda > 0$, (39) can be formally re-written as

$$u - R(\lambda, A)Bu = R(\lambda, A)f \tag{40}$$

and we can recover u provided the Neumann series

$$R(\lambda)f := \sum_{n=0}^{\infty}(R(\lambda, A)B)^n R(\lambda, A)f = \sum_{n=0}^{\infty} R(\lambda, A)(BR(\lambda, A))^n f \tag{41}$$

is convergent. There are three possible cases.

1. *Easy case.* If the spectral radius of $BR(\lambda, A)$ satisfies

$$\rho(BR(\lambda, A)) < 1, \tag{42}$$

for some $\lambda > s(A)$, where $s(A)$ is the spectral bound of A (and, since $R(\lambda, A) \geq 0$, for all larger ones), then

$$u = R(\lambda, A) \sum_{n=0}^{\infty} (BR(\lambda, A))^n f \in D(A)$$

is the solution to (39). Moreover, in Banach lattices, (42) is also necessary for the positive invertibility of (39), see [31] (it can be seen that Banach lattices and positive operators on them satisfy the more general assumption in *op. cit.*, see e.g. [4, Section 2.2.3]). Thus $A + B$ is a good candidate for the generator.
2. *Slightly less easy case.* If

$$\lim_{n\to\infty} (BR(\lambda, A))^n f = 0 \tag{43}$$

for any $f \in X$, then $R(\lambda) = R(\lambda, \overline{A + B})$, see [4, Proposition 4.7]. In other words,

$$(\lambda I - \overline{A + B})u = f, \quad f \in X, \lambda > 0. \tag{44}$$

Hence, (38) cannot be solved as it is, but we can hope to solve a modification of it with $A + B$ replaced by $\overline{A + B}$. As we have seen in Example 5.4, some essential features of the dynamics are preserved in such a case.
3. *Neither of them.* While, in general, the series in (41) may fail to converge or, even if it converges, $R(\lambda)$ may fail to be the resolvent of a densely defined operator, see e.g. [11], we are interested in cases when $R(\lambda) = R(\lambda, K)$ for some $K \supset A + B$, see e.g. Theorem 6.11.

Example 6.9 For the death problem (10), we have

$$R(\lambda, A)u = \left(\frac{u_1}{\lambda}, \frac{u_2}{\lambda + 3^2}, \ldots, \frac{u_n}{\lambda + 3^n}, \ldots\right)$$

and $(Bu)_n = 3^{n+1}u_{n+1}, n \geq 1$. Thus

$$((BR(\lambda, A))^k u)_n = u_{k+n} \prod_{j=n+1}^{n+k} \frac{3^j}{\lambda + 3^j}$$

and, for $u \in l_1$,

$$\|(BR(\lambda, A))^k u\|_{l_1} \le \sum_{n=1}^{\infty} |u_{k+n}| \prod_{j=n+1}^{n+k} \frac{3^j}{\lambda + 3^j} = \sum_{r=k+1}^{\infty} |u_r| \prod_{j=r-k+1}^{r} \frac{3^j}{\lambda + 3^j}$$
$$\le \sum_{r=k+1}^{\infty} |u_r|$$

so that

$$\lim_{k\to\infty} BR(\lambda, A))^k u = 0.$$

Thus, $\overline{A+B}$ is a plausible candidate for the generator of a (conservative) death semigroup. On the other hand, for each fixed k let us consider the sequence $u^r = (\delta_{jr})_{j\ge 1}$. Then

$$\|(BR(\lambda, A))^k\| \ge \sup_{r\ge 1} \|BR(\lambda, A))^k u^r\|_{l_1} = \lim_{r\to\infty} \prod_{j=r-k+1}^{r} \frac{3^j}{\lambda + 3^j} = 1,$$

as the product consists of only k terms. Hence $\rho(BR(\lambda, A)) = 1$. On the other hand, if $A + B$ was the generator of a positive semigroup, then $\lambda I - (A + B)$ would be positively invertible for large λ. Then, however, by the comment under (42), $\rho(BR(\lambda, A)) < 1$. Thus $A + B$ cannot be the generator.

Example 6.10 Consider now the birth problem (16). Here

$$(R(\lambda, A)u)_n = \frac{u_n}{\lambda + 3^n}, \qquad n \ge 1,$$

and $Bu = (0, 3u_1, 3^2u_2, \ldots)$, thus

$$((BR(\lambda, A))^k u)_n = \begin{cases} 0 & \text{for } 1 \le n \le k, \\ u_j \prod_{r=j}^{j+k-1} \frac{3^r}{\lambda+3^r} & \text{for } n = j + k. \end{cases}$$

Hence, for $u \ge 0$,

$$\sum_{n=1}^{\infty}((BR(\lambda, A))^k u)_n = \sum_{j=1}^{\infty} u_j \prod_{r=j}^{j+k-1} \frac{3^r}{\lambda + 3^r}$$

Now, if we take $u = (1, 0, \ldots)$, then for any k

$$\|(BR(\lambda, A))^k u\|_{l_1} = \prod_{r=1}^{k} \frac{3^r}{\lambda + 3^r} \geq \prod_{r=1}^{\infty} \frac{3^r}{\lambda + 3^r} > 0,$$

where the product is positive on account of the summability of $(3^{-r})_{r\geq 1}$. This is consistent with the result of Example 4.2 and (28) that ascertained that the relevant semigroup cannot be generated by $\overline{A + B}$.

While in the first example we have the resolvent and a candidate for the generator, in the second case we have neither and thus we need a tool for handling situations in which $\rho(B(R(\lambda, A)) = 1$ since clearly they occur in important applications and result in interesting dynamics.

Such a tool have been developed by employing the order structure of the underlying state space and has its origins in the fundamental paper [23], devoted to the Kolmogorov system equations (which birth-and-death problems are special case of) in l_1. The main ideas, however, can be applied to a much broader class of problems.

Let us recall that a Banach lattice is called a Kolomogorov–Banach space (a KB-space) if every norm bounded and nondecreasing sequence is norm convergent. We have already used this property in Sect. 4.1. All reflexive, as well as L_1, spaces are KB-spaces.

Theorem 6.11 ([4, Theorem 5.2]) *Let X be a KB-space. If*

- *(A1)* $\rho(BR(\lambda, A)) \leq 1$ *for some* $\lambda > 0$,
- *(A2)* $\langle u^*, (A + B)u\rangle \leq 0$ *for any* $u \in D(A)_+, u^* \in \mathcal{J}(u)_+$,

then there is an extension $(K, D(K))$ of $(A + B, D(A))$ generating a positive C_0-semigroup of contractions, say, $(G_K(t))_{t\geq 0}$. The generator K satisfies, for $\lambda > 0$,

$$R(\lambda, K)f = \sum_{k=0}^{\infty} R(\lambda, A)(BR(\lambda, A))^k f. \tag{45}$$

Main Ideas of the Proof As we mentioned above, the order structure plays a crucial role in the proof. For $0 \leq r < 1$ we define $K_r = A + rB$, $D(K_r) = D(A)$. By (A1), $\rho(rBR(\lambda, A)) \leq r < 1$ and hence

$$R(\lambda, K_r) = R(\lambda, A) \sum_{n=0}^{\infty} r^n (BR(\lambda, A))^n, \tag{46}$$

where the series converges absolutely and each term is positive. Let $u^* \in \mathcal{J}(u)_+$. Then, by (A2), for $u \in D(A)_+$ and $r < 1$,

$$\langle u^*, K_r u\rangle = \langle u^*, (A + B)u\rangle + (r - 1)\langle u^*, Bu\rangle \leq 0;$$

that is, K_r are dissipative, and thus

$$\|(\lambda I - K_r)u\| \geq \langle u^*, (\lambda I - K_r)u\rangle = \lambda\langle u^*, u\rangle - \langle u^*, K_r u\rangle \geq \lambda\|u\|,$$

for all $u \in D(A)_+$. We can rewrite the above inequality as

$$\|R(\lambda, K_r)y\| \leq \lambda^{-1}\|y\| \tag{47}$$

for all $y \in X_+$ and, because $R(\lambda, K_r)$ are positive, (47) can be extended to the whole space X. As in Theorem 6.7, all these properties can be extended to $\lambda > 0$. Since we are in a KB-space, for each $f \in X_+$ there is $R(\lambda)f \in X_+$ such that

$$\lim_{r\to 1^-} R(\lambda, K_r)f = R(\lambda)f$$

in X. It follows that $R(\lambda) \geq 0$ is the resolvent of a densely defined operator K that is an extension of $A + B$ and then the generation is a consequence of the Trotter-Kato theorem, e.g. [28, Theorem 3.4.4]. □

Remark 6.12 The main drawback of Theorem 6.11 is that it does not provide any constructive information about K. We can have $K = A + B$, $K = \overline{A + B}$, or K could be an extension of $\overline{A + B}$.

Theorem 6.11 is close to [28, Theorem 3.3.4] (see the reformulation in [4, Theorem 4.12] and [6, Theorem 3.6]) which also requires $A + rB$ to be dissipative for $r \in [0, 1]$ and allows for $\|BR(\lambda, A)\| = 1$. It requires, however, B^* to be densely defined but, in return, contrary to Theorem 6.11, provides the characterization $K = \overline{A + B}$.

If X is reflexive and B closable, then B^* is densely defined, see [28, Lemma 1.10.5], and thus in such a case [4, Theorem 4.12] is stronger than Theorem 6.11. Thus, though there are examples when even in reflexive spaces the former is not applicable but the latter works, see [6, Example 3.8], the real power of Theorem 6.11 is revealed in L_1 spaces.

So, let us assume that (Ω, μ) is a measure space with σ-finite measure μ and $X = L_1(\Omega, d\mu)$.

Corollary 6.13 *If for all* $u \in D(A)_+$

$$\int_\Omega (Au + Bu)d\mu \leq 0, \tag{48}$$

then the assumptions of Theorem 6.11 are satisfied.

The reason for this simplification is the additivity of the norm on the positive cone of X. Indeed, since $R(\lambda, A)$ is a surjection from X onto $D(A)$, for $x \in X_+$ we have $u = R(\lambda, A)x \in D(A)_+$. Integrating

$$(A + B)u = -x + BR(\lambda, A)x + \lambda R(\lambda, A)x$$

we get

$$-\int_\Omega x\, d\mu + \int_\Omega BR(\lambda, A)x\, d\mu + \lambda \int_\Omega R(\lambda, A)x\, d\mu \le 0; \tag{49}$$

that is,

$$\lambda \|R(\lambda, A)x\| + \|BR(\lambda, A)x\| - \|x\| \le 0, \qquad x \in X_+, \tag{50}$$

from which $\|BR(\lambda, A)\| \le 1$, that is, assumption (A1) is satisfied.

Remark 6.14 Though, as mentioned before, we do not have any explicit characterization of K, [4, Theorem 6.20] ensures that K obtained in Corollary 6.13 satisfies $K \subset K_{\max}$.

Example 6.15 The strength of Corollary 6.13 lies in the fact that (48) is checked on the domain of the minimal operator.

In particular, for (8), (48) takes the form (9),

$$\sum_{n=1}^{\infty} (-(d_n + b_n)u_n + d_{n+1}u_{n+1} + b_{n-1}u_{n-1}) = 0, \tag{51}$$

where the rearrangements of the terms in the summation is justified as in (27) and the existence of the solution semigroup follows immediately. Again, by Remark 6.14, the coordinates of the obtained (matrix) semigroup satisfy (coordinate-wise) (8).

Another important characterization of $(G(t))_{t\ge 0}$ is that it is a minimal semigroup in the following sense.

Proposition 6.16 *Let D be a core of A. If $(\bar{G}(t))_{t\ge 0}$ is another positive semigroup generated by an extension of $(A + B, D)$, then $\bar{G}(t) \ge G(t)$, $t \ge 0$.*

This result is used e.g. to clarify the relation between the solution to (10) obtained in Sect. 4.1 and the one constructed using Corollary 6.13—they coincide, see [4, Proposition 7.10].

7 Honesty of the Semigroup and Characterization of Its Generator

The presentation in this section is based on [26], [32, Sections 2.2& 2.3], [33] and [8, Sections 4.10.2& 4.10.3].

7.1 The Three Functionals

The way forward comes from the realization that (48) in fact means

$$\int_{\Omega} (Au + Bu)\, d\mu = -c(u) \leq 0, \qquad u \in D(A)_+, \tag{52}$$

where c is a nonnegative linear functional on $D(A)_+$ (defined in fact on $D(A)$). On the other hand, we can define

$$\int_{\Omega} Kud\mu = -\hat{c}(u) \quad u \in D(K).$$

Since, by Theorem 6.13, $(G(t))_{t\geq 0}$ is contractive, that is, K is dissipative, we have $\hat{c} \geq 0$.

Functional c extends to $D(K)$ by monotonic limits and by continuity in the graph norm of $D(K)$, [8, Theorem 4.10.6],

$$\bar{c}(R(\lambda, K)f) := \sum_{n=0}^{\infty} c\left(R(\lambda, A)(BR(\lambda, A))^n f\right) \tag{53}$$

and there is $\beta_\lambda \in X_+^*$ such that

$$\hat{c}(R(\lambda, K)f) - \bar{c}\,(R(\lambda, K)f) = \langle \beta_\lambda, f \rangle. \tag{54}$$

The functional $\bar{c}$ is independent of λ, see [26]. The crucial characterization theorem, combining results of [8, 26, 32] reads as follows.

Theorem 7.17 *The following are equivalent:*

1. $K = \overline{A + B}$;
2. $\beta_\lambda \equiv 0$ *on* X *for some/all* $\lambda > 0$;
3. $\bar{c} = \hat{c}$ *on* $D(K)$;
4. $\mathrm{Ker}(\lambda I - (A + B)^*) = \{0\}$ *for some/all* $\lambda > 0$.

This result has several important ramifications. First, let us provide a result related to Remark 6.14, given in [32, Theorem 2.3.4].

Theorem 7.18 *Let the assumptions of Theorem 6.13 be satisfied. Then*

(a) If $K = \overline{A+B}$, then $(G_K(t))_{t\geq 0}$ is the unique substochastic semigroup whose generator K is an extension of $A+B$.
(b) If $K \not\supseteq \overline{A+B}$, then there are infinitely many substochastic semigroups generated by extensions of $A+B$.

The construction in the case b) is based on the observation that the functional $0 \leq C := \hat{c} - \bar{c}$, defined on $D(K)$, vanishes on $D(\overline{A+B})$ but is nonzero for $u \in D(K) \setminus D(\overline{A+B})$. For any fixed $f_0 \in X_+ \setminus \{0\}$ with $\|f_0\| \leq 1$, the operator

$$\tilde{K}u = Ku + C(u)f_0, \quad f \in D(K),$$

is the generator of a substochastic semigroup. We observe that the domains of the generators are $D(K)$. However, since $K \subset K_{\max}$, we see that $\tilde{K} \not\subseteq K_{\max}$ and thus it cannot be constructed using Corollary 6.13.

Let us now specify the concept of honesty for the current situation.

Definition 7.19 Let $I \subseteq [0, \infty)$ be an interval and let $\mathring{u} \in D(K)_+$.

(a) We say that the trajectory $\{G(t)\mathring{u}\}_{t\geq 0}$ is honest on I if $G_K(t)\mathring{u}$ satisfies

$$\frac{d}{dt}\|G(t)\mathring{u}\| = -\bar{c}\,(G(t)\mathring{u})\,, \quad t \in I. \tag{55}$$

(b) The trajectory is called honest if it is honest on $[0, \infty)$.
(c) The semigroup $(G(t))_{t\geq 0}$ is honest if all its trajectories are honest.

The dishonesty, that is, the amount of mass lost, can be measured by the defect function which can be defined for any $f \in X_+$,

$$\begin{aligned} \mathscr{d}_{\mathring{u}}(s,t) &:= \|G(t)f\| - \|G(s)f\| + \bar{c}\left(\int_s^t G(r)f\right) \\ &= \bar{c}\left(\int_s^t G(r)f\,dr\right) - \hat{c}\left(\int_s^t G(r)f\,dr\right) \leq 0. \end{aligned}$$

Rephrasing Theorem 7.17 in terms of honesty, we obtain

Theorem 7.20 *$(G(t))_{t\geq 0}$ is honest if and only if $K = \overline{A+B}$.*

7.2 Structure of Honesty

Defining a dishonest semigroup as a semigroup that is not honest, we see that for a semigroup to be dishonest it is sufficient if only one of its trajectories is dishonest on an arbitrarily short time interval. Let us then have a closer look at the structure of honest and dishonest trajectories. We denote

$$H_I := \{f \in X_+ \ : \ \{G(t)f\}_{t\geq 0} \text{ is honest on } I\}$$

with $H := H_{[0,\infty)}$. It turns out that H_I has a nice order structure.

Lemma 7.21 *If $f \in H_I$ and $g \in X_+$ satisfies $g \leq f$, then $g \in H_I$.*

Proof Indeed,

$$0 \leq (\hat{c} - \bar{c})\left(\int_s^t G(r)gdr\right) \leq (\hat{c} - \bar{c})\left(\int_s^t G(r)fdr\right) = d_f(s,t) = 0.$$

Hence, $d_g(s,t) = 0$ for all $s, t \in I$, $s \leq t$, and $g \in H_I$. □

With some more work we arrive at the full order theoretic characterization of the honesty set.

Theorem 7.22 ([26], [8, Proposition 4.10.22]) *For any interval $I \subset [0,\infty)$, the set*

$$\mathcal{H}_I := Span H_I = H_I - H_I$$

is a projection band in X. If $I = [a,\infty)$ for some $a \geq 0$, then $\mathcal{H}_I$ is invariant under $(G(t))_{t\geq 0}$.

Then the characterisation of projection bands, [10, Proposition 10.15], yields

Corollary 7.23 *There is a measurable set $\Omega_1 \subset \Omega$ such that $\mathcal{H}_I = L_1(\Omega_1)$.*

Another immediate consequence of Theorem 7.22 is

Corollary 7.24 *Let $(G(t))_{t\geq 0}$ be an irreducible semigroup and $a \geq 0$. If $H_{[a,\infty)} \neq \{0\}$, then $H_{[a,\infty)} = X_+$.*

In other words, for irreducible semigroups, either all trajectories are dishonest, or all are honest on $[0,\infty)$. Otherwise, we have

$$X = \mathcal{H} \oplus \mathcal{H}^d = L_1(\Omega_1) \oplus L_1(\Omega_2) \tag{56}$$

for some measurable set Ω_2, that is, if $X_+ \ni f > 0$ on Ω' with $\mu(\Omega_2 \cap \Omega') > 0$, then $\{G(t)f\}_{t\geq 0}$ is dishonest on some interval I. In general, we can say very little about the behaviour of trajectories originating from such initial conditions:

(1) once a trajectory becomes dishonest, it cannot recover (but it can continue as a honest trajectory if it becomes supported in Ω_1),
(2) if $(G(t))_{t\geq 0}$ is dishonest, then there is an initial condition $g \in X_+$ such that the trajectory is immediately dishonest.

Proposition 7.25 ([26]) *Assume that $(G(t))_{t\geq 0}$ is not honest. Then any trajectory $\{G(t)f\}_{t\geq 0}$, where $f \in X_+$ is such that $f > 0$ a.e. on Ω_2, see (56), is immediately dishonest.*

Corollary 7.26 *Let $(G(t))_{t\geq 0}$ be a dishonest irreducible semigroup. Then all trajectories are dishonest and, moreover, the trajectories originating from positive a.e. initial conditions are immediately dishonest.*

A more detailed information on the trajectories can be obtained for specific applications using, in particular, probabilistic methods, see [12] for Markov chains, or [20, 21] for fragmentation equations.

7.3 A Hunt for Honest Semigroups

While the above results provide a nice theoretical framework, they do not give a working tool to determine whether a semigroup is honest. There are several approaches to this problem from which we present one based on the concept of operator extensions. First we note the following consequence of Theorem 7.17.

Theorem 7.27 *The semigroup $(G_K(t))_{t\geq 0}$ is honest if and only if for any $u \in D(K)_+$ we have*

$$\int_\Omega Ku\, d\mu \geq -\bar{c}(u). \tag{57}$$

The statement follows from the fact that, by (54), $\hat{c} \geq \bar{c}$, and thus (57) implies $\hat{c} = \bar{c}$, giving the honesty by Theorem 7.17, part 3.

Condition (57) may seem useless as a tool for determining the honesty of $(G(t))_{t\geq 0}$ since it requires the knowledge of K itself. We note, however, that if we can prove (57) for an extension of K (such as $K_{\max}$), then it will hold for K.

Corollary 7.28 *If there exists an extension $\mathcal{K}$ of K and $\tilde{c}$ of $\bar{c}$ from $D(K)$ to $D(\mathcal{K})$ such that*

$$\int_\Omega \mathcal{K}u\, d\mu \geq -\tilde{c}(u), \tag{58}$$

for all $u \in D(\mathcal{K})_+$, then $K = \overline{A+B}$.

To illustrate these results, we consider the birth-and-death Eq. (8), where we assumed $b_n \geq 0$ for $n \geq 1$ and $d_n \geq 0$ for $n \geq 2$. By Example 6.15, there is a unique minimal substochastic semigroup $(G(t))_{t\geq 0}$ solving (8) and, by (51), we have $\bar{c} \equiv 0$. Furthermore, since $K \subset K_{\max}$, for $u \in D(K)_+$ we have

$$\begin{aligned}\int_\Omega K u d\mu &= \sum_{n=0}^{\infty}(-(b_n + d_n)u_n + b_{n-1}u_{n-1} + d_{n+1}u_{n+1}) \\ &= \lim_{n\to\infty} \sum_{k=0}^{n}(-(b_k + d_k)u_k + b_{k-1}u_{k-1} + d_{k+1}u_{k+1}) \\ &= \lim_{n\to\infty} (-b_n u_n + d_{n+1}u_{n+1}) =: -\hat{c}(u).\end{aligned} \tag{59}$$

First we look again at the death and birth semigroups.

Example 7.29 It is immediately seen that the death semigroup is always honest as then $b_n = 0, n \geq 1$, and hence

$$\int_\Omega K u d\mu \geq 0.$$

On the other hand, for the birth semigroup

$$\int_\Omega K u \, d\mu = - \lim_{n\to\infty} b_n u_n.$$

The limit on the right-hand-side is negative if, for instance, $u = (u_n)_{n\geq 1} = (b_n^{-1})_{n\geq 1}$ and this, by Theorem 7.27, would suffice for showing dishonesty if we could prove that $u \in D(K)$. For this we use Lemma 6.8. Considering, for $\lambda > 0$,

$$\begin{aligned}\lambda u_1 &= -b_1 u_1, \\ \lambda u_n &= -b_n u_n + b_{n-1}u_{n-1}, \quad n \geq 2,\end{aligned} \tag{60}$$

we see that $\mathrm{Ker}(\lambda I - K_{\max}) = \{0\}$ and hence $D(K) = D(K_{\max})$. Now, $u \in D(K_{\max})$ is characterized by

$$\sum_{n=2}^{\infty} |u_n| < \infty, \quad \sum_{n=2}^{\infty} |b_n u_n - b_{n-1}u_{n-1}| < \infty.$$

Since for $u = (b_n^{-1})_{n\geq 1}$ the second sum is identically 0, the birth semigroup is honest if and only if $(b_n^{-1})_{n\geq 1} \notin l_1$, confirming the calculations of Examples 4.2 and 4.3.

A similar, but obviously more involved, argument can be used to prove the following fundamental results for the full birth-and-death equation, that is, (8) with $b_n > 0$ for $n \geq 1$ and $d_n > 0$ for $n \geq 2$. It goes back, in a slightly weaker form, to [29] and have been proved in [4, Section 7.4] by semigroups tools.

Theorem 7.30 $K = \overline{A + B}$ *if and only if*

$$\sum_{n=0}^{\infty} \frac{1}{b_n} \left(\sum_{i=0}^{\infty} \prod_{j=1}^{i} \frac{d_{n+j}}{b_{n+j}} \right) = +\infty \tag{61}$$

(where we put $\prod_{j=1}^{0} = 1$*).*

Theorem 7.31 $K \neq K_{\max}$ *if and only if*

$$\sum_{n=1}^{\infty} \frac{1}{d_n} \prod_{j=1}^{n-1} \frac{b_j}{d_j} \left(\sum_{i=0}^{n-1} \prod_{r=1}^{i} \frac{d_r}{b_r} \right) < +\infty, \tag{62}$$

where, as before, $\prod_{j=1}^{0} = 1$.

By playing with coefficients, we can construct all cases listed in Section 5, see [4, Section 7.4]. In particular, $\overline{K_{\min}} \not\subseteq K \not\subseteq K_{\max}$ (Case 5) occurs as a combination of (10) and (16), that is, for (8) with $b_n = 2 \cdot 3^n$ and $d_n = 3^n$. The full example has been thoroughly analysed in [12, Sections 2.4.10–16 & 3.4.2–6].

Next we use the results of Sect. 7.2 to provide a more precise description of the birth-and-death dynamics.

Proposition 7.32 *If the minimal birth-and-death semigroup* $(G(t))_{t \geq 0}$ *is dishonest, then all trajectories are dishonest and the trajectories emanating from strictly positive initial conditions* $\mathring{u}$ *are immediately dishonest. Moreover, for such initial conditions* $\mathring{u}$, $t \mapsto \|G(t)\mathring{u}\|$ *is strictly decreasing.*

Proof First we observe that the semigroup $(G(t))_{t \geq 0}$ is irreducible. Indeed, by e.g. [10, Example 14.11], it is sufficient to show that $(R(\lambda, K)f)_n > 0$ for some $\lambda > 0$ and all $n \geq 1$, whenever $0 \neq f \geq 0$. We use the representation (45) where we observe, similarly to Example 6.9, that

$$\begin{aligned} ((BR(\lambda, A))^k u)_n = {} & \frac{d_{n+1}}{\lambda + d_n + b_n} ((BR(\lambda, A))^{k-1} u)_{n+1} \\ & + \frac{b_{n-1}}{\lambda + d_n + b_n} ((BR(\lambda, A))^{k-1} u)_{n-1}, \quad n \geq 1, k \geq 1, \end{aligned}$$

where, recall, $d_1 = b_0 = 0$. Let $u^r = (\delta_{jr})_{j \geq 1}$ for some $r \geq 1$. Then, since

$$R(\lambda, K) \geq \sum_{k=0}^{N} R(\lambda, A)(BR(\lambda, A))^k$$

for any N, $(R(\lambda, K)u)_n > 0$ for all $n = \max\{1, r-N\}, \ldots r+N$. Since N is arbitrary and any $0 \neq f \geq 0$ satisfies $f \geq \alpha u^r$ for some $r \geq 1$ and $\alpha > 0$, the statement is proved.

Then the first two statements of the proposition follow from Corollaries 7.24 and 7.26. To prove the last one we observe that since $G(t) \geq G_A(t)$, where $(G_A(t))_{t\geq 0}$ is the semigroup generated by the diagonal operator A, for such $0 \neq \mathring{u} \geq 0$, $(G(t)\mathring{u})_n > 0$ for any $t \geq 0$ and thus the claim follows again from Corollary 7.26. □

8 Conclusion

In this paper we have discussed a functional analytic explanation of seemingly pathological properties of some dynamical systems, such as the breach of the conservation laws and the non-uniqueness of solutions, using a class of Kolmogorov equations as an example. In the text we mentioned that similar phenomena occur, and can be explained in the same framework, in fragmentation processes, see [4, 8]. These examples certainly do not exhaust the list of cases in which such a behaviour occur. The existence of multiple solutions to the Cauchy problem has been observed in parabolic problems in non-smooth domains, [1]. The breach of conservation laws in Markov process has been studied at least since the work of Feller [18, 19], and recently it has been given a thorough overview and update in [12]. It is not restricted, however, to Markov chains but occurs in transport equations, see [9], as well as diffusion problems, see [25]. The latter paper is mostly concerned with the behaviour of the semigroup in the space of continuous functions where, by duality, the semigroup is conservative if the constant function 1 is invariant under its action. It is worthwhile to note that this property has been extensively studied in the context of quantum dynamical semigroups, [14].

References

1. J. Banasiak, On L_2-solvability of mixed boundary value problems for elliptic equations in plane nonsmooth domains. J. Differ. Equ. **97**(1), 99–111 (1992)
2. J. Banasiak, On a non-uniqueness in fragmentation models. Math. Methods Appl. Sci. **25**(7), 541–556 (2002)
3. J. Banasiak, Population models with projection matrix with some negative entries—a solution to the Natchez paradox. Bull. South Ural State Univ. Ser. Math. Model. Program. Comput. Softw. **11**(3), 18–28 (2018)
4. J. Banasiak, L. Arlotti, *Perturbations of Positive Semigroups with Applications*. Springer Monographs in Mathematics (Springer, London, 2006)
5. J. Banasiak, A. Falkiewicz, P. Namayanja, Asymptotic state lumping in transport and diffusion problems on networks with applications to population problems. Math. Models Methods Appl. Sci. **26**(2), 215–247 (2016)

6. J. Banasiak, M. Lachowicz, Around the Kato generation theorem for semigroups. Stud. Math. **179**(3), 217–238 (2007)
7. J. Banasiak, M. Lachowicz, *Methods of Small Parameter in Mathematical Biology*. Modeling and Simulation in Science, Engineering and Technology (Birkhäuser/Springer, Cham, 2014).
8. J. Banasiak, W. Lamb, P. Laurençot, *Analytic Methods for Coagulation-Fragmentation Models*, vols. 1&2 (CRC Press, Boca Raton, 2019)
9. J. Banasiak, S.C. Oukouomi Noutchie, R. Rudnicki, Global solvability of a fragmentation-coagulation equation with growth and restricted coagulation. J. Nonlinear Math. Phys. **16**(suppl. 1), 13–26 (2009)
10. A. Bátkai, M.K. Fijavž, A. Rhandi, *Positive Operator Semigroups: From Finite to Infinite Dimensions*, vol. 257 (Birkhäuser, Basel, 2017)
11. A. Bobrowski, *Convergence of One-Parameter Operator Semigroups. In Models of Mathematical Biology and Elsewhere*. New Mathematical Monographs, vol. 30 (Cambridge University Press, Cambridge, 2016)
12. A. Bobrowski, *Generators of Markov Chains: From a Walk in the Interior to a Dance on the Boundary*. Cambridge Studies in Advanced Mathematics. (Cambridge University Press, Cambridge, 2020)
13. C. Cercignani, *The Boltzmann Equation and Its Applications*. Applied Mathematical Sciences, vol. 67 (Springer, New York, 1988)
14. A.M. Chebotarev, F. Fagnola, Sufficient conditions for conservativity of minimal quantum dynamical semigroups. J. Funct. Anal. **153**(2), 382–404 (1998)
15. M. Doumic Jauffret, P. Gabriel, Eigenelements of a general aggregation-fragmentation model. Math. Models Methods Appl. Sci. **20**(5), 757–783 (2010)
16. K.-J. Engel, R. Nagel, *One-Parameter Semigroups for Linear Evolution Equations*. Graduate Texts in Mathematics, vol. 194 (Springer, New York, 2000)
17. K.-J. Engel, R. Nagel, *A Short Course on Operator Semigroups*. Universitext (Springer, New York, 2006)
18. W. Feller, Boundaries induced by non-negative matrices. Trans. Am. Math. Soc. **83**, 19–54 (1956)
19. W. Feller, On boundaries and lateral conditions for the Kolmogorov differential equations. Ann. of Math. (2) **65**, 527–570 (1957)
20. B. Haas, Loss of mass in deterministic and random fragmentations. Stoch. Process. Appl. **106**(2), 245–277 (2003)
21. B. Haas, Appearance of dust in fragmentations. Commun. Math. Sci. **2**(suppl. 1), 65–73 (2004)
22. E. Hille, R.S. Phillips, *Functional Analysis and Semi-Groups* (American Mathematical Society, Providence, 1974). Third printing of the revised edition of 1957, American Mathematical Society Colloquium Publications, Vol. XXXI
23. T. Kato, On the semi-groups generated by Kolmogoroff's differential equations.J. Math. Soc. Jpn **6**, 1–15 (1954)
24. J.L. Lebowitz, S.I. Rubinow, A theory for the age and generation time distribution of a microbial population. J. Math. Biol. **1**(1), 17–36 (1974/1975)
25. G. Metafune, D. Pallara, M. Wacker, Feller semigroups on $\mathbf{R}^N$. Semigroup Forum **65**(2), 159–205 (2002)
26. M. Mokhtar-Kharroubi, J. Voigt, On honesty of perturbed substochastic C_0-semigroups in L^1-spaces. J. Operator Theory **64**(1), 131–147 (2010)
27. A. Okubo, S.A. Levin, *Diffusion and Ecological Problems: Modern Perspectives*. Interdisciplinary Applied Mathematics, vol. 14, 2nd edn. (Springer, New York, 2001)
28. A. Pazy, *Semigroups of Linear Operators and Applications to Partial Differential Equations*. Applied Mathematical Sciences, vol. 44 (Springer, New York, 1983)
29. G.E.H. Reuter, W. Ledermann, On the differential equations for the transition probabilities of Markov processes with enumerably many states. Proc. Camb. Philos. Soc. **49**, 247–262 (1953)
30. M. Rotenberg, Transport theory for growing cell populations. J. Theor. Biol. **103**(2), 181–199 (1983)

31. J. Voigt, On resolvent positive operators and positive C_0-semigroups on AL-spaces. Semigroup Forum **38**(2), 263–266 (1989). Semigroups and differential operators (Oberwolfach, 1988)
32. C.P. Wong, *Kato's Perturbation Theorem and Honesty Theory*. Ph.D. Thesis, University of Oxford, 2015
33. C.P. Wong, Stochastic completeness and honesty. J. Evol. Equ. **15**(4), 961–978 (2015)

Real Positive Maps and Conditional Expectations on Operator Algebras

David P. Blecher

Dedicated to the memory of E. G. Effros and Coenraad Labuschagne

Abstract Most of this article is an expanded version of our talk at the Positivity X conference. It is essentially a survey, but some part, like most of the lengthy Sect. 5, is comprised of new results whose proofs are unpublished elsewhere. We begin by reviewing the theory of real positivity of operator algebras initiated by the author and Charles Read. Then we present several new general results (mostly joint work with Matthew Neal) about real positive maps. The key point is that real positivity is often the right replacement in a general algebra A for positivity in C^*-algebras. We then apply this to studying contractive projections ('conditional expectations') and isometries of operator algebras. For example we generalize and find variants of certain classical results on positive projections on C^*-algebras and JB algebras due to Choi, Effros, Størmer, Friedman and Russo, and others. In previous work with Neal we had done the 'completely contractive' case; we focus here on describing the real positive contractive case from recent work with Neal. We also prove here several new and complementary results on this topic due to the author, indeed this new work constitutes most of Sect. 5. Finally, in the last section we describe a related part of some recent joint work with Labuschagne on what we consider to be a good noncommutative generalization of the 'characters' (i.e. homomorphisms into the scalars) on an algebra. Such characters are a special case of the projections mentioned above, and are shown to be intimately related to conditional expectations. The idea is to try to use these to generalize certain classical function algebra results involving characters.

Blecher was partially supported by a Simons Foundation Collaboration grant 527078.

D. P. Blecher (✉)
Department of Mathematics, University of Houston, Houston, TX, USA
e-mail: dpbleche@central.uh.edu

E. Kikianty et al. (eds.), *Positivity and its Applications*, Trends in Mathematics,
https://doi.org/10.1007/978-3-030-70974-7_5

Keywords Operator algebra · Jordan operator algebra · Contractive projection · Conditional expectation · Real positive · Completely positive · Noncommutative Banach– Stone theorem · JC*-algebra

Mathematics Subject Classification (2010) Primary 17C65, 46L05, 46L51, 46L70, 47L05, 47L07, 47L30, 47L70; Secondary 46H10, 46B40, 46L07, 46L30, 46L52, 47L75

1 Introduction: Real Positivity

Positivity plays a key role in physics, in fact one could say it is intrinsic to the structure of the (quantum) universe. For lack of a better name, we shall use the term *quantum positivity* to refer to the positivity found in the 'standard model' of quantum mechanics, that is, positivity for operators on a complex Hilbert space H, or positivity in algebraic systems comprised of Hilbert space operators. It is indeed absolutely fundamental and pervasive in quantum physics, modern analysis, noncommutative geometry, and related fields. The associated order on selfadjoint operators is sometimes called the *Löwner order*:

$$S \leq T \qquad \text{if and only if} \qquad \langle S\zeta, \zeta\rangle \leq \langle T\zeta, \zeta\rangle \text{ for all } \zeta \in H.$$

There are several other well known characterizations of the positive cone in this order, i.e. for $T \geq 0$, for example in terms of the spectrum, or the numerical range, or in terms of a metric inequality, or an algebraic identity ($T = S^*S$, or $T = R^2$ where $R = R^*$), etc. Here the $*$ is the usual adjoint on $B(H)$.

The latter characterizations all make sense in any *C^*-algebra*, that is, a selfadjoint (that is, closed under the adjoint operation $*$) norm closed subalgebra of the bounded linear operators on H, or, more abstractly, a Banach $*$-algebra A satisfying the C^*-identity $\|x^*x\| = \|x\|^2$. We recall that a *von Neumann algebra* is a weak* closed C^*-algebra, or abstractly a C^*-algebra with a Banach space predual. These two classes of algebras are typically regarded as, respectively, *noncommutative topology* and *noncommutative measure theory*. Simplistically one could say that these are the kind of noncommutative topology and measure theory needed for quantum physics. The positivity and order above are one of the main ingredients of the vast theory of C^*-algebras and von Neumann algebras. There is a sense in which, explicitly or implicitly, 'quantum positivity' underlies almost every proof in C^*-algebra theory (see for example the texts [7, 58, 68]).

In an ongoing program (see e.g. [6, 8, 8, 16, 17, 19, 20, 22–24, 26, 27]), we have been importing some of this vast panorama of C^*-algebraic positivity for use in more general algebras (Banach algebras, nonselfadjoint operator algebras, Jordan operator algebras, etc). The usual theory of 'quantum positivity' in operator algebras is so spectacular and powerful it makes sense to make many of these tools available elsewhere. To do this we use *real positivity* systematically. The main goal of the present paper is to describe some of the very recent updates in this program. We do not prove many of the results stated here, just those that are not proved elsewhere. For example, most of Sect. 5 is proved here for the first time.

In the present article, H, K will denote Hilbert spaces over the complex field. We write $B(H)$ for the algebra of bounded linear operators $T : H \to H$. This is just the $n \times n$ matrix algebra M_n if H is finite dimensional. An operator $T \in B(H)$ is *real positive* (or *accretive*) if Re $T = (T + T^*)/2 \geq 0$. Again there are several other equivalent characterizations of real positive operators, for example that the numerical range lies in the closed right half plane, or they may be characterized by a metric inequality, or an algebraic identity, etc. See e.g. [8, Lemma 2.4]. The latter characterizations all make sense in any unital Banach algebra A (by unital we mean that it has an identity element of norm 1). We write $\mathfrak{r}_A$ for the 'cone' of real positive $T \in A$. In [8, 19–21] we study real positivity in Banach algebras and L^p-operator algebras. In [26] (written after the present survey was submitted) we consider real positivity in real operator algebras, and real positive real linear maps. We could have also reported here on the last three cited papers. However for the sake of not becoming too dispersed, in the present paper all of our algebras will be *operator algebras* or *Jordan operator algebras*.

For us an operator algebra is a norm-closed associative (but not necessarily selfadjoint) subalgebra of $B(H)$. These were characterized abstractly in [25], and much of their general theory may be found in [13]. A Jordan operator algebra is a norm-closed linear subspace $A \subset B(H)$ which is closed under the 'Jordan product' $a \circ b = \frac{1}{2}(ab + ba)$ (or equivalently, with $a^2 \in A$ for all $a \in A$). The theory of Jordan operator algebras in this sense is quite recent, and may be found in [17, 18, 27, 28, 70]. There is a much older theory of *Jordan C^*-algebras* (also called *JC^*-algebras*). These are the Jordan operator algebras $A \subset B(H)$ which are also closed under the involution of $B(H)$. Indeed JC^*-algebras are historically essentially amongst the first examples of 'operator algebras'. An Annals paper of Jordan, von Neumann and Wigner from 1934 on these nonassociative algebras [48] begins with the line "One of us has shown that the statistical properties of the measurements of a quantum mechanical system assume their simplest form when expressed in terms of a certain hypercomplex algebra which is commutative but not associative". Their hope was that such algebras "would form a suitable starting point for a generalization of the present quantum mechanical theory". These days JC^*-algebras are often viewed within the larger theory of JB^*-algebras [29, 30, 42]. We will view them within the class of Jordan operator algebras defined above.

Of course every operator algebra is a Jordan operator algebra. The latter algebras turn out to be the correct most general setting for many of the results below. Thus we state such results for Jordan operator algebras; the reader who does not care about nonassociative algebras should simply restrict to the associative case. Indeed the statement of many of our results contain the phrase '(Jordan) operator algebra', this invites the reader to simply ignore the word 'Jordan'. However a few of our new results do apply only to (associative) operator algebras.

The main principle for us is that real positivity is often the right replacement in a general algebra A for positivity in C^*-algebras. To be honest, there are many C^*-

subtheories or results in which this approach does not work. We focus here on some of the many settings where it does give something interesting and behaves well. Another main subtheme is that of 'conditional expectation', as we shall see below together with the relations between this theme and real positivity.

Turning to the structure of our paper, in Sect. 2 we begin by recalling the basics of real positivity from our work with Charles Read (actually most results are stated in the later and more general setting from [27]). In Sect. 3 we survey some new and foundational results from [18] concerning real positive maps, generalizing some aspects of the basic theory of positive maps on C^*-algebras [64, 66]. Considering such maps, as opposed to the 'completely positive' or 'completely contractive' case (terms defined below), forces one into the more general setting of Jordan operator algebras. For example, the range of a positive projection (i.e. idempotent linear transformation) on a C^*-algebra need not be again be isomorphic to a C^*-algebra (consider $\frac{1}{2}(x + x^T)$ on M_2), but it is always a Jordan operator algebra. Also, as one sees already in Kadison's Banach–Stone theorem for C^*-algebras [50], isometries of C^*-algebras relate to Jordan $*$-homomorphisms and not necessarily to $*$-homomorphisms. We recall that a Jordan algebra homomorphism is a map satisfying $T(a \circ b) = T(a) \circ T(b)$ (or equivalently, with $T(a^2) = T(a)^2$) for all $a, b \in A$.

Indeed in Sect. 4 we also describe a new Banach–Stone type theorem from [18] for nonunital isometries between Jordan operator algebras; characterizing such isometries in the spirit of Kadison's Banach–Stone theorem mentioned above. This result is needed in the proof of a theorem stated towards the end of Sect. 5, the characterization of symmetric real positive projections. It requires an analysis of 'quasi-multipliers' of (Jordan) operator algebras, a nontrivial link between the latter and quasi-multipliers of any generated C^*-algebra, a little known C^*-algebra theorem about quasi-multipliers due to Akemann and Pedersen, and some theory of Jordan multiplier algebras. In Sect. 5 we study real positive conditional expectations, and contractive (i.e. norm ≤ 1) and bicontractive projections, on operator algebras or Jordan operator algebras. In earlier work [16] we had considered completely contractive and completely bicontractive projections; the focus in Sect. 5 is how much of this is still true with the word 'completely' removed. Some of the main questions here concern generalization of famous results of Tomiyama and Choi and Effros on a C^*-algebra A: The range of a positive contractive projection P from A onto a C^*-subalgebra is a *conditional expectation*, by which we mean that $P(aP(b)) = P(a)P(b) = P(P(a)b)$ for all $a, b \in A$. (We discuss why these are called conditional expectations in Sect. 6, and also review some aspects of the history and theory of classical (probabilistic) conditional expectations there.) And even if $P(A)$ is not a subalgebra, it is still a C^*-algebra in the canonical product $P(ab)$ defined by P (assuming P is completely positive, otherwise we are in the Jordan situation mentioned in the last paragraph), and with respect to this product P is still a 'conditional expectation'. This is the Choi and Effros result [34], and the latter product is often called the Choi-Effros product or P-product. Pioneering results about contractive projections on JB$*$-algebras may be found in [59, Theorem 2.21] and [49, Corollary 1]. In Sect. 5 we consider similar questions for a real positive contractive projection P on an operator algebra or Jordan operator algebra: is it a conditional expectation if the range is a subalgebra or Jordan subalgebra, in the general case is $P(A)$ again an operator algebra or Jordan operator algebra in the canonical product defined by P, and is P a 'conditional expectation'? Some of Sect. 5 surveys a selection of some results from Sections 4–6 of [18]. However, as we said earlier, most of Sect. 5 of the present paper consists of new results and proofs that do not appear elsewhere.

In Sect. 6 we discuss the concept and history of 'conditional expectation', and review very briefly some features of the important but difficult theory of normal conditional expectations of von Neumann algebras. In the last section we describe some joint work with Labuschagne (from [11, 12]) on what we consider to be a good noncommutative generalization of the 'characters' (i.e. homomorphisms into the scalars) on an algebra. Such characters are a special case of the projections mentioned above. The idea is to try to use these to generalize certain classical function algebra results involving characters. We focus here on some aspects of this work that relate to material and results from earlier in the present article. For example, we generalize the von Neumann algebraic setting of conditional expectations, surveyed in Sect. 6 which involves an inclusion $D \subset M$ of von Neumann algebras, to inclusions $D \subset A \subset M$ for a weak* closed subalgebra A. This is a setting where one can find a positive (in the usual sense) extension to M of a real positive map on A which does not increase the range of the map. (The example to bear in mind is from Hardy space theory: $\mathbb{C}1 \subset H^\infty(\mathbb{D}) \subset L^\infty(\mathbb{T})$.)

2 Some General Results on Real Positivity

For operator algebras or Jordan operator algebras the definition of 'real positive' (i.e. $T + T^* \geq 0$) does not depend of the particular representation of the algebras. Indeed as we said earlier there are nice equivalent definitions of 'accretive' which make this point clear. For example, an element x is real positive if and only if $\|1 - tx\| \leq 1 + t^2\|x\|^2$ for all $t > 0$. Here, the 1 above is well-defined, for every nonunital nonselfadjoint (even Jordan) operator algebra A. This is essentially 'Meyers theorem': in [27, Section 2.2] we established using a result of Meyer that every Jordan operator algebra A has a unitization A^1 which is unique up to isometric Jordan homomorphism.

It is useful that the 'real positive cone' $\mathfrak{r}_A$ above has a stronger subcone $\mathbb{R}_+\mathfrak{F}_A$, where $\mathfrak{F}_A = \{a \in A : \|1 - a\| \leq 1\}$. This is a proper cone (in the sense that 0 is the only element $x \in A$ with both x and $-x$ in this cone). Again, the 1 here is well-defined, as we explained above.

In [22–24] Charles Read and the author began using and systematically developing real positivity in operator algebras, proving results like the following. We will however state the more general analogous results for Jordan operator algebras from [27]. Thus throughout the rest of this section A is a Jordan operator algebra.

Proposition 2.1 $\mathfrak{r}_A = \overline{\mathbb{R}_+\mathfrak{F}_A}$.

Sometimes one first has to prove a positivity result using $\mathfrak{F}_A$, and then use the last Proposition to generalize to all real positive elements.

We write *cai* for short for an approximate identity which is contractive (that is, of norm ≤ 1). A *Jordan contractive approximate identity* for a Jordan operator algebra is a net (e_t) of contractions with $e_t \circ a \to a$ for all $a \in A$. If A is an associative operator algebra then one can show that the existence of a Jordan cai implies existence of a (two-sided) cai.

Theorem 2.2 *A has a (Jordan) cai if and only if A has a real positive (Jordan) cai, and if and only if the real positive elements in A span A.*

In the Jordan algebra case this follows from [27, Theorem 4.1]. The (earlier) operator algebra case is in [22]; this result suggests that the operator algebras with a cai are for some purposes the good generalization of C^*-algebras (recall that all C^*-algebras have a cai). We call a (Jordan) operator algebra with a cai *approximately unital*. We focus on this class here, although there are many things one can do for algebras without a cai.

We recall that an operator $T : A \to B$ between C^*-algebras (or *operator systems*, that is, selfadjoint unital subspaces of C^*-algebras) is *completely positive* if $T(A_+) \subset B_+$, and similarly at the matrix levels. That is T acting entrywise takes $M_n(A)_+$ into $M_n(B)_+$ for all $n \in \mathbb{N}$. Recall that $M_n(A)$ is also a C^*-algebra if A is a C^*-algebra. Similarly T is *completely contractive* if this entrywise action on $M_n(A)$ is contractive for all n. A linear map $T : A \to B$ between (Jordan) operator algebras or unital operator spaces is *real completely positive* if $T(\mathfrak{r}_A) \subset \mathfrak{r}_B$ and similarly at the matrix levels (i.e. $\mathfrak{r}_{M_n(A)}$ is taken to $\mathfrak{r}_{M_n(B)}$ for all $n \in \mathbb{N}$).

(Trace preserving) completely positive maps are the standard way to take a measurement in quantum information theory. They appear explicitly or implicitly in the important definitions of *quantum channels*, or *POVM's*, or *quantum instruments* in that subject. Completely positive maps are also often viewed as the correct 'quantum analogue' of positive maps between function spaces. Much of the present paper however focuses on positivity as opposed to complete positivity. So mostly we will consider maps/isometries/projections that are simply *real positive*, i.e. $\mathfrak{r}_A$ is taken to $\mathfrak{r}_B$. This will be less interesting for some (certainly for the author, who usually works on things related to operator spaces, see e.g. [13]). We justify this for the present article for two reasons. First, positive maps are quite interesting, for example in the Jordan approach to quantum physics or entanglement, etc, even if they are less important in many applications. Though they may be poorer relatives, they are often fine fellows, and can go places that the other cannot. Moreover even in the theory of complete positivity, maps that are positive but not completely positivity often make an appearance. In the physics literature this happens for example in the study of entanglement where positive but not completely positivity maps are needed (i.e. necessary and sufficient) to test if a fixed given mixed state is entangled or separable [47]. There are papers with titles like "Who's afraid of not completely positive maps?" (by Sudarshan et al). Positive but not completely positive maps are sometimes important in the theory. Thus the theory of general positive maps deserves to be worked out, and that is the setting of much of the present article. The second justification for looking at positivity as opposed to complete positivity here is that a main goal of the present paper is to describe the most recent updates, and recently we have been investigating what transpires if the word ‘completely’ is dropped.

Theorem 2.3 ([6, 22]) *A (not necessarily unital) linear map $T : A \to B$ between C^*-algebras or operator systems is completely positive in the usual sense if and only if it is real completely positive. Also, T is real positive if and only if it is positive in the usual sense.*

Thus one does not lose anything from the (completely) positive theory when considering real (completely) positive maps.

Theorem 2.4 (Extension and Stinespring-Type Result, [6, 22, 27]) *A linear map $T : A \to B(H)$ on an approximately unital Jordan operator algebra or unital operator space is real completely positive if and only if T has a completely positive (in the usual sense) extension $\tilde{T} : C^*(A) \to B(H)$. Here $C^*(A)$ is any C^*-algebra generated by A. This is equivalent to being able to write T as the restriction to A of $V^*\pi(\cdot)V$ for a $*$-representation $\pi : C^*(A) \to B(K)$, and an operator $V : H \to K$.*

This last result generalizes to a larger class of algebras both the original Stinespring theorem [63], and the Arveson extension theorem for completely positive maps [5, 57]. Of course the range of $\tilde{T}$ will typically be much larger than the range of T. In the final pages of this article we will describe a setting where the extension $\tilde{T}$ surprisingly has the same range as T.

Since it is easy to see that every real positive scalar valued functional on such a space A is real completely positive, it follows that such real positive functionals are simply the restriction to A of positive scalar multiples of states (that is, positive norm 1 linear functionals) on $C^*(A)$. Thus we obtain a generalization of the famous GNS theorem: A functional $\varphi : A \to \mathbb{C}$ is real positive if and only if there is a $*$-representation $\pi : C^*(A) \to B(K)$, and a vector $\zeta \in H$ with $\varphi(x) = \langle \pi(x)\zeta, \zeta \rangle$ for $x \in A$.

The ordering induced on A by the real positive cone is obviously $b \preccurlyeq a$ iff $a - b$ is real positive. One may then go ahead and follow the C^*-theory by studying for example the order theory in the unit ball. Order theory in the unit ball of a C^*-algebra, or of its dual, is crucial in C^*-algebra theory. A feature of the first such result below is that having the order theory is possible if and only if there is a contractive approximate identity around. On the other hand the following couple of results are fairly obvious for unital algebras.

Theorem 2.5 *Let A be an (Jordan) operator algebra which generates a C^*-algebra B, and let $\mathfrak{U}_A = \{a \in A : \|a\| < 1\}$. The following are equivalent:*

(1) *A is approximately unital.*
(2) *For any* positive $b \in \mathfrak{U}_B$ *there exists a real positive a with $b \preccurlyeq a \preccurlyeq 1$.*

(3) *For any pair* $x, y \in \mathfrak{U}_A$ *there exists a real positive contraction* a *with* $x \preccurlyeq a$ *and* $y \preccurlyeq a$.
(4) *For any* $b \in \mathfrak{U}_A$ *there exists a real positive contraction* a *with* $-a \preccurlyeq b \preccurlyeq a$.
(5) *For any* $b \in \mathfrak{U}_A$ *there exists a real positive contractions* x, y *with* $b = x - y$.
(6) $\mathfrak{r}_A$ *is a generating cone (that is,* $A = \mathfrak{r}_A - \mathfrak{r}_A$*).*

The real positive elements above may be chosen 'nearly positive' and in $\frac{1}{2}\mathfrak{F}_A$.

Corollary 2.6 *Thus if an approximately unital (Jordan) operator algebra* A *generates a* C^**-algebra* B*, then* A *is* order cofinal *in* B*: given* $b \in B_+$ *there exists* $a \in A$ *with* $b \preccurlyeq a$*. Indeed we can do this with* $b \preccurlyeq a \preccurlyeq \|b\| + \epsilon$.

We will use this result later.

We recall that the positive part of the open unit ball $\mathfrak{U}_B$ of a C^*-algebra B is a directed set, and indeed is a net which is a positive cai for B. The following generalizes this to operator algebras:

Corollary 2.7 *If* A *is an approximately unital (Jordan) operator algebra, then the real positive strict contractions, indeed the set of real positive elements* $\{a \in A : \|a\| < 1, \|1 - 2a\| \leq 1\}$*, is a directed set in the* $\preccurlyeq$ *ordering, and with this ordering this set is an increasing cai for* A.

It is also interesting to consider order theory in the dual of an (Jordan) operator algebra but we will not do so here (see e.g. [24]). In most of the remainder of the paper we turn to real positive maps and projections on operator algebras.

3 Real Positive Maps

We recall again that a linear map $T : A \to B$ is real positive if $T(\mathfrak{r}_A) \subset \mathfrak{r}_B$. We saw in Theorem 2.3 that the real positive maps on C^*-algebras or operator systems are just the positive maps in the usual sense. The following recent results are a sample from Section 2 in [18]. We have chosen to include several of these results partly because they are used in later theorems, and it is helpful to see how the theory builds on itself. We also remark that there are (historically earlier) operator space versions of most of the following results (with the word 'completely' added in many places), but that is not our focus here. The interested reader can find these in our papers referenced in the introduction, e.g. [16].

Proposition 3.1 *If* $T : A \to B$ *is a real positive linear map between unital (resp. approximately unital) Jordan operator algebras then* T *is bounded and* $\|T\| = \|T(1)\|$ *(resp.* $\|T\| = \sup_t \|T(e_t)\| = \|T^{**}(1)\|$*), if* (e_t) *is a (Jordan) cai for* A.

For us a contraction or contractive map means it has norm ≤ 1.

Proposition 3.2 *Contractive homomorphisms (resp. Jordan homomorphisms) between approximately unital operator algebras (resp. Jordan operator algebras) are real positive.*

A unital contraction between unital (Jordan) operator algebras is real positive.

Theorem 3.3 *Let A and B be approximately unital (Jordan) operator algebras, and write A^1 for a (unique, as we said at the start of Sect. 2) unitization of A. If A is unital choose $A^1 = A \oplus^\infty \mathbb{C}$ as usual (we do not necessarily make this requirement on B). A real positive contractive linear map $T : A \to B$ extends to a unital real positive contractive linear map from A^1 to B^1.*

Thus real positive contractions from A to B are just the restrictions of unital contractions from A^1 to B^1. This 'is a theorem' because it seems quite nontrivial. Indeed our proof uses the earlier nontrivial 'order theoretic' fact of cofinality in Corollary 2.6. Moreover this theorem is foundational and seems to be exceedingly useful, very often permitting a 'reduction to the unital case'. This trick may not have been noticed for example in the Jordan C^*-algebra literature, where sometimes complicated new arguments are used to deal with approximately unital algebras rather than a quick appeal to the unital case.

Remark 3.4 The last theorem is not true in general if A is not approximately unital. For example, consider $A_0(\mathbb{D})$, the continuous functions on the disk that are analytic inside the disk and which vanish at 0. If Re $f \geq 0$ and $f(0) = 0$ then Re f is identically 0 by the maximum modulus theorem for harmonic functions. Hence f is constant and zero. Thus the map $f \mapsto f'(0)$ is trivially real positive, and it is a contraction by the Schwarz inequality. However, the unital extension of the latter map is not a contraction, or equivalently is not positive. Indeed the states on $A(\mathbb{D}) = A_0(\mathbb{D})^1$ are integrals against probability measures μ on the circle. Nonetheless, the existence of a positive measure μ on the circle with $\int_{\mathbb{T}} z^n \, d\mu = 0$ for $n \geq 2$, and $\int_{\mathbb{T}} z \, d\mu = 1$, is ruled out by the solution to the well known trigonometric moment problem, since e.g. the 3×3 Toeplitz matrix whose entries are all 1 except for zeroes in the 1-3 and 3-1 entries is not positive. Thus the theorem fails in this case.

The following result, whose proof relies in an interesting way on Theorem 3.3, is useful for questions about real positivity because it shows that we can often get away with working with the simpler set $\mathfrak{F}_A = \{x \in A : \|1 - x\| \leq 1\}$.

Corollary 3.5 *A linear map $T : A \to B$ between approximately unital (Jordan) operator algebras is real positive and contractive if and only if $T(\mathfrak{F}_A) \subset \mathfrak{F}_B$.*

Corollary 3.6 *Let M be a unital weak* closed operator algebra, $\Phi : M \to M$ be a weak* continuous real positive contraction, and let M^Φ be the weak* closed unital subspace of fixed points of Φ. Then there exists a real positive contractive projection on M with range M^Φ.*

The space M^Φ is sometimes called a *Poisson boundary*. We assign M^Φ the new product $\Phi(xy)$ (or $\Phi(x \circ y)$ in the Jordan case). It follows from the idea in the next proof and results such as Theorems 5.5 and 5.9 and others in Sect. 5 below that M^Φ

is typically a Jordan operator algebra. Conversely, it may be an interesting question as to whether all Jordan operator algebras arise in this way as fixed points of real positive contractions.

Corollary 3.7 *Let M be a unital weak* closed operator algebra (resp. Jordan operator algebra), and let $\Phi : M \to M$ be a weak* continuous real completely positive complete contraction. Then there exists a real completely positive completely contractive projection on M with range M^Φ, and M^Φ with its new product is a unital dual operator algebra in the sense of [13, Section 2.7] (resp. is a unital Jordan operator algebra).*

Proof One may follow the proof in [37, Corollary 1.6], taking weak* limits in the unit ball of $CB(M, M) = (M\hat{\otimes}M_*)^*$ of averages of powers of Φ. One obtains a completely contractive projection P on M with range M^Φ. It is an exercise in weak* approximation that P is real positive, and similarly it is real completely positive. If M is a unital operator algebra then M^Φ is an operator algebra in the P-product by [16, Theorem 2.5], with identity $P(1)$. Since M^Φ consists of the fixed points of Φ it is weak* closed. Hence it is a dual operator space, thus a dual operator algebra in the sense of [13, Section 2.7], by Theorem 2.7.9 in that reference. The Jordan case follows similarly from [18, Theorem 4.18]. □

A similar result holds for weak* continuous complete contractions using Theorem 5.2 (1) in the proof in place of [16, Theorem 2.5].

The following is a nonselfadjoint analogue of the well known fact that the positive part of the kernel of a positive map T on a C^*-algebra B has the following 'ideal-like' property:

$$T(xy) = T(yx) = 0, \qquad y \in \text{Ker}(T)_+, \ x \in B.$$

Note that the entire kernel is rarely an ideal.

Lemma 3.8 *Suppose that A is an approximately unital operator algebra and that $T : A \to B(H)$ is a real positive map on A. If $x \in A$ and $y \in \mathfrak{r}_A \cap Ker(T)$ then xy and yx are in $Ker(T)$.*

The following, which is needed later e.g. for Theorems 5.5 and 5.20, is a sample corollary of this:

Corollary 3.9 *If A is an approximately unital Jordan operator algebra, $T : A \to B(H)$ is real positive, and if $J = Ker(T)$ is contained in the closed Jordan subalgebra generated by the real positive elements that J contains, then $Ker(T)$ is an approximately unital Jordan ideal in A.*

In fact one may replace 'Jordan subalgebra' in the last result with 'Jordan hereditary subalgebra'. A Jordan hereditary subalgebra of a Jordan operator algebra A is a closed approximately unital Jordan subalgebra D satisfying $dAd \subset D$ for all $d \in D$.

The following result, whose proof benefits from some ideas from Lemmas 2.8 and 4.6 in [18], will be also used later in Sect. 5.

Lemma 3.10 *Let A be a unital operator space (resp. approximately unital Jordan operator algebra), and let $T : A \to B(H)$ be a unital (resp. real positive) contraction. Suppose that e is a projection in A with $e \circ A \subset A$, such that $q = T(e)$ is a projection in $B(H)$. Then $T(eae) = qT(a)q$ and $T(a \circ e) = T(a) \circ q$ for all $a \in A$.*

Proof By Theorem 3.3 we can unitize if necessary. So we may assume that $T(1) = 1$ (since $e \circ A^1 \subset A^1$). Let $S = qT(\cdot)q$. Then S is real positive and $S(1) = q = S(e)$. By Lemma 2.8 in [18], $S(a) = qT(a)q = qT(eae)q$.

For any $a \in \mathrm{Ball}(A)$ we have $\|1 - e \pm eae\| \leq 1$, so that $\|1 - q \pm T(eae)\| \leq 1$. Hence $\|q^{\perp} \pm T(eae)\| \leq 1$. Since $q^{\perp}$ is an extreme point of $q^{\perp}B(H)q^{\perp}$ we see that $q^{\perp}T(eae)q^{\perp} = 0$. Looking at the matrix of $T(eae)$ with respect to $q^{\perp}$, and using $\|q^{\perp} \pm q^{\perp}T(eae)q^{\perp}\| \leq 1$, we also see that $T(eae) = qT(eae)q$. So we have proved that $T(eae) = qT(a)q$. Similarly, $T(e^{\perp}ae^{\perp}) = q^{\perp}T(a)q^{\perp}$.

The second identity follows from the facts in the last line, and from the identity $a \circ q = \frac{1}{2}(a + qaq - q^{\perp}aq^{\perp})$, and the similar identity with q replaced by e. □

The last few results are a sample of recent results on real positive maps (mostly from [18]). Almost all of these particular results are ingredients of proofs of results featured in the rest of the paper.

4 Quasimultipliers and a Banach–Stone Theorem

There are very many Banach–Stone type theorems in the literature. For example, we already mentioned Kadison's result that surjective linear isometries between C^*-algebras are precisely the maps $u\pi(\cdot)$ for a surjective Jordan $*$-isomorphism π and unitary multiplier u [50]. In particular linearly isometrically isomorphic C^*-algebras are Jordan $*$-isomorphic. By spectral theory (by a result of Harris [46, Proposition 3.4] if necessary, or see e.g. Proposition 3.4.4 in [29]) one can see that the converse is true, Jordan $*$-isomorphic JC^*-algebras are isometrically isomorphic.

Theorem 4.1 (Arazy-Solel [3]) *Surjective unital linear isometries between unital (Jordan) operator algebras are Jordan homomorphisms.*

A variant of this due to Arazy where the isometry need not take 1 to 1 may be found in [2, Theorem 3.1].

Theorem 4.2 ([27]) *An isometric surjection T between approximately unital (Jordan) operator algebras is real positive if and only if T is a Jordan algebra homomorphism.*

We now wish to extend Arazy and Solel's result above to the case of nonunital surjective isometries $T : A \to B$ between approximately unital (Jordan) operator algebras. This is needed in a theorem towards the end of the section, the characterization of symmetric real positive projections. It turns out that the 'obvious thing' is wrong in the Jordan operator algebra case. Although 'unitally linearly isometrically isomorphic' unital JC^*-algebras are Jordan $*$-isomorphic (this follows e.g. from the next theorem, although it is much older of course, due again to Harris [44]), there exist linearly isometric unital JC^*-algebras which are not Jordan isomorphic (see e.g. [29, Antitheorem 3.4.34 and Corollary 3.4.76]).

Proposition 4.3 *There exist linearly completely isometric unital JC^*-algebras which are not Jordan isomorphic. There exist unital JC^*-algebras which are Jordan $*$-isomorphic but not linearly completely isometric.*

Proof The second statement has a fairly obvious putative counterexample: take the map $d(x) = (x, x^{\mathsf{T}})$ from M_n to $M_n \oplus M_n$. The range is a unital JC^*-algebra and d is a Jordan $*$-isomorphism. Here is one way to see that the range R is not completely isometric to M_n. By way of contradiction, suppose that $u : M_n \to R$ is a linear complete isometry. Then it is easy to see that

$$\|[x_{ij}]\| = \|[u(x_{ij})]\| = \|[u(x_{ij})]^{\mathsf{T}}\| = \|[u(x_{ji})]\| = \|[x_{ji}]\| = \|[x_{ij}^{\mathsf{T}}]\|,$$

for $[x_{ij}] \in M_m(M_n)$). In other words, 'transpose' is a complete isometry on M_n. But this is well known to be false (take x_{ij} above to be the usual matrix unit basis of M_n).

For the first statement we just need to show that the two algebras in [29, Antitheorem 3.4.34] are actually completely isometric. These two algebras are two copies of a single Jordan $*$-subalgebra A of a C^*-algebra C, but the second copy B is given the Jordan product $\frac{1}{2}(xu^*y + yu^*x)$, and involution ux^*u, for a unitary u in C with u and u^* in A. Right multiplication by u^* on C restricts to a completely isometric Jordan $*$-isomorphism (since $(xu^*)^* = (ux^*u)u^*$) from B onto the Jordan $*$-subalgebra Au^* of C. (See also [18, Lemma 3.1].) Thus A and the latter Jordan $*$-subalgebra are completely isometric, but not Jordan isomorphic (since A and B are not Jordan isomorphic). □

Therefore Banach–Stone theorems for nonunital isometries between Jordan operator algebras are not going to look quite as one might first expect: one cannot expect the Jordan isomorphism appearing in the conclusion to map onto the second C^*-algebra exactly. Nonetheless we obtain a reasonable Banach–Stone type theorem for nonunital isometries between Jordan operator algebras. This Banach–Stone type theorem (from [18]) plays a crucial role in one of the theorems in the next section (Theorem 5.19). One of the main steps is to show that for T as above, $T^{**}(1)$ is in the Jordan multiplier algebra $JM(B) = \{x \in B^{**} : x \circ B \subset B\}$. But this is not at all clear, and requires an analysis of 'quasi-multipliers' of (Jordan) operator algebras, a nontrivial link between the latter and quasi-multipliers of any generated C^*-algebra, a little known C^*-algebra theorem about quasi-multipliers due to Akemann and Pedersen, and some theory of Jordan multiplier algebras. We define a *quasimultiplier* of B to be an element $w \in B^{**}$ with $bwb \in B$ for all

$b \in B$. Some of the steps in the proof of the next theorem are: showing that $T^{**}(1)$ gives rise to a quasimultiplier of B, and that such quasimultipliers are in $JM(B)$ (which uses several of the ingredients mentioned a few lines back).

Let $\Delta(A) = A \cap A^*$, sometimes called the 'diagonal' of A. One can show that this is a well defined C^*-algebra (or JC^*-algebra) independently of the particular representation of A on a Hilbert space [27]. If B is a C^*-algebra then $JM(B)$ is just the usual C^*-algebraic multiplier algebra $M(B)$. This follows from e.g. [30, Proposition 5.10.96]. We recall that for an approximately unital operator algebra D, the multiplier algebra $M(D)$ may be defined to be the unital operator algebra $\{x \in D^{**} : xD + Dx \subset D\}$.

We recall that a JW^*-algebra is a weak* closed JC^*-subalgebra of $B(H)$ (or of a von Neumann algebra). The bidual of a JC^*-algebra A is a JW^*-algebra. Indeed it is a weak* closed JC^*-subalgebra of B^{**} if A is a JC^*-subalgebra of a C^*-algebra B.

Theorem 4.4 *Suppose that $T : A \to B$ is an isometric surjection between approximately unital Jordan operator algebras. Suppose that B is a Jordan subalgebra of an (associative) operator algebra D, and that B generates D as an operator algebra. Then there exists a unitary $u \in \Delta(JM(B))$ which is also in $\Delta(M(D))$, and there exists an isometric Jordan algebra homomorphism $\pi : A \to D$, such that $\pi(A) = Bu^*$ is a Jordan subalgebra of D, and*

$$T = \pi(\cdot)u.$$

As is usual with noncommutative Banach–Stone theorems, we can also write the unitary u on the left: $T = u\,\theta(\cdot)$ (indeed simply set $\theta = u^*\pi(\cdot)u$).

In the unital case the following consequence for associative operator algebras also follows from [2, Theorem 3.1] (see also [31, Proposition 3.12]):

Corollary 4.5 *Suppose that $T : A \to B$ is an isometric surjection between approximately unital operator algebras. Then there exists a unitary $u \in \Delta(M(B))$ and there exists an isometric Jordan algebra homomorphism $\pi : A \to B$, such that $T = u\pi(\cdot)$.*

Of course if T is a complete isometry in the last results then so will be π.

5 Contractive Projections on Operator Algebras

The following theorem (essentially due to Effros and Størmer [37] in the case that $P(1) = 1$) shows what happens in the case of selfadjoint Jordan operator algebras (JC^*-algebras, see e.g. [29, 42]). The reader could take A to be a C^*-algebra if they wish. The new case of the theorem, i.e. the case that A is nonunital, or that $P(1) \neq 1$, can be dealt with by passing to the unitization by using Theorem 3.3.

Theorem 5.1 *If $P : A \to A$ is a positive contractive projection on a JC^*-algebra A then $P(A)$ is a JC^*-algebra in the new product $P(x \circ y)$, P is still positive as a map into the latter JC^*-algebra, and*

$$P(P(a) \circ P(b)) = P(a \circ P(b)) \ , \qquad a, b \in A \, .$$

If in addition $P(A)$ is a Jordan subalgebra of A then P is a Jordan conditional expectation: that is,

$$P(a \circ P(b)) = P(a) \circ P(b) \ , \qquad a, b \in A.$$

We remark that the variant of the last theorem with JB*- instead of JC*- was later proved in [33, Theorem 3.2]. The latter paper was in part inspired by our results in [17].

Our goal now is to try to generalize such results to more general algebras. The following result is from [17]:

Theorem 5.2 *Let A be a (Jordan) operator algebra, and $P : A \to A$ a completely contractive projection.*

(1) *The range of P with product $P(x \circ y)$, is completely isometrically Jordan isomorphic to a Jordan operator algebra.*
(2) *If A is an associative operator algebra then the range of P with product $P(xy)$, is completely isometrically algebra isomorphic to an associative operator algebra.*
(3) *If A is unital and $P(1) = 1$ then the range of P, with product $P(x \circ y)$, is unitally completely isometrically Jordan isomorphic to a unital Jordan operator algebra.*

By the *P-product* or *new product* on $P(A)$ we mean the bilinear map $P(x \circ y)$ (in the Jordan algebra case) or $P(xy)$ (in the algebra case) see also [39, 55] for some complementary results.

We do not recall having seen the following explicitly in the literature. In any case, in view of the present venue, we give a simple proof of it using Theorem 5.1.

Theorem 5.3 *If A is a JC^*-algebra which has a Banach space predual, then A is Jordan $*$-isomorphic, via a weak* homeomorphism, to a JW^*-algebra.*

Proof A JC^*-algebra with a predual has an identity (by e.g. the Krein–Milman theorem and [29, Theorem 4.2.36]). Now A^{**} is a JW^*-algebra as we showed above Theorem 4.4. Suppose that $E^* = A$. The canonical map $E \to E^{**} = A^*$ dualizes to give a weak* continuous contractive unital, hence positive and $*$-linear, surjection $\Phi : A^{**} \to A$. Regard A as a JC^*-subalgebra of A^{**}. It is easy to check that Φ extends the identity map on A, so that $\Phi \circ \Phi = \Phi$. Thus Φ is a weak* continuous 'conditional expectation' satisfying Theorem 5.1. Applying that result we have $x \circ y$ and $y \circ x$ are in $\mathrm{Ker}(\Phi)$ for any $x \in A$, $y \in \mathrm{Ker}(\Phi)$. It follows that for $x, y \in A$ we have $\Phi(x \circ y)$ equals

$$\Phi((x - \Phi(x)) \circ (y - \Phi(y)) + \Phi(\Phi(x) \circ (y - \Phi(y)) + \Phi(x \circ \Phi(y)),$$

which is just $\Phi(x \circ \Phi(y)) = \Phi(x) \circ \Phi(y)$. Hence Φ is a weak* continuous Jordan $*$-homomorphism. Thus $\mathrm{Ker}(\Phi)$ is a weak* closed selfadjoint two-sided ideal in A^{**}. By e.g. a variant of [30, Fact 5.1.10] (see also e.g. [27, Theorem 3.25]), there exists a central projection $p \in A^{**}$, with $\mathrm{Ker}(\Phi) = pA^{**}$. We have

$$\Phi((1-p)a) = \Phi(a) - \Phi(p)\Phi(a) = a, \qquad a \in A.$$

Thus Φ restricts to a surjective weak* continuous faithful Jordan $*$-homomorphism from the JW^*-algebra $(1-p)A^{**}$ onto A. □

Theorem 5.4 ([16, Theorem 2.5] and [18, Corollary 4.18]) *Let A be an approximately unital operator algebra (resp. Jordan operator algebra), and $P : A \to A$ a completely contractive completely real positive projection. Then $P(A)$ is an approximately unital operator algebra (resp. Jordan operator algebra) in the new product $P(xy)$ (resp. $P(x \circ y)$), P is still completely real positive as a map into the latter, and*

$$P(P(a)P(b)) = P(aP(b)) = P(P(a)b) \ , \qquad a, b \in A \, ,$$

(resp. $P(P(a) \circ P(b)) = P(a \circ P(b))$ for $a, b \in A$). If in addition $P(A)$ is a subalgebra (resp. Jordan subalgebra) of A then P is a conditional expectation (resp. Jordan conditional expectation).

The first step in the proof is to extend to the unitizations, using e.g. Theorem 3.3. This means that we may assume in the statement of the theorem that A is unital and $P(1) = 1$, and we may then discard the 'completely real positive' as being automatic. This is also the reason why P is still completely real positive as a map into the 'new algebra': this follows from Proposition 3.2 because P is (a restriction of) a unital complete contraction.

The above results are very satisfactory. As we said, the present goal is to try to generalize such results to more general algebras, and so we now mostly consider 'contractive real positive projection variants'. That is, we will see what happens e.g. to the parts of the last theorem when we drop the words 'completely'. Or in other words, we move away from the *operator space* (in the sense of e.g. [13, 57]) case. This is a very significant step, indeed loosely speaking one of the main points of operator space theory is that being the 'correct functional analysis' for many noncommutative problems, 'it makes things work'. We gave two justifications for attempting this step in the present paper above Theorem 2.3.

Thus one would not really expect very strong results about real positive projections on general Jordan operator algebras without a further hypothesis. An important first illustration of the breakdown is that $P(ab) \neq P(a)b$ in general for a unital operator algebra A and contractive unital (hence real positive) projection from A onto a subalgebra containing 1_A, and $a \in A, b \in P(A)$. This is not even true in general if A is commutative, which also rules out the Jordan variant $P(a \circ b) = P(a) \circ b$ (see [15, Corollary 3.6]). However we will see that the latter is true under further hypotheses, such as if either a or b is in $\Delta(A)$.

The next question, and one of the most important open questions here, is whether the range of a real positive contractive projection P on an approximately unital (Jordan) operator algebra is always again a Jordan operator algebra in the P-product? Probably the answer is in the negative. However we are able to prove this kind of result under various hypotheses, illustrated by some of the next several theorems.

Theorem 5.5 *Let A be an approximately unital operator algebra, and $P : A \to A$ a contractive real positive projection. Suppose that Ker(P) is densely spanned by the real positive elements which it contains. Then the range $P(A)$ is an approximately unital operator algebra with product $P(xy)$, and P is still real positive as a map into the latter algebra. Also*

$$P(ab) = P(aP(b)) = P(P(a)b) = P(P(a)P(b)), \qquad a, b \in A.$$

If further $P(A)$ is a subalgebra of A then P is a homomorphism with respect to the P-product on its range.

This last result is a simpler special case of [18, Theorem 4.10]. The 'densely spanned' hypothesis is a condition that is essentially always satisfied for positive projections on C^*-algebras, so it is not unnatural.

Proposition 5.6 *Let A be a (not necessarily approximately unital) Jordan operator algebra, and let $P : A \to A$ be a contractive real positive projection. The restriction of P to the JC^*-algebra $\Delta(A)$ is positive and satisfies*

$$P(\Delta(A)) = \Delta(P(A)) = \Delta(A) \cap P(A).$$

Proof Here $\Delta(A) = A \cap \Delta(A^1)$, and $\Delta(P(A))$ consists of the elements in $P(A)$ such that $x^* \in P(A)$ (so that $x \in \Delta(A)$, and we can take the last involution to be the one in $\Delta(A)$). The restriction of P to the JC^*-algebra $\Delta(A)$ is real positive, so it is positive and maps into $A \cap \Delta(A^1) = \Delta(A)$ by [18, Lemma 2.7] and the lines after [27, Corollary 2.5]. By the above, $\Delta(P(A)) \subset \Delta(A) \cap P(A)$. However $\Delta(A) \cap P(A) \subset P(\Delta(A))$ since if $x \in \Delta(A)$ with $x = P(x)$, then $x \in P(\Delta(A))$. Finally, $P(\Delta(A)) \subset \Delta(P(A))$, since if $x \in \Delta(A)$ with $x = x^*$, then $P(x) = P(x)^* \in \Delta(P(A))$. □

Lemma 5.7 *Suppose that A is an approximately unital Jordan operator algebra, and $P : A \to A$ is a contractive real positive projection such that $P(A)$ is a Jordan operator algebra with the P-product. Then this Jordan algebra is approximately unital, and $P^{**}(A^{**})$ is a unital Jordan operator algebra with the P^{**}-product.*

Proof Let B be $P(A)$ with the P-product. By basic functional analysis

$$P(A)^{\perp\perp} = (\,\mathrm{Ker}(P^*))^{\perp} = (I - P^*)(A^*)^{\perp} = \mathrm{Ker}((I - P^*)^*) = P^{**}(A^{**}),$$

since $\mathrm{Ker}(I - P^{**}) = P^{**}(A^{**})$. Thus $P(A)^{**} = B^{**} \cong P^{**}(A^{**})$, via the bidual of the canonical inclusion $i : P(A) \to A$. Now B^{**} is a Jordan operator algebra with separately weak* continuous product $m^{**}(\zeta, \eta) = \lim_s \lim_t P(a_s \circ b_t)$, assuming that $a_s \to \zeta$ and $b_t \to \eta$ weak* in B^{**}. Here $a_s, b_t \in B$. Now $i^{**}(P(a \circ b)) = P^{**}(i^{**}(a) \circ i^{**}(b))$ for $a, b \in B$. Hence $P^{**}(A^{**})$ is a Jordan operator algebra with product

$$\lim_s \lim_t i^{**}(P(a_s \circ b_t)) = \lim_s \lim_t P^{**}(i^{**}(a_s) \circ i^{**}(b_t)) = P^{**}(i^{**}(\zeta) \circ i^{**}(\eta)),$$

which is the P^{**}-product. These are weak* limits.

By the last assertion of [18, Lemma 4.9] applied in the bidual, $P^{**}(1)$ is an identity for the P^{**}-product. So B^{**} is a unital Jordan operator algebra, and hence B is approximately unital by [27, Lemma 2.6]. □

Theorem 5.8 *Let A be an approximately unital Jordan operator algebra, and let $P : A \to A$ be a contractive real positive projection. Then $\Delta(A) \cap P(A)$ is a JC^*-algebra in the P-product. If $P(A)$ is a Jordan operator algebra with the P-product then*

$$P(a \circ b) = P(P(a) \circ b),$$

for $a \in A$ and $b \in \Delta(A) \cap P(A)$; or for $a \in \Delta(A)$ and $b \in P(A)$. In particular, if A is unital then $P(1)$ is an identity for the P-product on $P(A)$.

Proof The restriction E of P to $\Delta(A)$ is a positive contractive projection by Proposition 5.6, with range $\Delta(A) \cap P(A)$. Thus by Theorem 5.1 we have that $\Delta(A) \cap P(A)$ is a JC^*-algebra in the new product $P(x \circ y)$ and the old involution, and E is still positive as a map into the latter JC^*-algebra. The case of the displayed equation when $a \in \Delta(A)$ and $b \in P(A)$ may be found in [18, Lemma 4.9].

We next show that if q is a projection in the latter JC^*-algebra then $P(qaq) = P(qP(a)q)$ and $P(a \circ q) = P(P(a) \circ q)$ for all $a \in A$. If q is an identity for A then these assertions are trivial. If not, by Theorem 3.3 we may assume that A and P are unital. So $q = P(q^2) \geq 0$. Claim: $P(q^n) = q$ for all $n \in \mathbb{N}$. In fact this is clear by the C^*-theory, but we give a short alternative proof. For $n = 1, 2$ this is clear. By Theorem 5.1 $P(q^{n+1}) = P(q \circ P(q^n))$, the claim follows by induction.

By functional calculus we can approximate $q^{\frac{1}{n}}$ appropriately by polynomials $p_m(q)$ in q, where p_n has no constant term. Then $P(p_m(q)) = p_m(1)\,q$ clearly. If this approximation is done carefully one sees that $P(q^{\frac{1}{n}}) = q$ for all $n \in \mathbb{N}$.

Let $Q = P^{**}$ and $M = A^{**}$. So M is unital and weak* closed and Q is weak* continuous and unital. Note that $\Delta(M)$ is then weak* closed (since if $a_t, a_t^* \in A$ and $a_t \to \eta$ weak* then $a_t^* \to \eta^* \in A^{**} = M$ weak*). Hence so is $Q(\Delta(M)) =$

$\Delta(M) \cap P(M)$ weak* closed. Note that $\Delta(M)$ contains the weak* closed algebra N in M generated by q and 1, a von Neumann algebra. Since $Q(q^{\frac{1}{n}}) = q$ we have $Q(s) = q$ where s is the support projection of q in M. Let R be the restriction of Q to the unital Jordan operator algebra sMs. Then R is real positive. By [18, Lemma 2.8] or the remark after it, we have $R(sas) = R(qsasq) = R(qaq)$ for all $a \in M$. On the other hand, let us now view Q as a map into $Q(M)$ with the P-product, which is a unital Jordan operator algebra by Lemma 5.7. Then we have $Q(sas) = Q(qQ(a)q)$ by Lemma 3.10. Hence

$$P(qaq) = Q(qaq) = R(qaq) = R(sas) = Q(sas) = Q(qQ(a)q) = P(qP(a)q).$$

Next note that $1 - q$ is a selfadjoint contraction with

$$P((1-q)^2) = 1 - 2q + P(q^2) = 1 - q.$$

Thus $P(q^\perp a q^\perp) = P(q^\perp P(a) q^\perp)$ too, by the last paragraph. By the argument at the end of the proof of Lemma 3.10 we deduce that $P(a \circ q) = P(a \circ P(q))$.

For the remaining case of the displayed equation, we may assume that A is weak* closed and P is weak* continuous (or else replace by M and Q above). This will work here because $\Delta(A) \cap P(A) \subset \Delta(A^{**}) \cap P^{**}(A^{**})$. By Theorem 5.3, $\Delta(A) \cap P(A)$ is a JW^*-algebra in the P-product. Hence $\Delta(A) \cap P(A)$ is spanned by the projections (in the P-product) which it contains (by essentially the same proof of the analogous fact for von Neumann algebras). Thus the second assertion yields the third (and centered) assertion. □

Corollary 5.9 *Let A be an approximately unital Jordan operator algebra, and $P : A \to A$ a real positive contractive projection with $P(A) \subset \Delta(A)$. Then $P(A)$ is a JC^*-algebra in the P-product, and the restriction of P to joa$(P(A))$ is a Jordan $*$-homomorphism onto this JC^*-algebra. In this case P is a Jordan conditional expectation with respect to the P-product:*

$$P(a \circ P(b)) = P(P(a) \circ P(b))$$

for a, b in A.

Proof By Proposition 5.6, the restriction of P to the JC^*-algebra $\Delta(A)$ is a real positive, hence positive, contractive projection onto $P(\Delta(A)) = P(A) \subset \Delta(A)$. It is also $*$-linear. So $P(A)$ is a JC^*-algebra in the P-product by Theorem 5.1.

By Remark 2 after Theorem 4.10 in [18], Ker$(P) \cap$ joa$(P(A))$ is densely spanned by the real positive elements which it contains. Hence P is a Jordan homomorphism from joa$(P(A))$ onto $P(A)$ with the P-product, by Theorem 4.10 in [18].

That P is a Jordan conditional expectation follows from the last assertion of Corollary 5.8, since $P(A)$ is selfadjoint. □

The following result of independent interest is contained in Theorem 3.5 of [61]. Indeed, the first paragraph in the lemma is the equivalence of (4) and (8) in that paper (taking $u = 1$ in (8)). Actually that paper is a revision of [60], and the first paragraph of our lemma may also be seen from Lemma 1.1 and the equivalence of (i) and (iv) in Corollary 2.7 of [60]. However we do not see some of it stated there in exactly the form below, except in the von Neumann algebra case, and so we thank Angel Rodríguez Palacios for allowing us to give here a direct proof avoiding nonassociative algebra. By a Banach algebra we mean an associative algebra with a complete norm that is submultiplicative: $\|xy\| \leq \|x\|\|y\|$.

Lemma 5.10 *Let B be a C^*-algebra with identity 1. The possible Banach algebra products on B with identity 1 are all C^*-algebra products. They are in a bijective correspondence with the central projections in the multiplier algebra of the commutator ideal of B.*

In addition, any Banach algebra product m on a unital JC^*-algebra is a C^*-algebra product such that $\frac{1}{2}(m(x, y) + m(y, x))$ is the original JC^*-algebra product.

Proof Let J be the commutator ideal of B, and let z be the support projection of J in B^{**}. Suppose that e is a central projection in $M(J)$. Then $ex = xe$ for all $x \in J$, hence for all $x \in J^{**}$. For $x \in B^{**}$ we have $ex = e(z + z^\perp)x = ezx = zxe$. Similarly $xe = zxe$, so that e is central in B^{**}. For $x, y \in B$ we have

$$exy + e^\perp yx = e(xy - yx) + yx \in J + B \subset B.$$

Then it is easy to check that $m(x, y) = exy + e^\perp yx$ defines a C^*-algebra product on B with identity 1.

If e, f are central projections in $M(J)$ such that $exy + e^\perp yx = fxy + f^\perp yx$ for all $x, y \in A$, then

$$e(xy - yx) = exy + e^\perp yx - yx = fxy + f^\perp yx - yx = f(xy - yx).$$

Hence $e = f$.

Suppose that m is a Banach algebra product on a JC^*-algebra B with identity 1. The hermitian elements in this Banach algebra are just the selfadjoint elements in the JC^*-algebra B, since Hermitians in a unital Banach algebra depend only on the norm and the identity element. These hermitians span B. Thus B with product m is a unital C^*-algebra by the Vidav–Palmer theorem (see e.g. [29, Theorem 2.3.32]). Then $\frac{1}{2}(m(x, y) + m(y, x))$ is a JC^*-algebra product on B which must be the original JC^*-algebra product by the Banach–Stone theorem (e.g. by Theorem 4.1).

Again suppose that C is a C^*-algebra and write C for B with product m. The identity map is a surjective unital isometry from C onto B, thus is a Jordan $*$-isomorphism θ by Kadison's Banach–Stone theorem for C^*-algebras [50]. By another result of Kadison in the same paper, there is a central projection $e \in C^{**}$ such that if $f = \theta^{**}(e)$ then $e\theta^{**}(\cdot)$ is a $*$-homomorphism and $f^\perp\theta^{**}(\cdot)$ is a $*$-anti-homomorphism. In our setting $e = f$ and

$$e\theta^{**}(m(a, b)) = e\theta^{**}(a)\theta^{**}(b), \quad e^\perp\theta^{**}(m(a, b)) = e^\perp\theta^{**}(b)\theta^{**}(a), a, b \in B.$$

That is,

$$em(a,b) = eab, \quad e^{\perp}m(a,b) = e^{\perp}ba, a, b \in B,$$

in the usual product on B^{**}. Thus $m(a,b) = eab + e^{\perp}ba$ for $a, b \in B$. (The facts in this paragraph also follow from Theorem 3.2 of [61].)

Note that for $x, y \in B$ we have

$$exy + e^{\perp}yx = e(xy - yx + yx) + e^{\perp}yx = ez(xy - yx) + yx = ezxy + (ez)^{\perp}yx.$$

Thus if we replace e by ez we may assume that e is a central projection in $J^{**} = zB^{**}$. We have

$$e(xy - yx) = (exy + e^{\perp}yx) - yx \in B \cap J^{**} = J, \qquad x, y \in J.$$

Thus $e \in M(J)$ and is central there. □

Remark 5.11 One may ask if B is a selfadjoint Jordan subalgebra of a unital C^*-algebra D containing 1, and if B has a Banach algebra product with identity 1, or equivalently (by the last result) a C^*-algebra product with identity 1, then is B an (associative) subalgebra of D? The answer is in the negative: consider $B = \{a \oplus a^{\intercal} : a \in M_2\} \subset M_4$.

Theorem 5.12 *Let A be an approximately unital (associative) operator algebra, and $P : A \to A$ a real positive contractive projection with $P(A) \subset \Delta(A)$. The P-product on $P(A)$ is associative if and only if*

$$P(P(a)P(b)) = P(aP(b)) = P(P(a)b), \qquad a, b \in A.$$

In this case $P(A)$ is a C^*-algebra with respect to the P-product, and P viewed as a map into this latter C^*-algebra is real completely positive and completely contractive.

Proof If the centered equation holds then

$$P(P(P(a)P(b))P(c)) = P(P(a)P(b)P(c)) = P(P(a)P(P(b)P(c))).$$

So the P-product on $P(A)$ is associative.

Suppose that the P-product on $P(A)$ is associative. By Theorem 3.3 we may assume that A is unital and $P(1) = 1$. (Note that if A is nonunital or $P(1) \neq 1$ then $\tilde{P}(A^1) \subset \Delta(A^1)$; and if the new product on $P(A)$ is associative, then so is the new product on $\tilde{P}(A^1)$. Here $\tilde{P}$ is the extension to A^1 from Theorem 3.3.) Write B for $P(A)$ in the P-product. This is a unital Banach algebra. By Corollary 5.9 $P(A)$ is also a unital JC^*-algebra. By the proof of Lemma 5.10, B is a unital C^*-algebra. By considering P^{**} and A^{**} we may suppose that B is a W^*-algebra.

Write $\cdot$ for the P-product. Claim: $a \cdot P(x) = P(ax)$ for all $a \in B, x \in A$. To prove this, as in the proof of [14, Lemma 3.2] it suffices to show that

$$p^{\perp} \cdot P(px) = 0, \qquad x \in X, \tag{1}$$

for all orthogonal projections $p \in B$. For then we get $p \cdot P(x) = P(px)$ as in that cited proof, and since B is densely spanned by its projections (as is any W^*-algebra), we conclude that $a \cdot P(x) = P(ax)$ for all $a \in B$.

To prove (1), we adjust the argument in the last cited proof. Let

$$y = P(px) + tP(p^{\perp}P(px)) = P(px) + tp^{\perp} \cdot P(px), \qquad t \in \mathbb{R}.$$

By associativity, $y = P(px + p^{\perp}P(p^{\perp}P(px)))$. Now

$$\|y\|^2 \leq \|px + tp^{\perp}P(p^{\perp}P(px))\|^2,$$

which as in the last cited proof is dominated by $\|px\|^2 + t^2\|P(p^{\perp}P(px))\|^2$. On the other hand, again writing $\cdot$ for the P-product, we have

$$p^{\perp} \cdot y = p^{\perp} \cdot P(px) + tp^{\perp} \cdot P(px) = (1 + t)p^{\perp} \cdot P(px).$$

It follows that

$$(1 + t)^2\|p^{\perp} \cdot P(px)\|^2 \leq \|px\|^2 + t^2\|p^{\perp} \cdot P(px)\|^2.$$

This implies that $2t\|p^{\perp} \cdot P(px)\|^2 \leq \|px\|^2$ for all $t > 0$, so that $p^{\perp} \cdot P(px) = 0$.

The Claim says that P is a left B-module map; similarly it is a right B-module map, giving the last assertion of our statement. It follows by a standard trick similar to e.g. 1.2.6 in [13] (see e.g. [57, Proposition 8.6]) that P is completely contractive. Since $P(1) = 1$, P is real completely positive by (the matrix versions of) considerations in the paragraph before Corollary 2.2 from [18]. □

Remark 5.13 By the proof above, one may relax the associativity condition in the last theorem to: $p \cdot (p \cdot b) = p \cdot b$ for $b \in P(A)$ and for $p \in P(A)$ which are projections in the P-product.

We also remark that a real positive completely contractive projection need not be real completely positive. Indeed $P(a, b) = (a, a^{\intercal}/2)$ on $M_2 \oplus M_2$ is a counterexample (we thank R. R. Smith for this). Note that this completely contractive positive projection does not even satisfy the Kadison-Schwarz inequality $P(a^*a) \leq P(a)^*P(a)$.

The C^*-algebra structure on $P(A)$ in the conclusion of the theorem may not induce the operator space structure on $P(A)$ inherited from A. To see this consider $P(a, b) = (a, a^{\intercal})$ on $M_2 \oplus M_2$.

The following result is inspired by the selfadjoint case due to Effros and Størmer (see e.g. [37, Lemma 1.4]). If A is a unital operator algebra, and P is a unital contractive or completely contractive projection on A, define

$$N = \{x \in A : P(xy) = P(yx) = 0 \text{ for all } y \in A\}.$$

If A or P is not unital, but P is also real positive, then we may extend P by Theorem 3.3 to a unital contractive projection on A^1, where A^1 is a unitization with $A^1 \neq A$, and set

$$N = \{x \in A : P(xy) = P(yx) = 0 \text{ for all } y \in A^1\}.$$

Then N is clearly a closed ideal in A, and is also a subspace of $\mathrm{Ker}(P)$. Define

$$B = \{x \in A : P(xy) = P(P(x)P(y)) \text{ and } P(yx) = P(P(y)P(x)) \text{ for all } y \in A\}.$$

Note that $N \subset B$ since e.g. if $x \in N \subset \mathrm{Ker}(P)$ then $P(xy) = 0 = P(P(x)P(y))$ for all $y \in A$. Note too that $1 \in B$ if A is unital and $P(1) = 1$.

Theorem 5.14 *If P is a real positive contractive projection on an approximately unital operator algebra A, and N, B are defined as above, then $P(A) \subset B$ if and only if*

$$P(P(a)b) = P(P(a)P(b)) = P(aP(b)) \textit{ for all } a, b \in A.$$

That is, if and only if P is a conditional expectation onto $P(A)$ with respect to the P-product. This is also equivalent to $B = P(A) + N$. If these hold then $P(A)$ with the P-product is isometrically isomorphic to an operator algebra, B is a subalgebra of A containing $P(A)$, and P is a homomorphism from B onto $P(A)$ with the P-product.

Proof The first 'if and only if' follows from the definition of B. This is also equivalent to $B = P(A) + N$, since $N \subset B$, and if $P(A) \subset B$ and $a \in B$ then

$$P((a - P(a))b) = P(P(a)P(b)) - P(P(a)b) = 0, \qquad b \in A.$$

Similarly, $P(b(a - P(a))) = 0$. Hence $a = a - P(a) + P(a) \in N + P(A)$.

In this case B is a subalgebra of A since e.g. if $a, b \in B$, $c \in A$ then

$$P(abc) = P(P(a)P(bc)) = P(P(a)P(P(b)P(c))) = P(P(a)P(bP(c)),$$

and by the centered equation in the theorem statement,

$$P(P(ab)P(c)) = P(abP(c)) = P(P(a)P(bP(c)).$$

Thus also the P-product on $P(A)$ is associative. Since $B/N \cong P(A)$ as in the proof of Lemma 4.4 in [18], and quotients of operator algebras are operator algebras [13, Proposition 2.3.4], we see that $P(A)$ with the P-product is isometrically isomorphic to an operator algebra. The last assertion again follows from the definition of B. □

Remark 5.15 Note that $N = \text{Ker}(P)$ if and only if $\text{Ker}(P)$ is an ideal. The latter holds (by the associative algebra variant of Lemma 4.4 in [18]) if and only if $B = A$, and then all of the conclusions of the last theorem hold.

If P is real completely positive and completely contractive then

$$P(P(a)b) = P(P(a)P(b)) = P(aP(b)), \qquad a, b \in A,$$

as is proved in [16, Section 2], so that the conclusions of the last theorem hold.

Corollary 5.16 *If A is an approximately unital (associative) operator algebra and $P : A \to A$ is contractive and real positive, and is a conditional expectation onto $P(A)$ equipped with the product $P(ab)$ (that is, if $P(P(a)b) = P(P(a)P(b)) = P(aP(b))$ for all $a, b \in A$), then $P(A)$ is an operator algebra with this product.*

This corollary shows that P being a conditional expectation for the P-product implies that the P-product is an operator algebra product. The converse is false as we observed below Theorem 5.3.

It is natural to ask if similar results hold for Jordan operator algebras, and in particular does P being a conditional expectation for the Jordan P-product imply that $P(A)$ is a Jordan operator algebras in the P-product. The latter is an interesting question that we have not yet been able to solve. If one goes through the proof above with a (without loss of generality) contractive unital projection P on a unital Jordan operator algebra A, one defines

$$N = \{x \in A : P(x \circ y) = 0 \text{ for all } y \in A\}.$$

Then N is a subspace of $\text{Ker}(P)$, and is a closed 'Jordan ideal' in

$$B = \{x \in A : P(x \circ y) = P(P(x) \circ P(y)) \text{ for all } y \in A\}.$$

Again $B = P(A) + N$, and P restricts to a contractive unital projection and 'Jordan homomorphism' from B onto $P(A)$ with kernel N. So $P(A) \cong B/N$ isometrically and Jordan isomorphically. However, two issues arise. First it is not clear that B is actually a Jordan (operator) algebra. Second, even if it were we do not known that a quotient of a Jordan operator algebra by a closed Jordan ideal is a Jordan operator algebra. This is a big open problem in the subject. If it is false then this suggests that perhaps the study of closed Jordan subalgebras of such quotients may be an interesting direction of research. The usual proof that the quotient of an operator algebra by a closed ideal is an operator algebra, uses the so-called BRS characterization of operator algebras (see [13, Theorem 2.3.2 and Proposition 2.3.4]). There is a somewhat similar theorem for Jordan operator algebras, namely [27, Theorem 2.1]. If one attempts the natural proof (analogous to the operator algebra quotient proof) one sees that it works for quotients of a Jordan operator algebra B by a closed Jordan ideal N such that B is contained in an operator algebra A with $B/N \subset A/[ANA]$ completely isometrically. Here $[ANA]$ is the closed associative ideal in A generated by N. That is, in this case B/N is a Jordan operator algebra. We do not know unfortunately when $B/N \subset A/[ANA]$. This would require $B \cap [ANA] = N$ at the very least. Perhaps there is a clever choice of A that will do this, perhaps even a C^*-algebra.

In particular, in the situation in the last paragraph, but now assuming in addition that $N = (0)$, one may ask if $B = P(A)$ a Jordan subalgebra of A, and if P is a Jordan conditional expectation?

Corollary 5.17 *Let A be an approximately unital (associative) operator algebra, and $P : A \to A$ a real positive contractive projection onto an associative subalgebra of A, with $P(A) \subset \Delta(A)$. Then P is completely contractive and real completely positive as a map into A, $D = P(A)$ is a C^*-algebra in the P-product, and P is a D-bimodule map: $P(P(a)P(b)) = P(aP(b)) = P(P(a)b)$ for all $a, b \in A$.*

The last result is a corollary of Theorem 5.12.

We now turn to bicontractive projections. The following result shows what happens in the case of selfadjoint Jordan operator algebras (JC^*-algebras). It is the 'solution to the bicontractive and symmetric projection problems' for JC^*-algebras, essentially due to deep work of Friedman and Russo, and Størmer [38, 40, 65]. We recall that P is *bicontractive* if $\|P\|, \|I - P\|$ are contractions, and *symmetric* if $\|I - 2P\| \leq 1$. Some of this hinges on Harris's Banach–Stone type theorem for J^*-algebras [46]. The following is essentially very well known (see the references above), but we do not know of a reference besides the work which we are surveying (see [17, Theorem 5.1]) which has all of these assertions, or is in the formulation we give:

Theorem 5.18 *If $P : A \to A$ is a projection on a JC^*-algebra A then P is bicontractive if and only if P is symmetric. Moreover P is bicontractive and positive if and only if there exists a central projection $q \in M(A)$ (indeed $q \circ a = qaq \in A$ for all $a \in A$) such that $P = 0$ on $q^\perp A q^\perp$, and there exists a Jordan $*$-automorphism θ of qAq of period 2 (i.e. $\theta \circ \theta = I$) so that $P = \frac{1}{2}(I + \theta)$ on qAq. Finally, $P(A)$ is a JC^*-subalgebra of A, and P is a Jordan conditional expectation.*

We remark that subsequently the variant of this theorem with JB^*- instead of JC^*-algebras has been proved in [33, Theorem 5.7].

We have a complete characterization of symmetric real positive projections on Jordan operator algebras, which relies on the very recent Banach–Stone theorem 4.4 above. The point is that if P is symmetric then it is very easy to show that $\upsilon = I - 2P$ is a surjective isometric isomorphism of A onto A which has period 2 (that is, $\upsilon \circ \upsilon = I_A$). Then one applies the Banach–Stone theorem 4.4 above. This is the main ingredient in the following result from [18]:

Theorem 5.19 *Let A be an approximately unital (Jordan) operator algebra, and $P : A \to A$ a symmetric real positive projection. Then the range of P is* *an approximately unital Jordan subalgebra of A and P is a Jordan conditional expectation. Moreover, $P^{**}(1) = q$ is a projection in $JM(A)$.*

Set $D = qAq$, the hereditary subalgebra ('corner') of A supported by q, which contains $P(A)$. There exists a period 2 surjective isometric Jordan homomorphism $\pi : D \to D$, such that

$$P = \frac{1}{2}(I + \pi) \quad \text{on } D,$$

and P is zero on the 'other three corners' of A (that is, on $q^{\perp}Aq^{\perp}+q^{\perp}Aq+qAq^{\perp}$).

The converse is true too, such an expression $P = \frac{1}{2}(I + \pi)$ is a symmetric real positive projection.

This is very close to the 'classical' selfadjoint characterization in Theorem 5.18 above. The main difference is q need not be central, and P symmetric is not equivalent to P bicontractive.

The form of the *bicontractive projection problem* that evolved in [16, 18] asks for conditions on a bicontractive real positive projection $P : A \to A$ so that $P(A)$ is a subalgebra of A? This is not always true, as [16, Corollary 4.8] shows.

In [16, Section 4] and the start of [18, Section 6] we gave a three step reduction that reduces the bicontractive projection problem to the case of a bicontractive projection $P : A \to A$ with $P(1) = 1$ and A is generated by $P(A)$. We will omit the details here, although we note that some of the ingredients in this reduction are taking the bidual, and then observing that for a bicontractive real positive projection on a unital operator algebra A, $P(1)$ is a projection. Also we use some facts about Jordan hereditary subalgebras.

Part of the following result is [18, Theorem 6.3].

Theorem 5.20 *Let A be a unital (Jordan) operator algebra, and let D be the elements in Ker(P) that are also in the closed Jordan subalgebra generated by $P(A)$. If $P : A \to A$ is a bicontractive unital projection on A, and if D is densely spanned by the real positive elements which it contains, or if $P(A) \subset \Delta(A)$, then $P(A)$ is a Jordan subalgebra of A.*

Proof We will just prove the result with hypothesis $P(A) \subset \Delta(A)$ here, for the other see [18, Theorem 6.3]. If $P(A) \subset \Delta(A)$ then as in the proof of Corollary 5.9, the restriction of P to the JC^*-algebra $\Delta(A)$ is a real positive, hence positive, bicontractive projection onto $P(\Delta(A)) = P(A) \subset \Delta(A) = P(A)$. By Theorem 5.18, $P(A)$ is a Jordan subalgebra of $\Delta(A)$, hence of A. □

(The hypotheses in the last theorem are conditions that are always satisfied for positive projections on C^*-algebras, so they are not unnatural.)

Finally, we remark that much of Sect. 5 was concerned with which results from [16] still have variants valid for *contractive real positive* projections on Jordan operator algebras. If one uses *completely contractive completely real positive* projections then essentially everything in [16] is valid for Jordan operator algebras, as is observed in [18].

6 What Are Conditional Expectations?

This section may be viewed as an appendix for nonexperts, giving some basic insights into the conditional expectation property in Sect. 5, and into some of the advanced considerations needed for the discussion in Sect. 7. Conditional expectations are no doubt also addressed in other articles in this conference proceedings, probably also with some discussion of their history. Thus there may be a very small amount of overlap here at the beginning of the present section. We believe that these other articles are located in the Banach lattice setting and are concerned with applications there. We will not focus on the lattice aspect at all, and our techniques are quite different.

The classical or 'probabilistic' conditional expectation may be taken to refer to a hugely important construction on a probability measure space $(K, \mathcal{A}, \mu)$ induced by a choice of a sub-σ-algebra $\mathcal{B}$ of the σ-algebra $\mathcal{A}$ on the set K. In this case one may identify $L^p(K, \mathcal{B}, \mu)$ isometrically with a subspace of $L^p(K, \mathcal{A}, \mu)$. Moreover dualizing these embeddings gives contractive positive unital projections $E_{\mathcal{B}}$ of $L^p(K, \mathcal{A}, \mu)$ onto the copy of $L^p(K, \mathcal{B}, \mu)$ (to get the expectation onto L^1 one takes the dual in the weak* topology of the L^∞ inclusions). These satisfy a list of beautiful properties often attributed to Kolmogorov (a great account of this list may be found e.g. in the Wikipedia article on conditional expectations). Writing $E_{\mathcal{B}}$ as E, the most notable of these is that

$$\int E(f)\, d\mu = \int f\, d\mu$$

(which in the quantum variant becomes the 'trace preserving' property). However this list also includes 'positivity': $E(f) \geq 0$ if $f \geq 0$, 'contractivity' ($|Ef| \leq E(|f|)$ which implies $\|E\| \leq 1$), the projection property $E \circ E = E$, and that $E(1) = 1$. It also includes the important 'module' property

$$E(fg) = E(f)g, \qquad g \in L^\infty(K, \mathcal{B}, \mu),$$

which defined what we called 'conditional expectation' in Sect. 5. It also has important continuity properties, like being weak* continuous on L^∞ (the latter is clear because as we said it is the dual of a map on L^1). Moreover $E_{\mathcal{B}}$ has a fundamental probabilistic interpretation. We will not rehearse this here since it is so well known. Indeed this interpretation is ubiquitous in scientific disciplines. It has been said that conditional expectations are the starting point of modern probability theory.

Most of this is still true if we start to mildly relax the condition that μ is a finite measure (although now $1 \notin L^1$). Historically, mathematical analysts tried to successively weaken the measure theoretic requirements on μ while still preserving a satisfactory theory of expectation. At some point beyond so-called 'localizable measures' (we warn the reader that there is ambiguity in the measure theory literature concerning this term) things break down and become pathological. It is interesting that when one tries to identify this point of breakdown, it seems to corroborate how perfectly von Neumann algebras capture the essence of these concepts. Namely, the breakdown occurs very slightly beyond the class of measures for which $L^\infty(K, \mu)$ is a von Neumann algebra. And in the latter case the rich theory of von Neumann algebraic conditional expectations applies.

Let μ be a localizable measure on $\mathcal{A}$. It is known that $L^\infty(K, \mathcal{B}, \mu)$ is a von Neumann subalgebra of $L^\infty(K, \mathcal{A}, \mu)$. If $\mathcal{B}$ is order-closed in $\mathcal{A}$, that is, closed under suprema in the associated measure algebra.

Conversely, any von Neumann subalgebra D of $L^\infty(K, \mathcal{A}, \mu)$ is of this form. Since we are not aware of a source in the literature we sketch a simple argument that $D = L^\infty(K, \mathcal{B}, \mu)$. Let $\mathcal{B} = \{B \in \mathcal{A} : \chi_B \in D\}$. It is an exercise to check that $\mathcal{B}$ is a σ-algebra. For example it is an algebra because $\chi_{B_1} \chi_{B_2} = \chi_{B_1 \cap B_2}$. Then if $B_1, B_2, \cdots$ are disjoint sets in $\mathcal{B}$, let (p_n) be the corresponding mutually orthogonal projections in D, let $F = \cup_n B_n$, and let $p = \sup_n p_n \in D$. Then there exists a set $B \in \mathcal{A}$ with $p = \chi_B$ μ-a.e.. Since $p_n p = p_n$ the set $B_n \setminus (B \cap B_n)$ is μ-null. Thus we may assume that $B_n \subset B$ for all n, so that $F \subset B$. On the other hand if $E \cap B_n = \emptyset$ for all n, then $\chi_E\, p_n = 0$ for all n, so that $\chi_E\, p = \chi_{E \cap B} = 0$ in M. We conclude that $\chi_F = p \in D$. Thus $F \in \mathcal{B}$, so that $\mathcal{B}$ is a σ-algebra. If $B \in \mathcal{B}$ then $\chi_B \in D$ by definition. Conversely, for any projection p in D there exists a set $B \in \mathcal{A}$ with $p = \chi_B$ μ-a.e., so that $B \in \mathcal{B}$ and $\chi_B \in L^\infty(K, \mathcal{B}, \mu)$. Using the fact that D and $L^\infty(K, \mathcal{B}, \mu)$ are both generated by their projections, we conclude that $D = L^\infty(K, \mathcal{B}, \mu)$.

In the discussion below we assume that μ is a probability measure for simplicity. It follows that for any von Neumann subalgebra D of $L^\infty(K, \mathcal{A}, \mu)$, there exists a canonical sub-σ-algebra $\mathcal{B}$ of $\mathcal{A}$ and a canonical contractive projection $E_{\mathcal{B}}$ from $L^\infty(K, \mathcal{A}, \mu)$ onto D. We call this the *probabilistic conditional expectation*. It is, as we said, 'trace preserving': $\int E_{\mathcal{B}}(f)\, d\mu = \int f\, d\mu$ for $f \in L^\infty(K, \mathcal{A}, \mu)$.

Some of the founders of modern probability theory and their students tried to characterize conditional expectations amongst the idempotent maps (i.e. projections) on $L^p(K, \mathcal{A}, \mu)$ (particularly in the case $p = 1$). See e.g. [35, 54] and references therein. There were early characterizations due to Moy [56] who characterizes conditional expectations in terms of operators on the positive measurable functions, and on L^p, obtaining particularly nice results for L^1. Later Douglas [36], Ando, Lacey and Bernau (see e.g. [54]), and others refined these results. It follows from this work that there are bijective correspondences between weak* continuous unital contractive projections P from $L^\infty(K, \mathcal{A}, \mu)$ onto a von Neumann subalgebra D, and density functions $h \in L^1(K, \mathcal{A}, \mu)_+$ with $E_{\mathcal{B}}(h) = 1$ (such h is called a 'weight function' for p). The correspondence is given by $P(x) = E_{\mathcal{B}}(hx)$, and h is the density of the normal state $x \mapsto \int P(x)\, d\mu$ (which may be written as $P_*(1)$, viewing $1 \in L^1$). If $\int P(f)\, d\mu = \int f\, d\mu$ for $f \in L^\infty$ then one sees that $\int hf = \int f$ for such f, which forces $h = 1$ and $P = E_{\mathcal{B}}$.

Summarizing the above discussion: we have seen at least in the setting above, the weak* continuous unital contractive projections P from L^∞ onto a von Neumann subalgebra D are simply the 'weightings' of the probabilistic conditional expectation $E_{\mathcal{B}}$, by the weights h above. That these projections have the important 'module' property $P(fg) = P(f)g$ for $g \in D$ (which defined what we called 'conditional expectation' in Sect. 5), may be viewed in this picture $P(x) = E_{\mathcal{B}}(hx)$ as coming immediately from the fact that the probabilistic conditional expectation $E_{\mathcal{B}}$ has this property. The probabilistic conditional expectation is characterized among all the weak* continuous unital contractive projections P onto D by the 'trace-preserving' condition $\int P(f) = \int f$.

So all weak* continuous unital contractive projections P from L^∞ onto a von Neumann subalgebra 'are' weighted probabilistic conditional expectations. In fact if P is faithful then it is a probabilistic conditional expectation, the probabilistic conditional expectation associated with a measure which we now describe. Indeed one may play the above game in reverse. Suppose that we are given a weak* continuous unital contractive projection P from $L^\infty(K, \mathcal{A}, \mu)$ onto D. Then $f \mapsto \int P(f)\, d\mu$ is a normal state of $L^\infty(K, \mathcal{A}, \mu)$, which is simply integration against a probability measure $d\nu = h\, d\mu$ on $(K, \mathcal{A})$. If P is faithful then h has full support and $L^\infty(K, \mathcal{A}, \mu) = L^\infty(K, \mathcal{A}, \nu)$. The measure space $(K, \mathcal{A}, \nu)$ can replace the role of $(K, \mathcal{A}, \mu)$ in the above discussion, to produce a probabilistic conditional expectation associated with ν. If there are no issues with the 'support' (that is, if h has full support), then one can see that this probabilistic conditional expectation is P.

Remark 6.1 One can somewhat generalize the discussion beyond the case that D is a subalgebra, to try to link up with the generalized 'conditional expectations' discussed in Sect. 5 with respect to the Choi-Effros product. For example, Douglas showed that the range of a contractive projection P on L^1 is isometric to an L^1 space [36], and investigated the relation of such P to the probabilistic conditional expectation. Let us give a quick proof of this range assertion. First assume that $\int P(f) = \int f$ for $f \in L^1$. (The latter is a much weaker condition than the condition $\int P^*(f) = \int f$ or $f \in L^\infty$ considered above, indeed $\int P(f) = \int f$ for $f \in L^1$ simply says that P^* is unital on L^∞, which is satisfied by all the projections in the earlier discussion.) Indeed note that in this case P^* is a contractive unital projection, and

$$(\mathrm{Ran}(P))^* \cong A^*/\mathrm{Ran}(P)^\perp = A^*/\mathrm{Ker}(P^*) \cong \mathrm{Ran}(P^*).$$

By the Choi and Effros result mentioned in the introduction, the range of P^* on L^∞ is a (commutative) C^*-algebra in the P^*-product. Since it is weak* closed it is a von Neumann algebra (by a result like Theorem 5.3 if necessary). By the uniqueness of von Neumann algebra preduals, $\mathrm{Ran}(P)$ is an L^1 space. There is a similar proof in the general case: by a result of Youngson [13, Theorem 4.4.9] which generalizes the Choi-Effros result used above, $P^*(xy^*z)$ is a (commutative) TRO product on $\mathrm{Ran}(P^*)$. As e.g. in the proof of [13, Theorem 4.4.9], any extreme point of the ball in this TRO is 'unitary', and any TRO with a unitary is isometric to a C^*-algebra. So again as above it is a von Neumann algebra and Ran(P) is an L^1 space. See [71] (and the Kirchberg result discussed and cited there) for a noncommutative generalization.

Let us now consider relaxing the condition that μ is a probability measure in the discussions above. We said earlier that there are serious pathologies for the most general kinds of measures, and suggested to simply consider measures with $L^\infty(K, \mathcal{A}, \mu)$ a von Neumann algebra. Going one step further, replace $L^\infty(K, \mathcal{A}, \mu)$ by a possibly noncommutative von Neumann algebra M. The measure μ, and associated integral, will be replaced by a certain 'trace' or 'weight' ν. Now we are in the setting of von Neumann algebraic conditional expectations, which gets into the deep taxonomy of von Neumann algebras, and the difficult theory of noncommutative integration [53, 69]. The case that we have dealt with above corresponds to the class of so-called 'finite' von Neumann algebras, where there exists a faithful normal tracial state τ on M. In this case it is a theorem that there exists a weak* continuous unital contractive projection E from M onto any von Neumann subalgebra D of M, and moreover there is a unique such E that is trace preserving (i.e. $\tau \circ E = \tau$). We may write this E as E_τ, and call this the 'probabilistic' (we should perhaps say 'tracial' here) conditional expectation. The other weak* continuous unital contractive projections E from M onto D are again the 'weightings' $E_\tau(hx)$ for densities $h \in L^1(M)_+$ which commute with D and which satisfy $E_\tau(h) = 1$. The reader should note the parallel with $E_{\mathcal{B}}$ above, the probabilistic conditional expectation. Again we see that all weak* continuous unital contractive projections from M onto a von Neumann subalgebra 'are' weighted 'probabilistic' conditional expectations. They may be viewed as a 'partial integral with respect to a noncommutative measure'.

A brief noncommutative history of conditional expectations: von Neumann, Dixmier, Nakamura and Turumaru, Umegaki, and others considered conditional expectations in the framework of von Neumann (or C^*-) algebras and established many properties of these objects (in particular the 'Kolmogorov list' above, especially in the context of von Neumann algebras with a finite trace (see e.g. [1] for references). Some of these works were aiming to generalize the Moy-Doob characterization mentioned above. Tomiyama added the modern perspective of conditional expectations in terms of norm one projections in C^*-algebras, his theorem stated in our introduction shows that all positive idempotents onto a C^*-algebra have the module property that leads to them being called conditional expectations. They are also completely positive and completely contractive as we said. See p. 132–133 in [7] for more on this and some other basic facts about conditional expectations. Others have generalized some of the work of Moy-Douglas-Ando-Lacey and Bernau, etc., that we described above, to positive contractive projections on noncommutative L^1 or L^∞ (i.e. on a von Neumann algebra M). Conditional expectations play a profound role in the classification of von Neumann algebras, e.g. in the structure theory of factors, or the fundamental work of Connes in which approximately finite von Neumann algebras are the amenable ones, and are the ones that are the range of a (not necessarily weak* continuous) conditional expectation on $B(H)$. Haagerup transferred conditional expectations to the powerful framework of operator valued weights and the extended positive part of a von Neumann algebra. The latter is the noncommutative version of $(L_0)_+$, the positive measurable functions, and consists of suprema of increasing sequences of elements of M_+. This gives the most general perspective, allows treatment of general noncommutative L^p spaces, etc. Conditional expectations are now a major and ubiquitous tool in the theory of C^*- and von Neumann algebras, and there are by now a huge number of important examples (see e.g. [51, 69] for more references).

Nonetheless, outside of the class of von Neumann algebras with a faithful normal tracial state, the existence of a weak* continuous conditional expectation onto a von Neumann subalgebra D is a difficult question (unless D is atomic). Indeed this question gets to the heart of, and uses the whole industry of the theory of noncommutative integration (due to Connes, Haagerup, Pedersen, Takesaki, and very many other brilliant operator algebraists). See [53, 69] for a taste of the latter. For a commutative von Neumann algebra $L^\infty(K, \mathcal{A}, \mu)$ again, but with μ not σ-finite one must use the theory of semifinite measures to construct a conditional expectation onto a von Neumann subalgebra. Now suppose that M is a noncommutative *semifinite von Neumann algebra*, for example $B(l^2)$. Then M has a faithful normal semifinite trace τ. However there need not exist any conditional expectation onto a fixed von Neumann subalgebra (e.g. it is known that there is no conditional expectation from $B(l^2)$ onto nonatomic von Neumann subalgebras). Indeed there exist a τ-preserving conditional expectation onto a von Neumann subalgebra D if and only if τ restricts to a semifinite trace on D. The one direction of this is [68, Proposition V.2.36]. For the other, if $0 \neq x \in D_+$ and $0 \neq y \in M_+$ with $\tau(y) < \infty$ and $y \leq x$, then $\tau(E(y)) = \tau(y) \in (0, \infty)$ and $E(y) \leq x$. Clearly $0 \neq E(y) \in D_+$. We have verified that the restriction of τ to D is semifinite.

For non-semifinite von Neumann algebras the situation is much more complicated, and gets into Haagerup's theory of operator valued weights (see [53, 69] for references). The conditions for existence of a weak* continuous conditional expectation onto a von Neumann subalgebra are much more intricate, such conditions involving the operator semigroup central to Tomita-Takesaki modular theory. See e.g. [69, Theorem 4.2]. For technical reasons and to avoid pathologies one usually insists that M possesses a faithful normal state ν (which is equivalent to M possessing a faithful state). This class of von Neumann algebras includes those on a separable Hilbert space, or with separable predual. Then there exists a ν-preserving conditional expectation E onto a von Neumann subalgebra D if and only if D is invariant under the modular automorphism group (σ_t^ν) of ν (see [68, Theorem IX.4.2]). Such a ν-preserving conditional expectation is again unique. We may write this E as E_ν, and again call this the 'probabilistic' conditional expectation (it depends on the fixed state ν, which can be thought of as a noncommutative probability integral). We will not go into further detail here.

In summary we have seen that under certain conditions on a von Neumann algebra M and a von Neumann subalgebra D, and on a positive functional or weight ν on M, conditions usually involving modular theory, there exists a unique ν-preserving weak* continuous unital contractive projection from M onto D. We call this a conditional expectation, and it is analogous to $E_{\mathcal{B}}$ above.

7 Noncommutative Characters on Noncommutative Function Algebras (Operator Algebras)

The last part of our Positivity X lecture was concerned with ongoing joint work with L. E. Labuschagne [11, 12], on a special case of the real positive projections considered in the last Sect. 5. This case we consider to be a good noncommutative generalization of the classical theory of 'characters' (i.e. homomorphisms into the scalars) of a function algebra (see e.g. [41]). Recall that if $A \subset C(K)$ is a function algebra or uniform algebra on compact set K, then the fundamental associated object is the set M_A of (scalar valued) characters on A.

A noncommutative function algebra for us is just an operator algebra in the earlier sense, a subalgebra A of a C^*-algebra C. We assume C unital and $1_C \in A$ for simplicity here. In the nonunital case we can unitize by the tricks in the early parts of Sects. 2 and 3 above. In this setting scalar valued characters are usually not so useful, however we have found that in the following setting one can generalize many of the classical function algebra character results. Namely, consider an inclusion $D \subset A \subset C$, where A, C are as before, and D is a C^*-subalgebra of A. A *D-character* is a unital contractive homomorphism $\Phi : A \to D$ which is also a D-bimodule map (or equivalently, is the identity map on D). The classical scalar valued characters χ on A fit into this setting by identifying χ with $\chi(\cdot)1_A$. We were motivated to study these because of their importance in (the definition of) *Arveson's subdiagonal algebras* [4, 10]. Arveson also gives very many good examples of such D-characters in that paper.

Note that these fall within the framework of Sect. 5, they are in fact automatically real completely positive completely contractive projections from A onto a subalgebra. That they are completely contractive follows from the standard trick mentioned at the end of the proof of Theorem 5.12. That they are real completely positive follows from e.g. Proposition 3.2. Recall also that by e.g. Theorem 5.4. a completely contractive unital projection onto a unital C^*-subalgebra D is automatically a D-bimodule map.

In this section we will for simplicity stick to the case of contractive unital characters. In the nonunital case one would consider (completely) contractive real positive homomorphisms from the operator algebra A onto a C^*-subalgebra. As in Sect. 5 these extend uniquely to (completely) contractive unital homomorphisms on A^1, and so we are back in the unital character case. Thus we may suppress discussion of real positivity in the next paragraphs: it is there but automatic.

In [11, 12], Labuschagne and the author consider several problems that arise when generalizing classical function algebra results involving characters. For the sake of the present article not becoming too scattered in theme we just mention briefly a couple of examples of these that use specific theorems from our earlier sections above. The first is a new noncommutative take on the classical theory of Gleason parts of function algebras. The Gleason relation ($\|\varphi - \psi\| < 2$) on characters of function algebras does not seem to have a $B(H)$ valued analogue suitable for our purposes, but we show that interestingly it does have a noncommutative variant for our D-characters.

We also use some concepts considered by Harris in e.g. [43, 44]:

$$T_x(y) = (1 - xx^*)^{-\frac{1}{2}}(x + y)(1 + x^*y)^{-1}(1 - x^*x)^{\frac{1}{2}}.$$

This makes sense for elements in the open unit ball in $B(H)$. For fixed such x the maps T_x are essentially exactly the biholomorphic self maps of the open unit ball in $B(H)$, or are *Möbius maps* of this open ball. The *hyperbolic distance* $\rho(x, y)$ is

$$\tanh^{-1} \|(1 - xx^*)^{-\frac{1}{2}}(x - y)(1 - x^*y)^{-1}(1 - x^*x)^{\frac{1}{2}}\| = \tanh^{-1} \|T_{-x}(y)\|.$$

Harris shows [45] that ρ is what is known as a *CRF pseudometric* on the open unit ball $\mathcal{U}_0$ and it satisfies the *Schwarz-Pick inequality*

$$\rho(h(x), h(y)) \leq \rho(x, y), \qquad x, y \in \mathcal{U}_0,$$

for any holomorphic $h : \mathcal{U}_0 \to \mathcal{U}_0$. We have equality here if h is biholomorphic.

We may also define an equivalence relation using the *real positive ordering*. If Φ, Ψ are maps from A into a C^*-algebra we write $\Phi \preceq \Psi$ if $\Psi - \Phi$ is a real positive map (in the sense of e.g. Sect. 3). Note that one may then show that in this situation it is real completely positive, and then apply Theorem 2.4. This permits us to define an equivalence relation on D-characters by the existence of strictly positive constants c, d with $\Phi \preccurlyeq c\Psi$ and $\Psi \preccurlyeq d\Phi$. The reasoning in the last few lines ties this equivalence relation with the famous notion of Harnack equivalence (see e.g. [67], and we thank Sanne ter Horst for this and many other references).

Using these ideas, and following classical methods, and results like Theorem 2.4 above, one may prove:

Theorem 7.1 *Consider inclusions $D \subset A \subset C$ as above. Suppose that D is represented nondegenerately on a Hilbert space H. Let $\Phi, \Psi : A \to D$ be D-characters. The following are equivalent:*

(1) $\|\Phi - \Psi\| < 2$.
(2) $\|\Phi_{|\,Ker\,\Psi}\| < 1$.
(3) *There is a constant $M > 0$ with $\rho(\Phi(a), \Psi(a)) \leq M$ for $\|a\| < 1, a \in A$.*
(4) *If $\|\Phi(a_n)\| \to 1$ for a sequence (a_n) in Ball(A), then $\|\Psi(a_n)\| \to 1$.*

The above conditions are implied by the equivalent conditions:

(5) *There are positive constants c, d with $\Phi \preccurlyeq c\Psi$ and $\Psi \preccurlyeq d\Phi$.*
(6) *There are positive constants c, d and completely positive $B(H)$-valued maps $\tilde{\Phi}, \tilde{\Psi}$ extending Φ, Ψ to C, with $\tilde{\Phi} \leq c\tilde{\Psi}$ and $\tilde{\Psi} \leq d\tilde{\Phi}$.*

If D is one dimensional then all the conditions here are equivalent.

At the time of writing we do not know if the conditions in the last theorem are equivalent in full generality. That is, we do not know if we have two distinct equivalence relations in the general case. Gleason parts are applied in [11] to the theory of Hankel and Fredholm Toeplitz operators.

Remark 7.2 If $\Phi : A \to A$ is a completely contractive unital projection then $\Phi(A)$ is an operator algebra in the Φ product and Φ is a '$\Phi(A)$-bimodule map' with respect to that operator algebra, by Theorem 5.4. If $\Phi(A) \subset \Delta(A)$ then by the idea in the proof of Corollary 5.9 but appealing to the Choi-Effros in our introduction instead of to Effros-Størmer, one sees that $\Phi(A)$ is selfadjoint, a C^*-algebra in the Φ product. This is not mentioned in [11] but many of the results in that paper, including parts of the theorem above, will go through for such maps.

Finally we discuss *noncommutative representing measures* for $\Phi : A \to D$, a D-character. A positive measure μ on a set K is called a *representing measure* for a character Φ of a function algebra A on K if $\Phi(f) = \int_K f \, d\mu$ for all $f \in A$. The functional $\widetilde{\Phi}(g) = \int_K g \, d\mu$ on $C(K)$ is a state on $C(K)$, and indeed representing measures for Φ are in a bijective correspondence with the extensions of Φ to a positive functional on $C(K)$. That is, representing measures for a character are just the Hahn-Banach extensions to $C(K)$ of that character.

Noncommutative representing measures will therefore be related somewhat to the earlier theorem 2.4 concerning positive extensions. Suppose that we are given a faithful representation of D on a Hilbert space H. The usual noncommutative analogue of a 'noncommutative representing measure' for say a D-character on A would be a $B(H)$-valued extension of Φ to a C^*-algebra B containing A, which is completely positive (or equivalently, in this case, completely contractive). Such noncommutative representing measures $\Psi : B \to B(H)$ always exist, by Theorem 2.4 (indeed by Arveson's extension theorem [5, Theorem 1.2.9]). However although these noncommutative notions are appropriate in many settings, they do not necessarily seem appropriate when generalizing some other important parts of the theory of uniform algebras. An intuitive reason we advance for now for this (other reasons will become clearer momentarily) is that $B(H)$ is too big, thus insensitive; in some settings one probably would not want to go too far from D in the range if one does not have to. We shall see below that for some purposes one should not have to.

The alternative *noncommutative representing measure* that we are proposing, again inspired by Arveson (but this time his noncommutative analyticity work [4]), is a completely positive extension to B that takes values in D (or possibly a weak* closure of D). Let us call these *tight noncommutative representing measures*.

Now however one has to face the problem of existence of such an extension. Such existence would in a real sense improve on Theorem 2.4 in the case of D-characters. This problem is dealt with by exploiting the C^*-algebraic or von Neumann algebraic theory of conditional expectations from B onto D.

In the following discussion, we have weak*-continuous unital inclusions $D \subset A \subset M$, where M is a von Neumann algebra, and A and D are unital weak* closed subalgebras, with D selfadjoint (hence a von Neumann subalgebra). We are also given a weak*-continuous D-character $\Phi : A \to D$. We seek a weak* continuous positive extension $\Psi : M \to D$. Now we can see that we are asking for something quite interesting in several ways. Firstly, we are asking for a generalization of the remarkable and deep theory of von Neumann algebra conditional expectations summarized briefly in Sect. 6. Indeed setting $A = D$, the question above becomes precisely the important question of the existence of a normal (i.e. weak* continuous) expectation of a fixed von Neumann algebra onto a von Neumann subalgebra. Second, it is interesting because weak* continuous positive extensions of weak* continuous *linear* unital contractive maps do not typically exist. Indeed saying 'positive' here is equivalent to saying 'contractive', and even in the case that the range is one dimensional (i.e. $D = \mathbb{C}1$) the Hahn-Banach theorem about extensions with the same norm usually fails drastically if all maps are supposed to be weak* continuous. This point is discussed early in [9], and we will end our paper with an example of such failure. It is important that Φ is a homomorphism for such a positive weak* continuous extension to exist. Third, this is precisely the setting of Arveson's famous paper [4] on noncommutative generalizations of Hardy spaces. In Arveson's approach to noncommutative analyticity/generalized analytic functions/Hardy spaces we have a normal conditional expectation $\Psi : M \to D$ extending a D-character Φ on A. In this 'generalized analytic function theory' it is very important that the representing measures are D-valued rather than $B(H)$-valued.

In the classical case if μ is a representing (probability) measure on a space K for a character θ of a function algebra A on K, we define $H^\infty(\mu)$ to be the weak* closure of A in $L^\infty(\mu)$. Similarly for $p < \infty$ define $H^p(\mu)$ to be the closure of A in $L^p(\mu)$. E.g. if A is the disk algebra or H^∞ of the disk, then $\theta(f) = f(0)$, and the important 'representing measure' is $\mu(f) = \int_{\mathbb{T}} f \, dm$, Lebesgue integration on the circle, which is a state on $C(\mathbb{T})$ and a weak* continuous state on $L^\infty(\mathbb{T})$. If A is a *Dirichlet* or *logmodular algebra*, and indeed much more generally, these Hardy spaces behave very similarly to the classical Hardy spaces of the disk. One obtains an F & M Riesz theorem, Beurling's theorem, Jensen and Szego theorems, Gleason-Whitney theorem, inner-outer factorization, and so on. Arveson was attempting a vast noncommutative generalization of all of this, using precisely the noncommutative representing measure of a D-character approach that we are describing.

Arveson gave many interesting examples, showing that his framework synthesized several theories that were emerging in the 1960's. Work on Arveson's spaces has continued over the decades by very many authors (see e.g. [10] for many references), being at present something of an international industry. His vision was realized in the case that $A + A^*$ is weak* dense in M (again, see e.g. [10]). We are trying to push this same noncommutative representing measure approach to operator algebras beyond the latter case.

Returning to tight noncommutative representing measures for a weak* continuous D-character $\Phi : A \to D$, the primary problem concerns their existence, which turns out to hold for quite subtle reasons. We seek a weak* continuous positive extension $\Psi : M \to D$ of Φ, where M is a fixed von Neumann algebra containing A unitally, and contains D as a von Neumann subalgebra. In [9] this is done if D is atomic (and it is explained there why this is a noncommutative generalization of an old theorem of Hoffman and Rossi).

Theorem 7.3 ([12]) *Consider weak*-continuous unital inclusions $D \subset A \subset M$, where M is a von Neumann algebra which is commutative, or which possesses a faithful normal tracial state, and A and D are unital weak* closed subalgebras, with D selfadjoint. If $\Phi : A \to D$ is a weak*-continuous D-character then Φ has a weak* continuous positive extension $\Psi : M \to D$.*

The main point is that this suggests that at least for some purposes one should not need, and probably should not use, general $B(H)$-valued extensions of D-characters. One in fact has the (surprising) existence of tight noncommutative representing measures. Indeed in the above theorem we have this existence for any von Neumann subalgebra D.

We also have much more general theorems giving existence of weak*-continuous representing measures [12]. They are similar to the last result: there exist weak* continuous positive extensions $\Psi : M \to D$ of weak*-continuous D-characters on weak* closed unital subalgebras of M for much more general classes of von Neumann algebras M (without a faithful normal tracial state). These results require extra conditions e.g. on modular automorphism groups in the same spirit as the second last paragraph of Sect. 6. That is, there exists a much more general, but considerably more technical, version of Theorem 7.3. Those familiar with the conditions from Tomita-Takesaki theory ensuring the existence of von Neumann algebraic conditional expectations (see e.g. [69, Theorem 4.2]) will be able to guess what the vague conditions are. Those not versed in modular theory would not be thankful for an explicit statement of these conditions! Basically we are saying that 'noncommutative representing measures' exist in many such settings, under basically the same conditions that von Neumann algebra conditional expectations exist. The proofs use, in additional to the arsenal of noncommutative integration theory alluded to above, techniques from [52, 53, 72] and elsewhere, as well as new ideas.

Example We end our paper with an example showing the necessity of using D-characters $\Phi : A \to D$ in the results above, even in the scalar valued case ($D = \mathbb{C}$). The suspicious reader might think that possibly the issue is that A needs to be an algebra, rather than that Φ needs to be a homomorphism. Take any subspace $\mathcal{S}$ of $L^\infty([0,1]$ possessing a norm 1 functional φ_1 with no weak* continuous Hahn Banach (state) extension to $N = L^\infty([0,1])$. For example, the polynomials of degree ≤ 1 with $\varphi_1(p) = p(1)$ will do. In [9, Proposition 2.6] we considered the four dimensional subalgebra A of $M_2(L^\infty([0, 1])$ consisting of upper triangular 2×2 matrices with constant functions on the diagonal, and an element from $\mathcal{S}$ in the 1-2 position. We will construct a weak* continuous state (hence a completely contractive D-module map onto D) on A with no weak* continuous Hahn-Banach extension to $M_2(N)$.

To do this let $s \in (0, 1)$. We will use the fact that the functional on the upper triangular 2×2 matrices which takes $E_{11} \mapsto s$, $E_{22} \mapsto 1 - s$, $E_{12} \mapsto \mu$, for $s \in [0, 1]$, is contractive (and hence is a state) if and only if $|\mu|^2 \leq s(1 - s)$. To see this note it is contractive if and only if it is a state. States on the upper triangular 2×2 matrices are easily seen to have unique state extensions to M_2. Indeed there is a bijectiion between these two state spaces. States on M_2 correspond to density matrices, that is positive matrices of trace 1. These are selfadjoint matrices with $s, 1 - s$ on the diagonal for $s \in [0, 1]$, and off diagonal entries coming from a scalar μ with $|\mu|^2 \leq s(1 - s)$.

We may scale: let $\psi = \sqrt{s(1 - s)}\,\varphi_1$. Then the functional φ on A defined by

$$\varphi(\begin{bmatrix} a & x \\ 0 & c \end{bmatrix}) = sa + (1 - s)c + \psi(x), \quad a, c \in \mathbb{C}, x \in X,$$

is a weak* continuous state on A. This uses the fact (clear from the formula (2.1) in [15]) that if z is the first matrix in the last displayed equation, then z has the same norm as the same matrix but with x replaced by $\|x\|$, and a and c replaced by their modulus. Since $\|\psi\| = \sqrt{s(1 - s)}$, we have by the last paragraph that

$$s|a| + (1 - s)|c| + |\psi(x)| = s|a| + (1 - s)|c| + \|\psi\|\|x\| \leq \|\begin{bmatrix} |a| & \|x\| \\ 0 & |c| \end{bmatrix}\| = \|z\|.$$

Thus $|\varphi(z)| \leq \|z\|$. We claim that φ has no weak* continuous Hahn-Banach extension to $M_2(N)$. Indeed if there were, then we obtain a weak* continuous Hahn-Banach extension to B, the set of upper triangular 2×2 matrices with constant functions on the diagonal, and an element from N in the 1-2 position. Thus we obtain a weak* continuous extension ξ of ψ to N, such that the functional

$$\begin{bmatrix} a & x \\ 0 & c \end{bmatrix} \mapsto sa + (1 - s)c + \xi(x), \quad a, c \in \mathbb{C}, x \in N,$$

is a weak* continuous state on B. If $x \in N$ with $\|x\| = 1$ then for any scalars a, b, c we have

$$|sa + (1 - s)c + b\xi(x)| \leq \|\begin{bmatrix} a & bx \\ 0 & c \end{bmatrix}\| = \|\begin{bmatrix} a & \|bx\| \\ 0 & c \end{bmatrix}\| = \|\begin{bmatrix} a & b \\ 0 & c \end{bmatrix}\|.$$

We have used again formula (2.1) in [15], twice. Thus by the last paragraph $|\xi(x)| \leq \sqrt{s(1-s)}$, so that $\|\xi\| \leq \sqrt{s(1-s)}$. This contradicts the fact that ψ has no weak* continuous Hahn Banach extension to N.

Acknowledgments We thank Louis Labuschagne and Matt Neal for very many discussions. We also thank Angel Rodríguez Palacios for many very useful comments and references. Some of our results on contractive projections, and some complementary results, he has subsequently extended in unpublished work [62] to the class called *Arazy algebras* and introduced in [31]; in particular to unital complete normed power-associative complex algebra satisfying von Neumann's inequality (we will not define all these terms). Some related theory of nonunital nonassociative algebras may be found in the last section of [32]. While writing this article we learned of the death of Ed Effros, many of whose beautiful and important ideas are featured here, and were struck again by his profound contributions to the subject. Similarly, we often fondly remembered Coenraad Labuschagne during this writing for his warm and kind personality, and for his fine work on the subject of conditional expectations.

References

1. L. Accardi, C. Cecchini, Conditional expectations in von Neumann algebras and a theorem of Takesaki. J. Funct. Anal. **45**, 245–273 (1982)
2. J. Arazy, Isometries of Banach algebras satisfying the von Neumann inequality. Math. Scand. **74**, 137–151 (1994)
3. J. Arazy, B. Solel, Isometries of nonselfadjoint operator algebras. J. Funct. Anal. **90**, 284–305 (1990)
4. W.B. Arveson, Analyticity in operator algebras. Am. J. Math. **89**, 578–642 (1967)
5. W.B. Arveson, Subalgebras of C^*-algebras. Acta Math. **123**, 141–224 (1969)
6. C.A. Bearden, D.P. Blecher, S. Sharma, On positivity and roots in operator algebras. Integr. Equ. Oper. Theory **79**, 555–566 (2014)
7. B. Blackadar, *Operator Algebras. Theory of C^*-Algebras and von Neumann Algebras.* Encyclopaedia of Mathematical Sciences, vol. 122 (Springer, Berlin, 2006)
8. D.P. Blecher, Generalization of C*-algebra methods via real positivity for operator and Banach algebras, in *Operator Algebras and Their Applications: A Tribute to Richard V. Kadison*, ed. by R.S. Doran, E. Park. Contemporary Mathematics, vol. 671 (American Mathematical Society, Providence, 2016), pages 35–66
9. D.P. Blecher, L.C. Flores, B.G. Zimmer, The Hoffman-Rossi theorem for operator algebras. Integr. Equ. Operator Theory **91**(2), 17 (2019), 7 pp.
10. D.P. Blecher, L.E. Labuschagne, Von Neumann algebraic H^p theory, in *Function Spaces: Fifth Conference on Function Spaces*. Contemp. Math. vol. 435 (Amer. Math. Soc. 2007), pp. 89–114
11. D.P. Blecher, L.E. Labuschagne, On vector-valued characters for noncommutative function algebras. Compl. Anal. Oper. Theory **14**, 31 (2020), 30 pp. https://doi.org/10.1007/s11785-020-00989-1
12. D.P. Blecher, L.E. Labuschagne, Von Neumann algebra conditional expectations with applications to generalized representing measures for noncommutative function algebras, preprint (2021)
13. D.P. Blecher, C. Le Merdy, *Operator Algebras and Their Modules—an Operator Space Approach* (Oxford University Press, Oxford, 2004)
14. D.P. Blecher, B. Magajna, Duality and operator algebras: automatic weak* continuity and applications. J. Funct. Anal. **224**, 386–407 (2005)

15. D.P. Blecher, B. Magajna, Duality and operator algebras II: Operator algebras as Banach algebras. J. Funct. Anal. **226**, 485–493 (2005)
16. D.P. Blecher, M. Neal, Completely contractive projections on operator algebras. Pac. J. Math. **283**(2), 289–324 (2016)
17. D.P. Blecher, M. Neal, Noncommutative topology and Jordan operator algebras. Math. Nachr. **292**, 481–510 (2019)
18. D.P. Blecher, M. Neal, Contractive projections and real positive maps on operator algebras. Stud. Mat. **256**, 21–60 (2021)
19. D.P. Blecher, N. Ozawa, Real positivity and approximate identities in Banach algebras. Pac. Math. J. **277**, 1–59 (2015)
20. D.P. Blecher, N.C. Phillips, L^p operator algebras with approximate identities I. Pac. J. Math. **303**, 401–457 (2019). https://doi.org/10.2140/pjm.2019.303.401
21. D.P. Blecher, N.C. Phillips, L^p operator algebras with approximate identities II. In preparation
22. D.P. Blecher, C.J. Read, Operator algebras with contractive approximate identities. J. Funct. Anal. **261**, 188–217 (2011)
23. D.P. Blecher, C.J. Read, Operator algebras with contractive approximate identities II. J. Funct. Anal. **264**, 1049–1067 (2013)
24. D.P. Blecher, C.J. Read, Order theory and interpolation in operator algebras. Stud. Math. **225**, 61–95 (2014)
25. D.P. Blecher, Z.-J. Ruan, A.M. Sinclair, A characterization of operator algebras. J. Funct. Anal. **89**, 188–201 (1990)
26. D.P. Blecher, W. Tepsan, Real operator algebras and real positive maps. Preprint July 2020, arXiv:2007.12259
27. D.P. Blecher, Z. Wang, Jordan operator algebras: basic theory. Math. Nachr. **291**, 1629–1654 (2018)
28. D.P. Blecher, Z. Wang, Jordan operator algebras revisited. Math. Nachr. **292**, 2129–2136 (2019)
29. M. Cabrera, A. Rodríguez, *Non-Associative Normed Algebras*. vol. 1, The Vidav-Palmer and Gelfand-Naimark Theorems. Encyclopedia of Mathematics and Its Applications, vol. 154 (Cambridge University Press, Cambridge, 2014)
30. M. Cabrera, A. Rodríguez, *Non-Associative Normed Algebras. Vol. 2, Representation Theory and the Zel'manov Approach*. Encyclopedia of Mathematics and Its Applications, vol. 167. (Cambridge University Press, Cambridge, 2018)
31. M. Cabrera, A. Rodríguez, The von Neumann inequality in complete normed non-associative complex algebras. Math. Proc. R. Ir. Acad. **118A**, 83–125 (2018)
32. M. Cabrera, A. Rodríguez, Multiplication algebras: algebraic and analytic aspects, in *Associative and Non-Associative Algebras and Applications, 3rd Moroccan Andalusian Meeting on Algebras and their Application* ed. by M. Siles, E.K. Laiachi, M. Louzari, L.M. Ben Yakoub, M. Benslimane. Springer Proceedings in Mathematics and Statistics, vol. 311 (Springer, Berlin, 2020), pp. 113–138
33. M. Cabrera, A. Rodríguez, Unit-free contractive projection theorems for C*-, JB*-, and JB-algebras. J. Math. Anal. Appl. **486**, 123–921 (2020)
34. M.-D. Choi, E.G. Effros, Injectivity and operator spaces. J. Funct. Anal. **24**, 156–209 (1977)
35. N. Dinculeanu, M.M. Rao, Contractive projections and conditional expectations. Multivar. Anal. **2**, 362–381 (1972)
36. R.G. Douglas, Contractive projections on an L^1 space. Pac. J. Math. **15**, 443–462 (1965)
37. E.G. Effros, E. Størmer, Positive projections and Jordan structure in operator algebras. Math. Scand. **45**, 127–138 (1979)
38. Y. Friedman, B. Russo, Conditional expectation without order. Pac. J. Math. **115**, 351–360 (1984)
39. Y. Friedman, B. Russo, Solution of the contractive projection problem. J. Funct. Anal. **60**, 56–79 (1985)
40. Y. Friedman, B. Russo, Conditional expectation and bicontractive projections on Jordan C^*-algebras and their generalizations. Math. Z. **194**, 227–236 (1987)
41. T.W. Gamelin, *Uniform Algebras,* 2nd edn. (Chelsea, New York, 1984)

42. H. Hanche-Olsen, E. Størmer, *Jordan Operator Algebras*. Monographs and Studies in Mathematics, vol. 21 (Pitman (Advanced Publishing Program), Boston, 1984)
43. L.A. Harris, Banach algebras with involution and Möbius transformations. J. Funct. Anal. **11**, 1–16 (1972)
44. L.A. Harris, *Bounded Symmetric Homogeneous Domains in Infinite Dimensional Spaces*. Lecture Notes in Math., vol. 364 (Springer, Berlin-New York, 1973)
45. L. A. Harris, Schwarz-Pick systems of pseudometrics for domains in normed linear spaces, in *Advances in holomorphy (Proc. Sem. Univ. Fed. Rio de Janeiro, Rio de Janeiro, 1977)*. North-Holland Math. Stud., vol. 34 (North-Holland, Amsterdam-New York, 1979), pp. 345–406
46. L.A. Harris, A generalization of C^*-algebras. Proc. Lond. Math. Soc. **42**, 331–361 (1981)
47. M. Horodecki, P. Horodecki, R. Horodecki, Separability of mixed states: necessary and sufficient conditions. Phys. Lett. A **223**, 1–8 (1996)
48. P. Jordan, J. von Neumann, E. Wigner, On an algebraic generalization of the quantum mechanical formalism. Ann. Math. **35**, 29–64 (1934)
49. W. Kaup, Contractive projections on Jordan $C*$-algebras and generalizations. Math. Scand. **54**, 95–100 (1984)
50. R.V. Kadison, Isometries of operator algebras. Ann. Math. **54**, 325–338 (1951)
51. R.V. Kadison, Non-commutative conditional expectations and their applications, in *Operator Algebras, Quantization, and Noncommutative Geometry*. Contemp. Math., vol. 365 (Amer. Math. Soc., Providence, 2004), pp. 143–179
52. L.E. Labuschagne, Invariant subspaces for H^2 spaces of σ-finite algebras. Bull. Lond. Math. Soc. **49**, 33–44 (2017)
53. L.E. Labuschagne, S. Goldstein, Notes on noncommutative L^p and Orlicz spaces (2020). Submitted ebook manuscript
54. H.E. Lacey, *The Isometric Theory of Classical Banach Spaces*. Die Grundlehren der Math. Wissenschaften, vol. 208 (Springer, New York-Heidelberg, 1974)
55. A.T. Lau, R.J. Loy, Contractive projections on Banach algebras. J. Funct. Anal. **254**, 2513–2533 (2008)
56. S.-T.C. Moy, Characterizations of conditional expectation as a transformation on function spaces. Pac. J. Math. **4**, 47–63 (1954)
57. V.I. Paulsen, *Completely Bounded Maps and Operator Algebras*. Cambridge Studies in Advanced Math., vol. 78 (Cambridge University Press, Cambridge, 2002)
58. G.K. Pedersen, C^*-algebras and their automorphism groups. (Academic Press, London, 1979)
59. A. Rodríguez, Non-associative normed algebras spanned by Hermitian elements. Proc. Lond. Math. Soc. **47**, 258–274 (1983)
60. A. Rodríguez, Isometries and Jordan isomorphisms onto $C*$-algebras. J. Oper. Theory **40**, 71–85 (1998)
61. A. Rodríguez, Approximately norm-unital products on C^*-algebras, and a non-associative Gelfand-Naimark theorem. J. Algebra **347**, 224–246 (2011)
62. A. Rodríguez, Conditional expectations in complete normed complex algebras satisfying the von Neumann inequality (2020). Unpublished work in progress
63. W.F. Stinespring, Positive functions on C^*-algebras. Proc. Am. Math. Soc. **6**, 211–216 (1955)
64. E. Størmer, Positive linear maps of operator algebras. Acta Math. **110**, 233–278 (1963)
65. E. Størmer, Positive projections with contractive complements on C^*-algebras. J. Lond. Math. Soc. **26**, 132–142 (1982)
66. E. Størmer, *Positive Linear Maps of Operator Algebras*. Springer Monographs in Mathematics (Springer, Berlin, 2013)
67. I. Suciu, I. Valusescu, On the hyperbolic metric on Harnack parts. Stud. Math. **55**, 97–109 (1975/1976)
68. M. Takesaki, *Theory of Operator Algebras I* (Springer, New York, 1979)
69. M. Takesaki, *Theory of Operator Algebras II*. Encyclopaedia of Mathematical Sciences, vol. 125. Operator Algebras and Non-Commutative Geometry, vol. 6 (Springer, Berlin, 2003)

70. Z. Wang, Theory of Jordan operator algebras and operator $*$-algebras. Ph.D. Thesis, University of Houston, 2019
71. P. Wong, N. Ozawa, A characterization of completely 1-complemented subspaces of noncommutative L^1-spaces. Pac. J. Math. **205**, 171–195 (2002)
72. Q. Xu, On the maximality of subdiagonal algebras. J. Oper. Theory **54**, 137–146 (2005)

Free Vector Lattices and Free Vector Lattice Algebras

Marcel de Jeu

Dedicated to the memory of Coenraad Labuschagne

Abstract We show how the existence of various free vector lattices and free vector lattice algebras can be derived from a theorem on equational classes in universal algebra. A discussion about free f-algebras over non-empty sets is given, where the main issues appear to be open. It is indicated how the existence results for free vector lattices and vector lattice algebras can be used for easy proofs of existence results for free Banach lattices and free Banach lattice algebras. A detailed exposition of the necessary material from universal algebra is included.

Keywords Free vector lattice · Free vector lattice algebra · Equational class

Mathematics Subject Classification (2010) Primary 46A40; Secondary 06F25

1 Introduction and Overview

In recent years, there has been a growing interest in free Banach lattices. Definitions have been given of a free Banach lattice over a set (see [12]), over a Banach space (see [4, 5, 22]), and over a lattice (see [3]). These objects have been shown to exist, and properties beyond their mere existence have been studied.

The starting point for the existence proofs in these papers is a concrete model for a free object that has been obtained earlier. The most basic of these concrete models appears to be the usual model for the free vector lattice over a non-empty

M. de Jeu (✉)
Mathematical Institute, Leiden University, Leiden, The Netherlands

Department of Mathematics and Applied Mathematics, University of Pretoria, Pretoria, South Africa
e-mail: mdejeu@math.leidenuniv.nl

E. Kikianty et al. (eds.), *Positivity and its Applications*, Trends in Mathematics,
https://doi.org/10.1007/978-3-030-70974-7_6

set S as a sublattice of $\mathbb{R}^{\mathbb{R}^S}$; see [6] or [10], for example. In [12], this concrete model is then used in the construction of the free Banach lattice over a non-empty set S (see [12, Definition 4.4]). Likewise, it is an ingredient in the construction of the free Banach lattice over a Banach space (see [4, beginning of the proof of Theorem 2.5]). In [3, p. 583], the existence of the free Banach lattice over a non-empty set, as established in [12], is used to construct the free Banach lattice over a vector lattice. This construction is, therefore, in the end also based on the usual concrete model for the free vector lattice over a non-empty set. In [22], this model is used to construct the free Banach lattice over non-empty sets again (simplifying the existence proof in [12]); this, in turn, is used to construct the free Banach lattice over a Banach space (simplifying the existence proof in [4]).

It seems to have escaped notice so far that there is an alternative and, as we believe, simpler way to obtain the existence of such free functional analytic objects. The general strategy is to start with the mere existence—a concrete model is not needed—of a corresponding free object in an algebraic context and then, almost as an afterthought, add the norm to the picture by using a few standard constructions. Let us give a detailed example on how to construct a free Banach lattice over a Banach space along these lines.

Suppose that X is a (real) Banach space. By Theorem 6.2, below, there exist a vector lattice E and a map $j : X \to E$ with the property that, for every vector lattice Y and for every linear map $\varphi : X \to Y$, there exists a unique vector lattice homomorphism $\overline{\varphi}$ such that the diagram

$$\begin{array}{ccc} X & \xrightarrow{\ j\ } & E \\ & \searrow^{\varphi} & \downarrow{\scriptstyle \overline{\varphi}} \\ & & Y \end{array}$$

is commutative. Such a vector lattice E is called a free vector lattice over the vector space X. It is easy to see that E is generated, as a vector lattice, by its subset $j(X)$. For $e \in E$, set

$$\rho(e) := \sup\left\{\|\overline{\Psi}(e)\| : Y \text{ is a Banach lattice and } \Psi : X \to Y \text{ is a contraction}\right\}.$$

Using the correspondence between lattice seminorms on E and vector lattice homomorphisms from E into Banach lattices, one easily sees that one can, equivalently, set

$$\rho(e) := \sup\left\{\sigma(e) : \sigma \text{ is a lattice seminorm on } E \text{ and } \sigma \circ j \text{ is contractive on } X\right\},$$

thereby avoiding possible set-theoretical subtleties.

For $x \in X$, it is clear that $\rho(j(x)) \leq \|x\|$. Since the subset of E on which q is finite is easily seen to be a vector sublattice of E, and since $j(X)$ generates E as a vector lattice, we conclude that ρ is a lattice seminorm on E.[1] The kernel of ρ is an order ideal in E. We let $q : E \to E/\ker\rho$ denote the quotient map. On setting $\|q(e)\| := \rho(e)$ for $e \in E$, the vector lattice $E/\ker\rho$ becomes a normed vector lattice. It is then immediate that $q \circ j : X \to E/\ker\rho$ is contractive.

Let Y be a Banach lattice and let $\varphi : X \to Y$ be a bounded linear map. Suppose that $e \in E$ is such that $\rho(e) = 0$. We claim that then also $\overline{\varphi}(e) = 0$. This is clear if $\varphi = 0$. When $\varphi \neq 0$, so that $\varphi/\|\varphi\|$ is a contraction, this follows from the definition of ρ. Hence there exists a unique vector lattice homomorphism $\overline{\overline{\varphi}}$ such that the diagram

$$\begin{array}{ccccc} X & \xrightarrow{\;j\;} & E & \xrightarrow{\;q\;} & E/\ker\rho \\ & \searrow^{\varphi} & \downarrow{\scriptstyle\overline{\varphi}} & \swarrow_{\overline{\overline{\varphi}}} & \\ & & Y & & \end{array} \tag{1.1}$$

is commutative.

We claim that $\overline{\overline{\varphi}}$ is bounded and that $\|\overline{\overline{\varphi}}\| = \|\varphi\|$. We may suppose that $\varphi \neq 0$, so that $\varphi/\|\varphi\|$ is a contraction. For $e \in E$, we then have

$$\begin{aligned} \left\|\overline{\overline{\varphi}}(q(e))\right\| &= \left\|\overline{\varphi}(e)\right\| \\ &= \|\varphi\| \left\| \overline{\left(\frac{\varphi}{\|\varphi\|}\right)}(e) \right\| \\ &\leq \|\varphi\|\, \rho(e) \\ &= \|\varphi\|\, \|q(e)\|. \end{aligned}$$

Hence $\overline{\overline{\varphi}}$ is bounded and $\|\overline{\overline{\varphi}}\| \leq \|\varphi\|$. On the other hand, the fact that $\varphi = \overline{\overline{\varphi}} \circ (q \circ j)$, combined with the fact that $q \circ j$ is contractive, shows that $\|\varphi\| \leq \|\overline{\overline{\varphi}}\|$. Hence $\|\overline{\overline{\varphi}}\| = \|\varphi\|$, as claimed.

There exists an isometric linear map from X into a Banach lattice. Indeed, the canonical embedding of X into the bounded real-valued functions on the unit ball of its dual is such a map. Take such an isometric linear embedding for φ. Then

[1] It is a non-trivial fact that ρ is actually a lattice norm on E; this follows from [22, Theorem 3.1]. For the present construction to go through this is, however, not needed.

$\|\overline{\overline{\varphi}}\| = \|\varphi\| \leq 1$. Since we already know that $q \circ j$ is contractive, we have, for $x \in X$,

$$\begin{aligned}
\|x\| &= \|\varphi(x)\| \\
&= \left\| [\overline{\overline{\varphi}} \circ (q \circ j)](x) \right\| \\
&\leq \|\overline{\overline{\varphi}}\| \, \|(q \circ j)(x)\| \\
&= \|\varphi\| \, \|(q \circ j)(x)\| \\
&\leq \|(q \circ j)(x)\| \\
&\leq \|x\|.
\end{aligned}$$

We conclude that $q \circ j$ is isometric.

We note that, since $(q \circ j)(X)$ generates $E/\ker\rho$ as a vector lattice, $\overline{\overline{\varphi}} : E/\ker\rho \to Y$ is uniquely determined as a vector lattice homomorphism by the requirement that $\varphi = \overline{\overline{\varphi}} \circ (q \circ j)$.

We have thus found a normed vector lattice $E/\ker\rho$ and an isometric linear map $q \circ j : X \to E/\ker\rho$ with the property that, for every Banach space Y, and for every bounded linear map $\varphi : X \to Y$, there exists a unique vector lattice homomorphism $\overline{\varphi} : E/\ker\rho \to Y$ such that the diagram

$$\begin{array}{ccc}
X & \xrightarrow{\;q\circ j\;} & E/\ker\rho \\
& \searrow^{\varphi} & \downarrow{\scriptstyle \overline{\varphi}} \\
& & Y
\end{array}$$

is commutative. Moreover, $\overline{\varphi}$ is a bounded linear operator, and $\|\overline{\overline{\varphi}}\| = \|\varphi\|$.

Let F be the norm completion of $E/\ker\rho$, and set $j_F := q \circ j$, seen as a map from X into F. We see, removing the construction from the notation, that there exist a Banach lattice F and an isometric linear map $j : X \to F$ with the property that, for every Banach lattice Y, and for every bounded linear map $\varphi : X \to Y$, there exists a unique vector lattice homomorphism $\varphi : F \to Y$ with $\|\overline{\varphi}\| = \|\varphi\|$ such that the diagram

$$\begin{array}{ccc}
X & \xrightarrow{\;j\;} & F \\
& \searrow^{\varphi} & \downarrow{\scriptstyle \overline{\varphi}} \\
& & Y
\end{array}$$

is commutative. Using only the existence of a free vector lattice over a vector space as a starting point, we have thus retrieved the existence of a free Banach lattice over a Banach space as was first established in [4].

The type of arguments in the preceding construction have been used earlier in the papers cited above, but this was always done using the setting in which a concrete model for the free object was constructed. The existence of the free object and the proof that a particular object was a concrete model for it came at the same time. In the above line of reasoning, however, nothing specific is used. Once one has the existence of a free vector lattice over a vector space, the rest is an argument that takes place in the category of vector lattices and in that of Banach spaces.

It is clear that this abstract approach can be used in other situations. One starts with a free object in the algebraic context, introduces an appropriate seminorm on it, divides out its kernel, and completes. For example, the free Banach lattice over a lattice from [3] can also be constructed along these lines, with the existence of the free vector lattice over a lattice (see Theorem 7.1, below) as a starting point. In fact, all free objects in the algebraic context in Theorems 6.2 and 7.1, below, can be used to construct their functional analytic counterparts. We intend to report separately on this in the future. In particular, it will then be seen that free Banach lattice algebras over non-empty sets exist, as well as unitisations of Banach lattice algebras; this solves Problems 13 and 15 in Wickstead's list [23]. In the case of Banach lattice algebras, it is then necessary to incorporate certain bounds into the construction. The final example in Examples 1, below, may serve to make this plausible.

We hasten to add that this general abstract approach will only lead to an easy existence proof for certain free functional analytic objects. It will not inform us how to find a concrete model that enables a further study of its structure. For example, as an algebraic analogue, it is not difficult to see that free vector lattices over non-empty sets exist (this follows from a general result in universal algebra), but it requires creativity as in [10] to show that they can be realised as lattices of functions. It is only then that it becomes clear that they are Archimedean. Our approach can, therefore, not replace the papers cited above where the structure of various free functional analytic objects is studied. The virtue of the general abstract approach is that it provides a smooth standard route to the basic existence result, after which the actual work can begin.

The present paper is intended to provide an algebraic basis for future existence results for free Banach lattices and Banach lattice algebras. As will become clear in the sequel, it is actually quite easy to prove that, for example, free vector lattice algebras with positive identity elements over non-empty sets exist. One merely needs to show that these are a so-called equational class, which is not too difficult, and then invoke a general theorem from universal algebra that holds for such classes to conclude the proof. It is even easier to show the existence of free vector lattices over non-empty sets along these lines.

It seems, however, as if this possibility of invoking a ready-to-use result from universal algebra in the context of vector lattices and vector lattice algebras was once known, but has later faded into the background. In his 1973 paper [10] already mentioned above, Bleier showed that the free vector lattice over a set has the usual concrete model as a vector lattice of functions. It is instrumental for him to know a priori that such a free vector lattice exists, but he does not even find it necessary to give a reference for this existence result. There is a simple remark *'If S is a* *non-empty set, then, since the class vector lattices is equationally definable, there exists a (unique up to isomorphism) vector lattice F which is free on S'* (see [10, p.74]). A reference is then given to the general theory of free abstract algebras in [9, p. 143–144]. Apparently, this was sufficient in that period of time.

On the other hand, the Problems 13 and 15 on Banach lattice algebras in [23], already mentioned above, were posed at a 2014 workshop at the Lorentz Center in Leiden where various senior researchers in the field of positivity were present. None of them was aware of the fact that, at the algebraic level, the existence of free vector lattice algebras over non-empty sets and of the unitisations of vector lattice algebras can easily be derived from one single theorem in universal algebra.

Therefore, apart from providing an algebraic basis for future existence results for free Banach lattices and Banach lattice algebras, this paper is also intended to revitalise this knowledge of universal algebra in the specific context of lattices, and in such a way that it can easily be used in other situations. The main theorem we need can be found in textbooks on universal algebra, but there it is among much more material that is not relevant for our purposes, and it may take some effort to isolate what one actually needs. Furthermore, vector lattice algebras, for example, are hard to find in such books—if at all—and we really need to recognise them as abstract algebras where certain identities are satisfied. It is not directly obvious how one can capture a partial ordering and the existence of suprema and infima in identities, but the fact that this is nevertheless possible (see Lemma 4.2, below) is crucial. We have, therefore, included the details for everything we need; the paper is self-contained. Our coverage of the material on universal algebra, culminating in the existence of free objects of equational classes over non-empty sets (see Theorem 5.4, below), is an exposition of parts of a known theory. It is, however, a very selective one, aiming for the one result we need and nothing more, and tailored to the context of lattices. We also need some basic known facts about free objects in a categorical language. The remainder of the paper, as well as the blend of category theory, universal algebra and vector lattices and vector lattice algebras, appears to be new.

This paper is organised as follows.

In Sect. 2, we introduce the notion of a free object. Some examples are given, and preparations for later sections are made.

In Sect. 3, we start our exposition on universal algebra. The goal is the existence theorem for a free (abstract) algebra of a given type over a non-empty set; see Theorem 3.10, below.

Section 4 is concerned with a crucial point: capturing the partial ordering and the existence of infima and suprema in a lattice in identities. Proposition 4.5, below, shows that the unital vector lattice algebras are precisely the abstract algebras of a certain type where a list of identities are satisfied.

After Sect. 4, there is a need to formalise the notion of an abstract algebra 'satisfying identities'. This is done in Sect. 5, where we continue our exposition on universal algebra and equational classes are introduced. Theorem 5.4, below, is the result we are after. It guarantees the existence of free abstract algebras in equational classes over non-empty sets.

In Sect. 6, the harvest is brought in. We start by inferring the existence of free unital vector lattice algebras over non-empty sets. From this one existence result, the existence of a host of other free objects is established. Together with their interrelations, they are collected in Theorem 6.2, below.

In Sect. 7, the existence of a number of free objects over (not necessarily distributive) lattices is shown; see Theorem 7.1, below. Although an independent approach is also possible, we have chosen to derive the results in this section from those in Sect. 6.

Finally, in Sect. 8, we consider various types of free f-algebras over non-empty sets. We can neither prove nor disprove that they exist. It is motivated how the observation that the free vector lattice over a non-empty set is, in fact, Archimedean can lead one to wonder whether free vector lattice algebras are, in fact, f-algebras.

We conclude this section by mentioning some terminology and conventions.

All vector spaces are over the real numbers. An *algebra* is an associative algebra. An algebra need not be unital. An algebra homomorphism between two unital algebras need not be unital. When convenient, a unital algebra will be denoted by A^1. A *vector lattice algebra*, also called a Riesz algebra in the literature, is a vector lattice that is also an algebra such that the product of two positive elements is positive. The identity element of a unital vector lattice algebra need not be positive. A *vector lattice algebra homomorphism* between vector lattice algebras is a lattice homomorphism that is also an algebra homomorphism. When convenient, a unital vector lattice algebra with a positive identity element will be denoted by A^{1+}. A *bi-ideal* in a vector lattice algebra is a linear subspace that is an order ideal as well as a two-sided algebra ideal.

We let $\mathbb{N}_0 = \{0, 1, 2 \ldots\}$.

2 Free Objects

In this section, we review basic facts about free objects. This notion, to be defined below, can be introduced whenever a category is a subcategory of another category. In our case, the main six categories of interest are:

- Set: the sets with the maps as morphisms;
- VS: the vector spaces, with the linear maps as morphisms;
- VL: the vector lattices, with the vector lattice homomorphisms as morphisms;
- VLA: the vector lattice algebras, with the vector lattice algebra homomorphisms as morphisms;
- VLA^1: the unital vector lattice algebras, with the unital vector lattice algebra homomorphisms as morphisms;
- VLA^{1+}: the unital vector lattice algebras that have a positive identity element, with the unital vector lattice algebra homomorphisms as morphisms.

There will also be an appearance of Lat, the category of not necessarily distributive lattices, with lattice homomorphisms as morphism. There are two ways to define lattices. The fact that these two are equivalent (see Lemma 4.2) is essential for this paper.

The main six categories of our interest can be ordered in the following chain. With the exception of Set, each is a subcategory of the one to the left of it:

$$\mathsf{Set} \supset \mathsf{VS} \supset \mathsf{VL} \supset \mathsf{VLA} \supset \mathsf{VLA}^1 \supset \mathsf{VLA}^{1+}. \tag{2.1}$$

Except for $\mathsf{VLA}^1 \supset \mathsf{VLA}^{1+}$, all these subcategories are non-full subcategories.

All in all, there are 15 instances of a category and a subcategory of it associated to this chain. For each of these, there is a notion of free objects. We shall now define this.

Definition 2.1 Suppose that Cat_1 and Cat_2 are categories, and that Cat_2 is a subcategory of Cat_1. Take an object O_1 of Cat_1. Then a *free object over* O_1 *of* Cat_2 is a pair $(j, \mathrm{F}_{\mathsf{Cat}_1}^{\mathsf{Cat}_2}[O_1])$, where $\mathrm{F}_{\mathsf{Cat}_1}^{\mathsf{Cat}_2}[O_1]$ is an object of Cat_2 and $j : O_1 \to \mathrm{F}_{\mathsf{Cat}_1}^{\mathsf{Cat}_2}[O_1]$ is a morphism in Cat_1, with the property that, for every object O_2 of Cat_2 and every morphism $\varphi : O_1 \to O_2$ of Cat_1, there exists a unique morphism $\overline{\varphi} : \mathrm{F}_{\mathsf{Cat}_1}^{\mathsf{Cat}_2}[O_1] \to O_2$ of Cat_2 such that the diagram

$$\begin{array}{ccc} O_1 & \xrightarrow{\ j\ } & \mathrm{F}_{\mathsf{Cat}_1}^{\mathsf{Cat}_2}[O_1] \\ & \searrow^{\varphi} & \downarrow{\scriptstyle \overline{\varphi}} \\ & & O_2 \end{array}$$

in Cat_1 is commutative.

Suppose that $(j', \mathrm{F}_{\mathsf{Cat}_1}^{\mathsf{Cat}_2}[O_1]')$ is another pair with this property. The unique morphism $\overline{j'} : \mathrm{F}_{\mathsf{Cat}_1}^{\mathsf{Cat}_2}[O_1] \to \mathrm{F}_{\mathsf{Cat}_1}^{\mathsf{Cat}_2}[O_1]'$ such that $j' = \overline{j'} \circ j$ and the unique morphism $\overline{j} : \mathrm{F}_{\mathsf{Cat}_1}^{\mathsf{Cat}_2}[O_1]' \to \mathrm{F}_{\mathsf{Cat}_1}^{\mathsf{Cat}_2}[O_1]$ such that $j = \overline{j} \circ j'$ are then such that $\overline{j} \circ \overline{j'}$ is the identity morphism of $\mathrm{F}_{\mathsf{Cat}_1}^{\mathsf{Cat}_2}[O_1]$ and $\overline{j'} \circ \overline{j}$ is the identity morphism of $\mathrm{F}_{\mathsf{Cat}_1}^{\mathsf{Cat}_2}[O_1]'$. Hence $\mathrm{F}_{\mathsf{Cat}_1}^{\mathsf{Cat}_2}[O_1]$, if it exists, is uniquely determined up to a unique compatible isomorphism. When convenient, we shall, therefore, simply say that $\mathrm{F}_{\mathsf{Cat}_1}^{\mathsf{Cat}_2}[O_1]$ *exists* when there exists a pair $(j, \mathrm{F}_{\mathsf{Cat}_1}^{\mathsf{Cat}_2}[O_1])$ as above, and let $\mathrm{F}_{\mathsf{Cat}_1}^{\mathsf{Cat}_2}[O_1]$ stand for any realisation of it, the accompanying map j being understood.

Examples 1

(1) Let Grp denote the category of groups with the group homomorphisms as morphisms, and let AbGrp denote its full subcategory of abelian groups. Take a group G. Then $\mathrm{F}_{\mathsf{Grp}}^{\mathsf{AbGrp}}[G]$ exists and is the quotient of G modulo its commutator subgroup.
(2) Take a non-empty set S. Then $\mathrm{F}_{\mathsf{Set}}^{\mathsf{VS}}[S]$ exists and is the vector space of real-valued functions on S with finite supports.
(3) Let Alg denote the category of algebras with the algebra homomorphisms as morphisms, and let Alg^1 denote its subcategory of unital algebras with the unital algebra homomorphisms as morphisms. Take an algebra A. Set $A^1 := \mathbb{R} \oplus A$,

as a vector space direct sum, and supply it with the usual structure of a unital algebra by setting $(\lambda, a) \cdot (\mu, b) := (\lambda\mu, \lambda b + \mu a + ab)$ for $\lambda, \mu \in \mathbb{R}$ and $a, b \in A$. Then $\mathrm{F}_{\mathsf{Alg}}^{\mathsf{Alg}^1}[A]$ exists and equals A^1.

(4) Take a vector lattice algebra A. Set $A^1 := \mathbb{R} \oplus A$ as a vector lattice direct sum, and supply it with the structure of a unital associative algebra as above. Then A^1 is a unital vector lattice algebra. It is, however, not generally the free unital vector lattice algebra over A. The problem is that the natural factoring map $\overline{\varphi}$, although a unital algebra homomorphism, need not be a lattice homomorphism. It is nevertheless true that the unitisation $\mathrm{F}_{\mathsf{VLA}}^{\mathsf{VLA}^1}[A]$ of A exists; see Theorem 6.2.

(5) Take a non-empty set S. Contrary to the previous example, in this case there does not even appear to be a natural (flawed) Ansatz for the free unital vector lattice algebra $\mathrm{F}_{\mathsf{Set}}^{\mathsf{VLA}^1}[S]$ over S. The fact that it nevertheless exists (see Theorem 6.2) is the foundation on which the other existence results in this paper are built.

(6) Let Met be the category of metric spaces with continuous maps, and let ComMet be its full subcategory of complete metric spaces. Take a metric space M. Then $\mathrm{F}_{\mathsf{Met}}^{\mathsf{ComMet}}[M]$ exists and is the metric completion of M.

(7) Let BA be the category of Banach algebras with the continuous algebra homomorphisms as morphisms. Consider a set $\{s\}$ with one element. Then $\mathrm{F}_{\mathsf{Set}}^{\mathsf{BA}}[\{s\}]$ does not exist. To prove this, suppose, to the contrary, that there exist a Banach algebra $\mathrm{F}_{\mathsf{Set}}^{\mathsf{BA}}[\{s\}]$ and a map $j : \{s\} \to \mathrm{F}_{\mathsf{Set}}^{\mathsf{BA}}[\{s\}]$ such that, for every Banach algebra A and every map $\varphi : \{s\} \to A$, there exists a unique continuous algebra homomorphism $\overline{\varphi}$ such that the diagram

$$\begin{array}{ccc} \{s\} & \xrightarrow{\ j\ } & \mathrm{F}_{\mathsf{Set}}^{\mathsf{BA}}[\{s\}] \\ & \searrow^{\varphi} & \downarrow \overline{\varphi} \\ & & A \end{array}$$

is commutative. For A, we take the Banach algebra of the real numbers and, for every $x > 0$ in $\mathbb{R}$, we define the map $\varphi_x : \{s\} \to \mathbb{R}$ by setting $\varphi(s) := x$. Take $x > 0$. Then, for every $n \geq 0$, we have

$$\begin{aligned} x^n &= \|[\varphi_x(s)]^n\| \\ &= \|[\overline{\varphi}_x(j(s))]^n\| \\ &= \|\overline{\varphi}_x([j(s)]^n)\| \\ &\leq \|\overline{\varphi}_x\| \, \|[j(s)]^n\| \\ &\leq \|\overline{\varphi}_x\| \, \|j(s)\|^n. \end{aligned}$$

On letting n tend to infinity, we see that we must have $\|j(s)\| \geq x$. Since $x > 0$ is arbitrary, this is impossible.

We shall suppose for the remainder of this paper that the objects of categories are sets.

Remark 2.3 Suppose, in the setting of Definition 2.1, that $\mathrm{F}_{\mathsf{Cat}_1}^{\mathsf{Cat}_2}[O_1]$ exists and that, for each pair of different elements $x, y \in O_1$, there exists an object O_2 of Cat_2 and a morphism $\varphi : O_1 \to O_2$ of Cat_1 such that $\varphi(x) \neq \varphi(y)$. Then the map $j : O_1 \to \mathrm{F}_{\mathsf{Cat}_1}^{\mathsf{Cat}_2}[O_1]$ is clearly injective. The converse is obviously true because $\mathrm{F}_{\mathsf{Cat}_1}^{\mathsf{Cat}_2}[O_1]$ is an object of Cat_2 and the injective morphism j of Cat_1 then separates the elements of O_1 all at once. We shall often use this observation and collect a number of elementary facts in this vein, where actually one injective map $\varphi : O_1 \to O_2$ already separates all elements of O_1.

Lemma 2.4

(1) *Let S be a non-empty set. There exists a vector space V and an injective map $\varphi : S \to V$.*
(2) *Let V be a vector space. There exist a vector lattice E and an injective linear map $\varphi : V \to E$.*
(3) *Let E be a vector lattice. There exist a commutative vector lattice algebra A^{1+} with a positive identity element and an injective vector lattice homomorphism $\varphi : E \to A^{1+}$.*
(4) *Let A be a vector lattice algebra. There exist a unital vector lattice algebra A^{1+} with a positive identity element and an injective vector lattice algebra homomorphism $\varphi : A \to A^{1+}$.*

The parts of Lemma 2.4 can be combined to see that, for example, there is always an injective linear map from a given vector space into a commutative vector lattice algebra with a positive identity element. A number of (combined) inclusions of categories $\mathsf{Cat}_1 \supset \mathsf{Cat}_2$ from the chain Eq. (2.1) can thus be 'reversed' in the sense that every object of Cat_1 embeds, via a morphism in Cat_1, into an object of Cat_2. Since it is not true that the identity of every unital vector lattice algebra is positive, there is no general 'reversal' for the inclusion $\mathsf{VLA}^1 \supset \mathsf{VLA}^{1+}$.

Proof of Lemma 2.4 For part (1), we take for V the real-valued functions on S, together with the canonical map from S into V.

For part (2), we let $V^\#$ denote set of all linear functionals on V. For E we take the vector lattice of all real-valued functions on $V^\#$. The canonical map from V into E is linear and injective.

For part (3), we first make E into a commutative vector lattice algebra by supplying it with the zero multiplication. Subsequently, we apply the usual unitisation procedure to that vector lattice algebra. All in all, we take the vector lattice direct sum $A^{1+} := \mathbb{R} \oplus E$, supplied with the multiplication $(\lambda, x) \cdot (\mu, y) := (\lambda\mu, \lambda y + \mu x)$ for $\lambda, \mu \in \mathbb{R}$ and $x, y \in E$, together with the canonical map from E into A^{1+}.

For part (4), we take the vector lattice direct sum $A^{1+} := \mathbb{R} \oplus A$, supplied with the usual multiplication to make it into a unital vector lattice algebra with a positive identity element, together with the canonical map from A into A^{1+}. □

Remark 2.5 In our situations of interest, the objects of the category Cat_2 in Definition 2.1 are sets with operations. This implies that $\mathrm{F}_{\mathsf{Cat}_1}^{\mathsf{Cat}_2}[O_1]$, if it exists, must be generated, in the sense of Cat_2, by $j(O_1)$. Take a non-empty set S, for example. If $\mathrm{F}_{\mathsf{Set}}^{\mathsf{VL}}[S]$ exists, then it must be generated, as a vector lattice, by its subset $j(S)$. The reason is simply that the vector sublattice that is generated by $j(S)$ and the restricted factoring map $\overline{\varphi}$ obviously also have the required universal property. The essential uniqueness of such a pair then implies that this vector sublattice must coincide with $\mathrm{F}_{\mathsf{Set}}^{\mathsf{VL}}[S]$. As another example, take a vector space V. If $\mathrm{F}_{\mathsf{VS}}^{\mathsf{VLA}^{1+}}[V]$ exist, then it must be generated, as a unital vector lattice algebra, by (its identity element and) its subspace $j(V)$. We shall often use this observation.

Remark 2.6 Let $\mathsf{Cat}_1 \supset \mathsf{Cat}_2 \supset \mathsf{Cat}_3$ be a chain of categories. Take an object O_1 of Cat_1, and suppose that $\mathrm{F}_{\mathsf{Cat}_1}^{\mathsf{Cat}_2}[O_1]$ exists in Cat_2, with accompanying map $j_{12} : O_1 \to \mathrm{F}_{\mathsf{Cat}_1}^{\mathsf{Cat}_2}[O_1]$. Suppose that $\mathrm{F}_{\mathsf{Cat}_2}^{\mathsf{Cat}_3}\big[\mathrm{F}_{\mathsf{Cat}_1}^{\mathsf{Cat}_2}[O_1]\big]$ exists in Cat_3, with accompanying map $j_{23} : \mathrm{F}_{\mathsf{Cat}_1}^{\mathsf{Cat}_2}[O_1] \to \mathrm{F}_{\mathsf{Cat}_2}^{\mathsf{Cat}_3}\big[\mathrm{F}_{\mathsf{Cat}_1}^{\mathsf{Cat}_2}[O_1]\big]$. It is easy to see that $\mathrm{F}_{\mathsf{Cat}_1}^{\mathsf{Cat}_3}[O_1]$ then also exists. In fact, one can take $\mathrm{F}_{\mathsf{Cat}_1}^{\mathsf{Cat}_3}[O_1] := \mathrm{F}_{\mathsf{Cat}_2}^{\mathsf{Cat}_3}\big[\mathrm{F}_{\mathsf{Cat}_1}^{\mathsf{Cat}_2}[O_1]\big]$ and $j_{13} := j_{23} \circ j_{12}$ as accompanying map $j_{13} : O_1 \to \mathrm{F}_{\mathsf{Cat}_1}^{\mathsf{Cat}_3}[O_1]$.

Remark 2.7 Let Cat_1 be a category, and let Cat_2 be a subcategory. Suppose that $\mathrm{F}_{\mathsf{Cat}_1}^{\mathsf{Cat}_2}[O_1]$ exists for every object O_1 of Cat_1. Since $\mathrm{F}_{\mathsf{Cat}_1}^{\mathsf{Cat}_2}[O_1]$ is not uniquely determined, there is no natural functor that assigns 'the' free object $\mathrm{F}_{\mathsf{Cat}_1}^{\mathsf{Cat}_2}[O_1]$ of Cat_2 to O_1. This can be remedied to some extent, as follows. Suppose that, for each object O_1 of Cat_1, a free object $\mathrm{F}_{\mathsf{Cat}_1}^{\mathsf{Cat}_2}[O_1]$ of Cat_2 over O_1 has been chosen, together with its accompanying map $j : O_1 \to \mathrm{F}_{\mathsf{Cat}_1}^{\mathsf{Cat}_2}[O_1]$. Suppose that O_1' is an object of Cat_1, and that $\varphi : O_1 \to O_1'$ is a morphism in Cat_1. For the chosen free object $\mathrm{F}_{\mathsf{Cat}_1}^{\mathsf{Cat}_2}[O_1']$ and accompanying map $j' : O_1' \to \mathrm{F}_{\mathsf{Cat}_1}^{\mathsf{Cat}_2}[O_1']$, there exists a unique morphism $\overline{\varphi} : \mathrm{F}_{\mathsf{Cat}_1}^{\mathsf{Cat}_2}[O_1] \to \mathrm{F}_{\mathsf{Cat}_1}^{\mathsf{Cat}_2}[O_1']$ of Cat_2 such that $(j' \circ \varphi) = \overline{\varphi} \circ j$. Then an actual functor from Cat_1 to Cat_2 is defined by sending an object O_1 to the chosen free object $\mathrm{F}_{\mathsf{Cat}_1}^{\mathsf{Cat}_2}[O_1]$ of Cat_2, and a morphism $\varphi : O_1 \to O_1'$ of Cat_1 to its associated morphism $\overline{\varphi} : \mathrm{F}_{\mathsf{Cat}_1}^{\mathsf{Cat}_2}[O_1] \to \mathrm{F}_{\mathsf{Cat}_1}^{\mathsf{Cat}_2}[O_1']$ of Cat_2.

Remark 2.8 There appears to be no general agreement about the terminology for the objects $\mathrm{F}_{\mathsf{Cat}_1}^{\mathsf{Cat}_2}[O_1]$ from Definition 2.1. A different way to look at the pairs $(j, \mathrm{F}_{\mathsf{Cat}_1}^{\mathsf{Cat}_2}[O_1]$, leading to a different terminology, is as follows. Take an object O_1 of Cat_1, and consider the pairs (φ, O_2), where O_2 is an object of Cat_2 and $\varphi : O_1 \to O_2$ is a morphism in Cat_1. We form a new category that consists of all such pairs, and where a morphism from a pair (φ, O_2) to a pair (φ', O_2') is a morphism ψ of Cat_2 such that the diagram

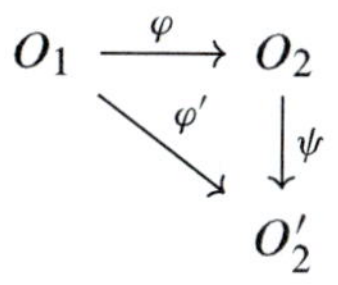

is commutative. The pairs $(j, \mathrm{F}_{\mathsf{Cat}_1}^{\mathsf{Cat}_2}[O_1])$ from Definition 2.1 are then precisely the universally repelling objects (also known as the initial objects) of this new category. From this viewpoint, it is natural to speak of a *universal* object over O_1 of Cat_2. This term is used in several places in the literature; see [19, p.83] or [2, p.153], for example. In the terminology of [15, p.179], $(j, \mathrm{F}_{\mathsf{Cat}_1}^{\mathsf{Cat}_2}[O_1])$ is called a Cat_2-*reflection of* O_1. In the terminology of [17, Definition 2.10], $\mathrm{F}_{\mathsf{Cat}_1}^{\mathsf{Cat}_2}[O_1]$ is a *free object with base* O_1. The terminology in [1, Definition 8.22] agrees with ours. The same is true for the overview paper by Pestov [20] and many titles in its biography; see [20]. It is clear from this source [20] that, in a topological or analytical context, 'free' is the prevailing term. Since an analytical context is, in the end, our main motivation for the present work, and since it is also used in the papers [3–5, 10, 12] that are directly related to the present paper, we have chosen to adapt this too.

3 Universal Algebra: Part I

In this section, we review the first part of the material from universal algebra that we need. It is largely based on the exposition in [7]. Our treatment is slightly different in the sense that we prefer to speak of constants instead of 0-ary operations, and that we have singled them out in definitions. There is then no longer any need for conventions to be in force when a definition 'degenerates' for an 'operation' that does not have variables at all. We also speak of an 'abstract algebra' rather than of an 'algebra', since we want to keep our convention in force that the latter term refers to an associative algebra over the real numbers. Since both notions do actually occur in one context, it seems unavoidable to make such a distinction.

Definition 3.1 Suppose that $\mathcal{F}$ is a non-empty (possibly infinite) set, and that $\rho : \mathcal{F} \to \mathbb{N}_0$ is a map. Then the pair $(\mathcal{F}, \rho)$ is called a *type*. Let A be a non-empty set and suppose that, for each $f \in \mathcal{F}$, the following is given:

(1) when $\rho(f) = 0$: an element f^A of A;
(2) when $\rho(f) \geq 1$: a map $f^A : A^{\rho(f)} \to A$.

We set $\mathcal{F}^A := \{ f^A : f \in \mathcal{F} \}$. The pair $\langle A, \mathcal{F}^A \rangle$ is then called an *abstract algebra of type* $(\mathcal{F}, \rho)$. The elements of $\mathcal{F}$ are called *operation symbols*. The elements f^A of A for those $f \in \mathcal{F}$ such that $\rho(f) = 0$ are called the *constants of* A, and the $\rho(f)$-ary maps $f^A : A^{\rho(f)} \to A$ for those f such that $\rho(f) \geq 1$ are called the *operations on* A.

When everything else is clear from the context, we shall also simply refer to A as an abstract algebra, the rest being tacitly understood.

Suppose B is a non-empty subset A that contains the constants of A and such that $f^A(B^{\rho(f)}) \subseteq B$ for all $f \in \mathcal{F}$ such that $\rho(f) \geq 1$. Supplied with the constants of A and the restricted operations on A, B is then called an *abstract subalgebra* of A. It is of the same type $(\mathcal{F}, \rho)$ as A.

Let $(\mathcal{F}, \rho)$ be a type. Suppose that I is a non-empty index set and that, for each $i \in I$, A_i is an abstract algebra of type $(\mathcal{F}, \rho)$. Then the product $\prod_{i\in I} A_i$ becomes an abstract algebra of type $(\mathcal{F}, \rho)$ in the obvious coordinate-wise way; it is then called the *abstract product algebra of the* A_i.

Example 3.2 Take $\mathcal{F} = \{f_0, f_1, f_2\}$ for some symbols f_0, f_1, and f_2. Set $\rho(f_0) := 0$, $\rho(f_1) := 1$, and $\rho(f_2) := 2$. Let G be a group.

(1) Set $f_0^G := e$, where e is the identity element of G; set $f_1^G(x) = x^{-1}$ for $x \in G$; and set $f_2^G(x, y) := xy$ for $x, y \in G$. Then $\langle G, \{f_0^G, f_1^G, f_2^G\}\rangle$ is an abstract algebra of type $(\mathcal{F}, \rho)$.
(2) Take an element x_0 of G. Set $\tilde{f}_0^G := x_0$; set $\tilde{f}_1^G(x) = x^7$ for $x \in G$; and set $\tilde{f}_2^G(x, y) := x^2yx^{-1}y^3x^2$ for $x, y \in G$. Then $\langle G, \{\tilde{f}_0^G, \tilde{f}_1^G, \tilde{f}_2^G\}\rangle$ is an abstract algebra of type $(\mathcal{F}, \rho)$.

The notation as used in part (1) of Example 3.2 is not very suggestive. Given a group G, it would be more natural to simply speak of the associated abstract algebra $\langle G, \{e, {}^{-1}, \cdot\}\rangle$, where the type $(\mathcal{F}, \rho)$ with an underlying set $\mathcal{F}$ of cardinality 3, and the map $\rho : \mathcal{F} \to \{e, {}^{-1}, \cdot\}$ (the set containing the constant and the two actual operations on G) understood to be evident from the context. Given two groups G_1 and G_2, it would then, strictly speaking, be necessary to write $\langle G_1, \{e^{G_1}, {}^{-1^{G_1}}, \cdot^{G_1}\}\rangle$ and $\langle G_2, \{e^{G_2}, {}^{-1^{G_2}}, \cdot^{G_2}\}\rangle$. When working with concrete examples we shall omit these superscripts. For example, let V be a vector space. Then there is a naturally associated abstract algebra $\langle V, \{0, +, \mathrm{ADDINV}, \{m_\lambda : \lambda \in \mathbb{R}\}\}\rangle$. The unspecified set $\mathcal{F}$ is now uncountable, and to its elements correspond a constant 0 of V, an obvious binary operation $+$, a unary operation ADDINV that sends $x \in V$ to $-x$, and, for every $\lambda \in \mathbb{R}$, a unary operation m_λ that sends $x \in V$ to λx. It is then also clear what the function $\rho : \mathcal{F} \to \mathbb{N}_0$ is; it takes the values 0,1, and 2. When W is another vector space, we denote its associated abstract algebra by $\langle W, \{0, +, \mathrm{ADDINV}, \{m_\lambda : \lambda \in \mathbb{R}\}\}\rangle$.

Not every abstract algebra $\langle V, \{0, +, \mathrm{ADDINV}, \{m_\lambda : \lambda \in \mathbb{R}\}\}\rangle$, with a constant 0, a binary operation +, a unary operation ADDINV, and unary operations m_λ for $\lambda \in \mathbb{R}$ becomes a vector space when one attempts to introduce the vector space operations in the obvious way. For this, certain relations between the constants and the operations have to hold, such as $m_{\lambda_1\lambda_2}(x) = m_{\lambda_1}(m_{\lambda_2}(x))$ for $\lambda_1, \lambda_2 \in \mathbb{R}$, and $x \in A$, and $x + (\mathrm{ADDINV}(x)) = 0$ for all $x \in A$. This need not always be the case. In Lemma 5.2, below, it will become clear how one can always pass to an abstract quotient algebra (to be defined below) of $\langle V, \{0, +, \mathrm{ADDINV}, \{m_\lambda : \lambda \in \mathbb{R}\}\}\rangle$ that *is* a vector space.

Definition 3.3 Let $\langle A, \mathcal{F}^A\rangle$ and $\langle B, \mathcal{F}^B\rangle$ be abstract algebras of the same type $(\mathcal{F}, \rho)$. Suppose that $h : A \to B$ is a map. Then h is an *abstract algebra homomorphism* when the following are both satisfied:

(1) $h(f^A) = f^B$ for all $f \in \mathcal{F}$ such that $\rho(f) = 0$;
(2) $h\big(f^A(a_1, \ldots, a_{\rho(f)})\big) = f^B\big(h(a_1), \ldots, h(a_{\rho(f)})\big)$ for all $f \in \mathcal{F}$ such that $\rho(f) \geq 1$.

The inclusion map from an abstract subalgebra to the abstract super-algebra is an abstract algebra homomorphism. The projections from an abstract product algebra to its factors are abstract algebra homomorphisms.

Example 3.4

(1) Let G_1 and G_2 be groups. The abstract algebra homomorphism $h : \langle G_1, \{e, {}^{-1}, \cdot\}\rangle \to \langle G_2, \{e, {}^{-1}, \cdot\}\rangle$ are maps between the underlying sets that are unital, preserve the inverse of one element, and preserve the product of two elements. Since G_1 and G_2 are actually groups, this is equivalent to preserving the product of two elements. Thus the abstract algebra homomorphisms between the associated abstract algebras are in a natural bijection with the group homomorphisms between the groups in the usual meaning of the word.
(2) Let G_1 and G_2 be groups. For G_2, take operations as in the second part of Example 3.2. In that case, the abstract algebra homomorphisms $h : \langle G_1, \{e, {}^{-1}, \cdot\}\rangle \to \langle G_2, \{\tilde{f}_1^{G_2}, \tilde{f}_2^{G_2}, \tilde{f}_3^{G_2}\}\rangle$ are the maps $h : G_1 \to G_2$ such that $h(e) = x_0$, $h(x^{-1}) = x^7$ for $x \in G_1$, and $h(xy) = x^2yx^{-1}y^3x^2$ for $x, y \in G_1$. Besides not being obviously natural or useful, it may well be the case that, for certain combinations of G_1, G_2, and x_0, such abstract algebra homomorphisms h do not exist.
(3) Let V and W be vector spaces, and let $\langle V, \{0^V, +^V, \text{ADDINV}^V, \{m_\lambda^V : \lambda \in \mathbb{R}\}\}\rangle$ and $\langle W, \{0^W, +^V, \text{ADDINV}^W, \{m_\lambda^W : \lambda \in \mathbb{R}\}\}\rangle$ denote the naturally associated abstract algebras of the same unspecified type $(\mathcal{F}, \rho)$. Then the abstract algebra homomorphisms between the associated abstract algebras are in a natural bijection with the linear maps between the vector spaces.

Definition 3.5 Let A and B be abstract algebras, and let $h : A \to B$ be a map. Then the *kernel of* h, denoted by $\ker h$, is defined as

$$\ker h := \Big\{(x, y) \in A^2 : h(x) = h(y)\Big\}.$$

Note that $\ker h$ is not a subset of A. In many practical contexts, however, it can be described in terms of a subset of A that will then be called the kernel of h in the pertinent context. For example, let G_1 and G_2 be groups, and let $h : \langle G_1, \{e, {}^{-1}, \cdot\}\rangle \to \langle G_2, \{e, {}^{-1}, \cdot\}\rangle$ be an abstract algebra homomorphism. Set $N := \{x \in G_1 : h(x) = e\}$. Then $\ker h = \{(x, y) \in G_1^2 : xy^{-1} \in N\}$. As another example, let V and W be vector spaces, and let $h : V \to W$ be an abstract algebra homomorphism between the two associated abstract algebras. Set $L := \{x \in V : h(x) = 0\}$. Then $\ker h = \{(x, y) \in V^2 : x - y \in L\}$.

When $\theta \subseteq A^2$ is a binary relation on A, and $x, y \in A$, then we shall write $x\,\theta\,y$ for $(x, y) \in \theta$. The kernels of abstract algebra homomorphisms turn out to be precisely the binary relations on A that we shall now define.

Definition 3.6 Let $\langle A, \mathcal{F}^A\rangle$ be an abstract algebra of type $(\mathcal{F}, \rho)$. Then a binary relation $\theta \subseteq A^2$ on A is called a *congruence relation on* A when the following are both satisfied:

(1) θ is an equivalence relation on A;
(2) when $f \in \mathcal{F}$ is such that $\rho(f) \geq 1$, then

$$f^A(x_1, \ldots, x_{\rho(f)}) \;\theta\; f^A(y_1, \ldots, y_{\rho(f)})$$

whenever $x_1, \ldots, x_{\rho(f)} \in A$ and $y_1, \ldots, y_{\rho(f)} \in A$ are such that $x_i \,\theta\, y_i$ for $i = 1, \ldots, \rho(f)$.

It is clear that A^2 is the largest congruence relation on A, and that $\{ (x, x) : x \in A \}$ is the smallest. The intersection of an arbitrary non-empty collection of congruence relations on A is again a congruence relation on A. Suppose that $S \subseteq A^2$ is an arbitrary subset. The intersection of all congruence relations on A that contain S is the smallest congruence relation on A that contains S; it is called the *congruence relation on A that is generated by S*.

It is immediate from the definitions that the kernel of an abstract algebra homomorphism between abstract algebras of the same type is a congruence relation on the domain. All congruence relations on an abstract algebra occur in this fashion, as will become clear from the following construction of abstract quotient algebras.

Let A be an abstract algebra of type $\langle \mathcal{F}, \rho\rangle$. Suppose that θ is a congruence relation on A. Let A/θ denote the set of equivalence classes in A with respect to θ, and let $q_\theta : A \to A/\theta$ denote the canonical map. When $f \in \mathcal{F}$ is such that $\rho(f) = 0$, we set

$$f^{A/\theta} := q_\theta(f^A).$$

When $f \in \mathcal{F}$ is such that $\rho(f) \geq 1$, then, for $x_1, \ldots, x_{\rho(f)} \in A$, we set

$$f^{A/\theta}\big(q_\theta(x_1), \ldots, q_\theta(x_{\rho(f)})\big) := q_\theta\big(f^A(x_1, \ldots, x_{\rho(f)})\big).$$

Since θ is a congruence relation on A, the maps $f^{A/\theta}$ are well defined. Thus $\langle A/\theta, \{ f^{A/\theta} : f \in \mathcal{F} \}\rangle$ is an abstract algebra of type $(\mathcal{F}, \rho)$. By its construction, the map $q_\theta : A \to A/\theta$ is an abstract algebra homomorphism, and $\ker q_\theta = \theta$.

The following result, which is [7, Exercise 1.26.8], is an immediate consequence of the definitions.

Lemma 3.7 *Let A and B be abstract algebras of the same type, and let $h : A \to B$ be an abstract algebra homomorphism. Suppose that θ is a congruence relation on A such that $\theta \subseteq \ker h$. Then there exists a unique map $\overline{h} : A/\theta \to B$ such that $h = \overline{h} \circ q_\theta$, and this map $\overline{h}$ is an abstract algebra homomorphism.*

We now come to the main point of this section. Let $(\mathcal{F}, \rho)$ be a type. The abstract algebras of type $(\mathcal{F}, \rho)$, together with the abstract algebra homomorphisms between them, form a subcategory $\mathsf{AbsAlg}_{(\mathcal{F},\rho)}$ of Set. Take a non-empty set S. Does there exists a free abstract algebra of type $(\mathcal{F},\rho)$ over S? The answer is affirmative, and we shall now construct such an object $\mathbf{F}_{\mathsf{Set}}^{\mathsf{AbsAlg}(\mathcal{F},\rho)}[S]$ of $\mathsf{AbsAlg}_{(\mathcal{F},\rho)}$. It will be denoted by $T_{(\mathcal{F},\rho)}(S)$.

The idea is quite easy. One starts by defining a set $T_{(\mathcal{F},\rho)}(S)$ of words that reflect the concept of applying maps (symbolised by the elements of $\mathcal{F}$) to their appropriate numbers of variables (as prescribed by ρ), and keep repeating combining the outcomes to get new maps of an ever increasing degree of complexity (measured by what will be called the 'height', below). The symbols in these words that reflect the concept of variables are taken from S. This set of words is then made into an abstract algebra of type $(\mathcal{F},\rho)$ in a natural way, with concatenation reflecting the concept of combining outcomes of operations as input for another operation. Furthermore, when A is any abstract algebra of type $(\mathcal{F},\rho)$, and $h : S \to A$ is any map, then there is a natural abstract algebra homomorphism $\overline{h}$ from $T_{(\mathcal{F},\rho)}(S)$ into A that extends h. This map $\overline{h}$ simply replaces each symbol f from $\mathcal{F}$ in a word in $T_{(\mathcal{F},\rho)}(S)$ by the concrete operation (or constant) f^A in the context of A, and replaces each symbol s from S by the concrete element $h(s)$ of A. All in all, this map $\overline{h}$ takes in a word from $T_{(\mathcal{F},\rho)}(S)$ and then applies the 'actual map that the word stands for in the context of A' to values of its arguments that are the pertinent given elements $h(s)$ of A.

The details are as follows; they are taken from [7, p. 95–96] (where the case where $S=\emptyset$ but $\{\, f \in \mathcal{F} : \rho(f)=0 \,\} \neq \emptyset$ is also included). The structure of the proof of Theorem 3.10, which we include for the convenience of the reader, is also taken from that source; cf. [7, Theorem 4.32].

Definition 3.8 Let $(\mathcal{F},\rho)$ be a type. Let S be a non-empty (possibly infinite) set that is disjoint from $\mathcal{F}$. We recursively define a set of words in symbols f for $f \in \mathcal{F}$ and symbols s for $s \in S$, as follows. We set

$$T_0(S) := \{\, s : s \in S \,\} \cup \{ f \in \mathcal{F} : \rho(f) = 0 \}\}$$

and, for $n = 1, 2, \ldots$, we set

$$T_{n+1}(S) := T_n(S) \cup \{\, f t_1 \ldots t_{\rho(f)} : f \in \mathcal{F},\ \rho(f) \geq 1,\ t_1, \ldots, t_{\rho(f)} \in T_n(S) \,\}.$$

We define $T_{(\mathcal{F},\rho)}(S) := \bigcup_{n\geq 0} T_n(S)$, and refer to elements of $T_{(\mathcal{F},\rho)}(S)$ as *terms of type* $(\mathcal{F},\rho)$ *over* S. For a term $t \in T_{(\mathcal{F},\rho)}(S)$, the smallest n such that $t \in T_n(S)$ is called the *height* of t.

The terms of height zero, i.e., the elements of $T_0(S)$, can come in two kinds. Since S is non-empty, there are always terms in $T_0(S)$ that consist of a single symbol s from S. These can be thought of as 'variables'. The other words in $T_0(S)$ are the symbols f from $\mathcal{F}$ such that $\rho(f) = 0$. There need not be any such f, but when there are, then the corresponding terms can be thought of as 'constants' or, perhaps even better with an eye towards applications, as 'distinguished elements'.

Example 3.9 By way of (a rather finite) example, we consider the case where $\mathcal{F} = \{f_0, f_1, f_2\}$, $\rho(f_0) = 0$, $\rho(f_1) = 1$, $\rho(f_2) = 2$, and where $S = \{x, y, z\}$. Then the set $T_0(S)$ of terms of height 0 consists of the 'variables' x, y, and z, together with the 'constant' f_0. The set $T_1(S)$ consists of all terms in $T_0(S)$ (all of height 0); the terms (all of height 1) f_1z, f_1y, f_1z, and f_1f_0; and $4 \cdot 4 = 16$ terms (all of height 1) of the form $f_2\xi_1\xi_2$, where each of ξ_1, ξ_2 can be taken from $T_0(S) = \{x, y, z, f_0\}$. The term f_2xy of height 1, secretly translated into $f_2(x, y)$, reflects the concept of applying an operation that depends on two variables. The term f_2xf_0, translated into $f_2(x, f_0)$, reflects the concept of applying a map that depends on two variables with the second one fixed at the value f_0. The subset $T_2(S)$ of $T_{(\mathcal{F},\rho)}(S)$ contains terms such as $f_2f_2f_0yf_1z$ and $f_2f_1xf_2yx$. After translating these terms of height 2 into their more readable forms $f_2(f_2(f_0, y), f_1(z))$ and $f_2(f_1(x), f_2(y, x))$, respectively, it becomes clear which concepts they reflect.

We shall now supply the set $T_{(\mathcal{F},\rho)}(S)$ with the structure of an abstract algebra of type $(\mathcal{F}, \rho)$. If $f \in \mathcal{F}$ is such that $\rho(f) = 0$, then we set

$$f^{T_{(\mathcal{F},\rho)}(S)} := f, \tag{3.1}$$

and when $f \in \mathcal{F}$ is such that $\rho(f) \geq 1$, then we use concatenation of words to set

$$f^{T_{(\mathcal{F},\rho)}(S)}(t_1, \ldots, t_{\rho(f)}) := ft_1 \ldots t_{\rho(f)} \tag{3.2}$$

for all $t_1, \ldots, t_{\rho(f)} \in T_{(\mathcal{F},\rho)}(S)$.

Theorem 3.10 *Let $(\mathcal{F}, \rho)$ be a type. For every abstract algebra A of type $(\mathcal{F}, \rho)$ and every map $h : S \to A$, there is a unique abstract algebra homomorphism $\overline{h} : T_{(\mathcal{F},\rho)}(S) \to A$ such that $\overline{h}(s) = h(s)$ for all $s \in S$:*

$$\begin{array}{ccc} S & \subset & T_{(\mathcal{F},\rho)}(S) \\ & \searrow^{h} & \downarrow \overline{h} \\ & & A \end{array}$$

That is, $F_{\mathsf{Set}}^{\mathsf{AbsAlg}(\mathcal{F},\rho)}[S]$ exists and is equal to $T_{(\mathcal{F},\rho)}(S)$; the accompanying inclusion map j is injective.

Proof The construction of the map $\overline{h}$ and the proof of its uniqueness can be given simultaneously, using induction on the height of a term.

Take a term $t \in T_{(\mathcal{F},\rho)}(S)$ of height 0. If t is a word that consists of a single symbol from S, then $\overline{h}(s)$ is prescribed, and we define $\overline{h}(s) := h(s)$ accordingly. If t is a symbol f from $\mathcal{F}$ such that $\rho(f) = 0$, then, since $\overline{h}$ is supposed to be an abstract algebra homomorphism, Eq. (3.1) implies that we must have $\overline{h}(f) = h(f^{T_{(\mathcal{F},\rho)}(S)}) = f^A$. Hence we define $\overline{h}(f) := f^A$ accordingly.

Suppose that, for some $n \geq 0$, the uniqueness of $\overline{h}(t)$ has already been shown for all terms $t \in T_{(\mathcal{F},\rho)}(S)$ of height at most n, and that $\overline{h}(t)$ has already been defined accordingly for such t. Take a term t of height $n+1$. Then $t = ft_1 \dots t_{\rho(f)}$ for a unique $f \in \mathcal{F}$ such that $\rho(f) \geq 1$ and unique terms $t_1, \dots, t_{\rho(f)} \in T_{(\mathcal{F},\rho)}(S)$ of height at most n. Since $\overline{h}$ is supposed to be an abstract algebra homomorphism, Eq. (3.2) implies that we must have

$$\begin{aligned}\overline{h}(t) &= \overline{h}(ft_1 \dots t_{\rho(f)})\\ &= \overline{h}(f^{T_{(\mathcal{F},\rho)}(S)}(t_1, \dots, t_{\rho(f)}))\\ &= f^A(\overline{h}(t_1), \dots, \overline{h}(t_{\rho(f)})).\end{aligned}$$

As a consequence of the induction hypotheses, this shows that $\overline{h}(t)$ is also uniquely determined. Since, also as a consequence of the induction hypothesis, $\overline{h}(t_1), \dots, \overline{h}(t_{\rho(f)})$ have already been defined, we can now define

$$\overline{h}(t) := f^A(\overline{h}(t_1), \dots, \overline{h}(t_{\rho(f)}))$$

accordingly. This completes the induction step.

We have now shown that $\overline{h}$ is uniquely determined as well as explicitly constructed the only possible candidate. It is immediate from this construction and the definitions in Eqs. (3.1) and (3.2) that this candidate is indeed an abstract algebra homomorphism.

The final sentence of the statement is then clear. □

Remark 3.11 It is evident from its construction that $T_{(\mathcal{F},\rho)}(S)$ is generated by S, in the sense that it equals its smallest abstract subalgebra that contains S. Of course, Remark 2.5 also makes clear that this must be the case.

4 Various Lattices as Abstract Algebras Satisfying Identities

Before proceeding with the general theory from universal algebra that we need, we pause to discuss structures that involve a partial ordering.

It is clear that, for a vector space, the validity of the vector space axioms can be expressed in terms of identities that involve the constant 0 and the operations of the naturally associated abstract algebra. For a unital vector lattice algebra, for example, it is, however, far less clear that there is an associated abstract algebra for which is possible. After all, the axioms of a unital vector lattice algebra also involve *in*equalities and the assumption of the existence of the infimum and supremum of two elements. At first sight, it may seem counterintuitive that these can also be described in terms of constants, operations, and identities. Nevertheless, this is possible. As we shall see, it is precisely this fact that lies at the heart of the existence proof for the free unital vector lattice algebra over a non-empty set. Later on, we shall then use this one existence result to obtain the existence of all other fourteen free objects in Theorem 6.2, below.

We start with the classical observation that the partial ordering in a lattice can equivalently be formulated in terms of operations and identities; see [7, Definition 1.7 and p.23], for example. Since this is crucial, and since we strive to keep this paper self-contained, we include the details for this. The ad-hoc terminology in the following definition is ours. We refrain from claiming any other originality here.

Definition 4.1 Let S be a non-empty set.

(1) Suppose that $\leq$ is a partial ordering on S. Then the partially ordered set $(S, \leq)$ is a *partially ordered lattice* if, for all $x, y \in S$, the supremum $x \vee y$ and the infimum $x \wedge y$ exist in S.
(2) Suppose that S is supplied with binary operations $\owedge$ and $\ovee$. Then the abstract algebra $(S, \owedge, \ovee)$ is an *algebraic lattice* if, for all $x, y, z \in S$,

$$\begin{aligned} x \owedge (y \owedge z) &= (x \owedge y) \owedge z, & x \ovee (y \ovee z) &= (x \ovee y) \ovee z, \\ x \owedge x &= x, & x \ovee x &= x, \\ x \owedge y &= y \owedge x, & x \ovee y &= y \ovee x, \\ x \owedge (x \ovee y) &= x, \quad \text{and} & x \ovee (x \owedge y) &= x. \end{aligned}$$

Let us mention explicitly that distributivity is not supposed.

Lemma 4.2 *Let S be a non-empty set.*

(1) *Suppose that S is supplied with a partial ordering such that the partially ordered set $(S, \leq)$ is a partially ordered lattice. For $x, y \in S$, set*

$$x \owedge y := x \wedge y$$

and

$$x \ovee y := x \vee y.$$

Then the abstract algebra $(S, \ovee, \owedge)$ is an algebraic lattice.
(2) *Suppose that S is supplied with two binary operations $\owedge$ and $\ovee$ such that the algebra $(S, \owedge, \ovee)$ is an algebraic lattice. For $x, y \in S$, say that $x \leq y$ if and only if*

$$x \owedge y = x.$$

Then $\leq$ is a partially ordering on S, and the partially ordered set $(S, \leq)$ is a partially ordered lattice. Moreover, for $x, y \in S$, we have

$$x \wedge y = x \owedge y$$

and

$$x \vee y = x \ovee y,$$

where $x \wedge y$ and $x \wedge y$ refer to the infimum and the supremum, respectively, in the partial ordering $\leq$.

Proof It is completely routine to verify the statement in part (1).

We turn to part (2). It is an easy consequence of the first three identities in the left column in Definition 4.1 that $\leq$ is a partially ordering on S.

Take $x, y \in S$. We claim that $x \wedge y$ exists in the partially ordered set $(S, \leq)$ and that, in fact, $x \wedge y = x \owedge y$. Since $(x \owedge y) \owedge y = x \owedge (y \owedge y) = x \owedge y$, we have $x \owedge y \leq y$. Since $(x \owedge y) \owedge x = x \owedge (x \owedge y) = (x \owedge x) \owedge y = x \owedge y$, we have $x \owedge y \leq x$. Then also $x \owedge y = y \owedge x \leq x$. Take $z \in S$, and suppose that $z \leq x$ and $z \leq y$, i.e., suppose that $z \owedge x = z$ and $z \owedge y = y$. Then $z \owedge (x \owedge y) = (z \owedge x) \owedge y = z \owedge y = z$. Hence $z \leq x \owedge y$ and the proof of the claim is complete.

Before turning to the supremum, we note that, for $x, y \in S$, the fact that $x \owedge y = x$ is equivalent to the fact that $x \ovee y = y$. It is here that the two identities in the fourth line of identities in Definition 4.1 come in. Indeed, suppose that $x \owedge y = x$. Then $x \ovee y = (x \owedge y) \ovee y = y \ovee (x \owedge y) = y \ovee (y \owedge x) = y$. Conversely, suppose that $x \ovee y = y$. Then $x \owedge y = x \owedge (x \ovee y) = x$.

Now take $x, y \in S$. We claim that $x \vee y$ exists in the partially ordered set $(S, \leq)$ and that, in fact, $x \vee y = x \ovee y$. Since $x \owedge (x \ovee y) = x$, we have $x \leq x \ovee y$. Since $y \owedge (x \ovee y) = y \owedge (y \ovee x) = y$, we have $y \leq x \ovee y$. Take $z \in S$ and suppose that $x \leq z$ and $y \leq z$, i.e., suppose that $x \owedge z = x$ and $y \owedge z = y$. By what we have just established in the intermezzo, this is equivalent to supposing that $x \ovee z = z$ and $y \ovee z = z$. Then $(x \ovee y) \ovee z = x \ovee (y \ovee z) = x \ovee z = z$. Again by the intermezzo, it follows that $(x \ovee y) \owedge z = x \ovee y$. Hence $x \ovee y \leq z$ and the proof of the claim is complete. Part (2) has now been established.

□

The following is now clear from Lemma 4.2.

Proposition 4.3 *The constructions in the parts (1) and (2) of Lemma 4.2 yield mutually inverse bijections between the category of partially ordered lattices (with the lattice homomorphisms as morphisms) and the category of abstract algebras that are algebraic lattices (with the abstract algebra homomorphisms as morphisms). Under this isomorphism, the underlying sets and the maps that are the morphisms are kept.*

With Lemma 4.2 available, it is not so difficult to describe the unital vector lattice algebras as the abstract algebras (of a common unspecified type) where certain identities are satisfied.

Lemma 4.4 *Let A be an abstract algebra with (not necessarily different) constants 0 and 1, a binary map $\oplus$, a unary map $\ominus$, a unary map m_λ for every $\lambda \in \mathbb{R}$, a binary map $\odot$, and binary maps $\owedge$ and $\ovee$. Suppose that all of the following are satisfied:*

(1) $(x \oplus y) \oplus z = x \oplus (y \oplus z)$ *for all $x, y, z \in A$;*
(2) $x \oplus 0 = x$ *for all $x \in A$;*
(3) $x \oplus (\ominus x) = 0$ *for all $x \in A$;*
(4) $x \oplus y = y \oplus x$ *for all $x, y \in A$;*

(5) $m_\lambda(x \oplus y) = m_\lambda(x) \oplus m_\lambda(y)$ *for all* $\lambda \in \mathbb{R}$ *and* $x, y \in A$;
(6) $m_{\lambda+\mu}(x) = m_\lambda(x) \oplus m_\mu(x)$ *for all* $\lambda, \mu \in \mathbb{R}$ *and* $x \in A$;
(7) $m_{\lambda\mu}(x)) = m_\lambda(m_\mu(x))$ *for all* $\lambda, \mu \in \mathbb{R}$ *and* $x \in A$;
(8) $m_1(x) = x$ *for all* $x \in A$;
(9) $(x \odot y) \odot z = x \odot (y \odot z)$ *for all* $x, y, z \in A$;
(10) $x \odot (y \oplus z) = (x \odot y) \oplus (x \odot z)$ *for all* $x, y, z \in A$;
(11) $(x \oplus y) \odot z = (x \odot z) \oplus (y \odot z)$ *for all* $x, y, z \in A$;
(12) $m_\lambda(x \odot y) = m_\lambda(x) \odot y = x \odot m_\lambda(y)$ *for all* $\lambda \in \mathbb{R}$ *and* $x, y \in A$;
(13) $1 \odot x = x \odot 1 = x$ *for all* $x \in A$;
(14) $x \owedge (y \owedge z) = (x \owedge y) \owedge z$ *and* $x \ovee (y \ovee z) = (x \ovee y) \ovee z$ *for all* $x, y, z \in A$;
(15) $x \owedge x = x$ *and* $x \ovee x = x$ *for all* $x \in A$;
(16) $x \owedge y = y \owedge x$ *and* $x \ovee y = y \ovee x$ *for all* $x, y \in A$;
(17) $x \owedge (x \ovee y) = x$ *and* $x \ovee (x \owedge y) = x$ *for all* $x, y \in A$;
(18) $x \oplus (y \owedge z) = (x \oplus y) \owedge (x \oplus z)$ *for all* $x, y, z \in A$;
(19) $m_\lambda(0 \owedge x) = 0 \owedge (m_\lambda(x))$ *for all* $\lambda \in \mathbb{R}_{\geq 0}$ *and* $x \in A$;
(20) $0 \owedge ((x \owedge (\ominus x)) \odot (y \owedge (\ominus y))) = 0$ *for all* $x, y \in A$.

Set

(a) $x + y := x \oplus y$ *for* $x, y \in A$;
(b) $\lambda x := m_\lambda(x)$ *for* $\lambda \in \mathbb{R}$ *and* $x \in A$;
(c) $xy := x \odot y$ *for* $x, y \in A$.

Supplied with the operations as defined under (a), (b), *and* (c), *A is an associative algebra over the real numbers with zero element 0 and identity element 1.*

For $x, y \in A$, *say that* $x \leq y$ *when* $x \owedge y = x$. *Then* $\leq$ *is a partial ordering on A that makes A into a partially ordered lattice. Moreover, for* $x, y \in A$, *we have* $x \wedge y = x \owedge y$ *and* $x \vee y = x \ovee y$, *where* $\wedge$ *and* $\vee$ *refer to the supremum resp. infimum in the partial ordering* $\leq$.

Supplied with the partial ordering $\leq$ *and with the operations as defined under* (a), (b), *and* (c), *A is a unital vector lattice algebra with zero element 0 and identity element 1.*

Proof The identities in (1)–(4) show that A becomes an abelian group under $+$ with identity element 0. Those in (5)–(8) show that A becomes a vector space with zero element 0 when the scalar multiplications as in (b) are added, and those in (9)–(13) guarantee that A becomes an associative algebra with zero element 0 and identity element 1 when the multiplication as in (c) is added.

It becomes more interesting when the partial ordering is brought in. In view of Lemma 4.2, the identities in (14)–(17) guarantee that $\leq$ is a partial ordering on A that makes A into a partially ordered lattice where the infimum $x \wedge y$ and supremum $x \vee y$ of two elements x, y are given by $x \owedge y$ and $x \ovee y$, respectively.

It remains to be shown that the partial ordering $\leq$ is a vector space ordering, and also that the product of two positive elements of A is again positive.

We start with the vector space ordering. Take $y, z \in A$ and suppose that $y \leq z$, i.e, suppose that $y \owedge z = y$. Take $x \in A$. Using the identity in (18), we have

$$(x + y) \wedge (x + z) = (x \oplus y) \owedge (x \oplus z) = x \oplus (y \owedge z) = x \oplus y = x + y.$$

Hence $x + z \leq y + z$. Take $x \in A$ and $\lambda \in \mathbb{R}_{\geq 0}$. Suppose that $0 \leq x$, i.e, suppose that $0 \owedge x = 0$. Using the identity in (19) (and, in the final equality, the fact that we already know that A is a vector space), we have

$$0 \wedge (\lambda x) = 0 \owedge (m_\lambda(x)) = m_\lambda(0 \owedge x) = m_\lambda(0) = \lambda 0 = 0.$$

Hence $0 \leq \lambda x$,

We turn to the product of two positive elements of A. Since we know by now that A is a vector lattice, we can equivalently formulate the identity in (20) as the fact that $0 \wedge (|x||y|) = 0$ for all $x, y \in A$. This implies (and is equivalent to) the fact that the product of two positive elements of A is again positive. □

Obviously, any unital vector lattice algebra gives rise, in a natural way, to an abstract algebra with constants and operations as in Lemma 4.4 where all identities in (1)–(20) in Lemma 4.4 are satisfied. As for partially ordered lattices and algebraic lattices, the two constructions are mutually inverse. Moreover, the unital vector lattice algebra homomorphisms correspond to the abstract algebra homomorphisms. We therefore have the following analogue of Proposition 4.3.

Proposition 4.5 *The category* VLA^1 *of unital vector lattice algebras, with the unital vector lattice algebra homomorphisms as morphism, is isomorphic to the category of abstract algebras with constants and operations as in Lemma 4.4 where the identities (1)–(20) in Lemma 4.4 are satisfied, with the abstract algebra homomorphisms as morphisms. Under this isomorphism, the underlying sets and the maps that are the morphisms are kept.*

Obviously, there are isomorphisms similar to those in Propositions 4.3 and 4.5 for VL and VLA. Once one notices that one can express the positivity of an identity element 1 of a vector lattice algebra by requiring that $0 \owedge 1 = 0$, it becomes clear that there is also a similar isomorphism for VLA^{1+}. For many categories of algebraic structures (groups, abelian groups, vector spaces, rings with identity elements, algebras, commutative algebras, ...), where there is no partial ordering that needs to be 'equationalised', the existence of a similar isomorphism with a category of abstract algebras is immediate from the axioms for these structures.

The existence of a 'similar' isomorphism can be made precise by saying that all these categories are isomorphic to an equational class of abstract algebras. This brings us to the next section.

5 Universal Algebra: Part II

We have seen in Propositions 4.3 and 4.5 how two categories from the partially ordered realm are isomorphic to categories of abstract algebras. In the abstract algebraic side of the picture, the objects of the category are those abstract algebras (all of a common type) 'where certain identities are satisfied'. We shall now formalise this concept of 'identities being satisfied'.

Let A be an abstract algebra of type $(\mathcal{F}, \rho)$. By way of example, suppose that one of the operations on A is a binary operator $\oplus$. We want to express the fact that this operation is associative, i.e., that $x_1 \oplus (x_2 \oplus x_3) = (x_1 \oplus x_2) \oplus x_3$ for all $x_1, x_2, x_3 \in A$. A first attempt would be to take a set $S = \{s_1, s_2, s_3\}$ of three elements, take the terms (rewritten in a legible way) $(s_1 \oplus s_2) \oplus s_3$ and $s_1 \oplus (s_2 \oplus s_3)$ in $T_{(\mathcal{F},\rho)}(S)$, and require that $(s_1 \oplus s_2) \oplus s_3 = s_1 \oplus (s_2 \oplus s_3)$ 'is satisfied in A'. The problem is that this does not make sense. Firstly, the left and the right hand sides are not elements of A. They are elements of $T_{(\mathcal{F},\rho)}(S)$ and, secondly, they are *not* equal in $T_{(\mathcal{F},\rho)}(S)$. There are two ways to get further.

The first one is to take the terms $(s_1 \oplus s_2) \oplus s_3$ and $s_1 \oplus (s_2 \oplus s_3)$, and assign to them the ternary operations on A that send a triple $(x_1, x_2, x_3) \in A^3$ to $x_1 \oplus (x_2 \oplus x_3)$ and $(x_1 \oplus x_2) \oplus x_3$, respectively. The associativity can then be expressed by saying that these two maps from A^3 to A are equal. This is the approach that is taken for the general case in [7, Definitions 4.31 and 4.35]. Here, in order to be able to accommodate identities in an arbitrarily large number of variables, one takes a countably infinite set S. Given two terms $t_1, t_2 \in T_{(\mathcal{F},\rho)}(S)$, one associates operations t_1^A and t_2^A with them, and requires that these be equal as maps from A^n (where n is appropriate) to A. There are some formalities to be taken care of then, however. For example, it could be the case that the 'natural' number of arguments of t_1^A differs from that of t_2^A. One could want to express the fact that $x_1 \oplus x_2 = (x_3 \oplus x_2) \oplus x_1$ for all $x_1, x_2, x_3 \in A$, but the natural domain for the map for the left hand side is A^2, whilst for the right hand side this is A^3.

The second one, and the one we shall take, is the following. Take a three-point set S again. For all $x_1, x_2, x_3 \in A$, there exists a map $h_{x_1,x_2,x_3} : S \to A$ such that $h_{x_1,x_2,x_3}(s_i) = x_i$ for $i = 1, 2, 3$. By Theorem 3.10, such a map extends uniquely to an abstract algebra homomorphism $\overline{h}_{x_1,x_2,x_3} : T_{(\mathcal{F},\rho)}(S) \to A$. Then $\overline{h}_{x_1,x_2,x_3}(s_1 \oplus (s_2 \oplus s_3)) = x_1 \oplus (x_2 \oplus x_3)$ and $\overline{h}_{x_1,x_2,x_3}((s_1 \oplus s_2) \oplus s_3) = (x_1 \oplus x_2) \oplus x_3$. It follows from this that the associativity of $\oplus$ in A can equally well be expressed by requiring that $\overline{h}(s_1 \oplus (s_2 \oplus s_3)) = \overline{h}((s_1 \oplus s_2) \oplus s_3)$ for every map $h : S \to A$, where, as usual, $\overline{h} : T_{(\mathcal{F},\rho)}(S) \to A$ is the abstract algebra homomorphism that extends h. Since every abstract algebra homomorphism from $T_{(\mathcal{F},\rho)}(S)$ to A is the unique extension of its restriction to S, one can equally well (with a change in notation) require that $h(s_1 \oplus (s_2 \oplus s_3)) = h((s_1 \oplus s_2) \oplus s_3)$ for every abstract algebra homomorphism $h : T_{(\mathcal{F},\rho)}(S) \to A$. Of course, one wants to be able to do this with an arbitrarily large number of variables involved. This leads to the following definition, as in [8, Definition 9.4.1] and [16, Section 2.8]

Fix, once and for all, a countably infinite set $S_{\aleph_0} = \{s_1, s_2, \dots\}$.

Definition 5.1 Let $(\mathcal{F}, \rho)$ be a type. Take two terms $t_1, t_2 \in T_{(\mathcal{F},\rho)}(S_{\aleph_0})$. Let A be an abstract algebra of type $(\mathcal{F}, \rho)$. Then A *satisfies* $t_1 \approx t_2$ when $h(t_1) = h(t_2)$ for every abstract algebra homomorphism $h : T_{(\mathcal{F},\rho)}(S_{\aleph_0}) \to A$. For a subset Σ of $T_{(\mathcal{F},\rho)}(S_{\aleph_0}) \times T_{(\mathcal{F},\rho)}(S_{\aleph_0})$, A *satisfies* Σ when A satisfies $t_1 \approx t_2$ for every pair $(t_1, t_2) \in \Sigma$.

This definition depends on the choice of $S_{\aleph_0}$ because t_1 and t_2 are elements of the set $T_{(\mathcal{F},\rho)}(S_{\aleph_0})$ that depends on this choice. A moment's thought shows that the equalities of operations on A—which is what we are after—that is equivalent to the satisfaction of the identities from Σ is independent of this choice. It is for this reason that there is no harm in fixing a particular choice for $S_{\aleph_0}$ as we have done.

For $\Sigma \subseteq T_{(\mathcal{F},\rho)}(S_{\aleph_0}) \times T_{(\mathcal{F},\rho)}(S_{\aleph_0})$, the class of all abstract algebras of type $(\mathcal{F}, \rho)$ satisfying Σ is called the *equational class defined by* Σ. Together with the abstract algebra homomorphism between them, it forms the subcategory $\mathsf{AbsAlg}_{(\mathcal{F},\rho);\Sigma}$ of $\mathsf{AbsAlg}_{(\mathcal{F},\rho)}$.

It is an important point that we can force identities to be satisfied by passing to an abstract quotient algebra.

Lemma 5.2 *Let A be an abstract algebra of type $(\mathcal{F}, \rho)$, and let θ be a congruence relation on A. Take $t_1, t_2 \in T_{(\mathcal{F},\rho)}(S_{\aleph_0})$. Then A/θ satisfies $t_1 \approx t_2$ if and only if $(h(t_1), h(t_2)) \in \theta$ for all abstract algebra homomorphisms $h : T_{(\mathcal{F},\rho)}(S_{\aleph_0}) \to A$.*

Proof Let $q_\theta : A \to A/\theta$ be the quotient map. It is a consequence of the surjectivity of the abstract algebra homomorphism q_θ and the universal property of $T_{(\mathcal{F},\rho)}(S_{\aleph_0})$ that the abstract algebra homomorphisms $h_\theta : T_{(\mathcal{F},\rho)}(S_{\aleph_0}) \to A/\theta$ are precisely the compositions $q_\theta \circ h$ for the abstract algebra homomorphisms $h : T_{(\mathcal{F},\rho)}(S_{\aleph_0}) \to A$. The statement in the lemma is an immediate consequence of this. □

Lemma 5.3 *Let A and B be abstract algebras of the same type $(\mathcal{F}, \rho)$, and let $h : A \to B$ be an abstract algebra homomorphism. Take $t_1, t_2 \in T_{(\mathcal{F},\rho)}(S_{\aleph_0})$. If B satisfies $t_1 \approx t_2$, then $(h'(t_1), h'(t_2)) \in \ker h$ for every abstract algebra homomorphism $h' : T_{(\mathcal{F},\rho)}(S_{\aleph_0}) \to A$.*

Proof Take an abstract algebra homomorphism $h' : T_{(\mathcal{F},\rho)}(S_{\aleph_0}) \to A$. Then $h \circ h' : T_{(\mathcal{F},\rho)}(S_{\aleph_0}) \to B$ is an abstract algebra homomorphism. Hence $(h \circ h')(t_1) = (h \circ h')(t_2)$, showing that $(h'(t_1), h'(t_2)) \in \ker h$. □

We can now construct a free object of an equational class over a non-empty set. Let $(\mathcal{F}, \rho)$ be a type, and take a subset $\Sigma \subseteqq T_{(\mathcal{F},\rho)}(S_{\aleph_0}) \times T_{(\mathcal{F},\rho)}(S_{\aleph_0})$. Let S be a non-empty set. We then take the abstract term algebra $T_{(\mathcal{F},\rho)}(S)$, which contains S as a subset, and we let θ be the smallest congruence relation on $T_{(\mathcal{F},\rho)}(S)$ that contains the pairs $(h'(t_1), h'(t_2))$ for all $(t_1, t_2) \in \Sigma$ and all abstract algebra homomorphisms $h' : T_{(\mathcal{F},\rho)}(S_{\aleph_0}) \to T_{(\mathcal{F},\rho)}(S)$. Let $q_\theta : T_{(\mathcal{F},\rho)}(S) \to T_{(\mathcal{F},\rho)}(S)/\theta$ denote the quotient map, and let $q_\theta|_S$ denote its restriction to S.

We see from Lemma 5.2 that $T_{(\mathcal{F},\rho)}(S)/\theta$ satisfies Σ, i.e., it is an object of $\mathsf{AbsAlg}_{(\mathcal{F},\rho);\Sigma}$.

Furthermore, we claim that the pair $(q_\theta|_S, T_{(\mathcal{F},\rho)}(S)/\theta)$ is a free object of $\mathsf{AbsAlg}_{(\mathcal{F},\rho);\Sigma}$ over the object S in Set. To see this, let $A \in \mathsf{AbsAlg}_{(\mathcal{F},\rho);\Sigma}$, and let $h : S \to A$ be a map. By Theorem 3.10, there exists a unique abstract algebra homomorphism $\overline{h} : T_{(\mathcal{F},\rho)}(S) \to A$ such that $\overline{h}(s) = h(s)$ for all $h \in S$.

Take $(t_1, t_2) \in \Sigma$, and let $h' : T_{(\mathcal{F},\rho)}(S_{\aleph_0}) \to T_{(\mathcal{F},\rho)}(S)$ be an arbitrary abstract algebra homomorphism. Since A satisfies $t_1 \approx t_2$, Lemma 5.3 shows that $(h'(t_1), h'(t_2)) \in \ker \overline{h}$. Thus the congruence relation $\ker f$ on $T_{(\mathcal{F},\rho)}(S)$ contains the generators of θ, and we conclude that $\theta \subseteq \ker \overline{h}$. Lemma 3.7 then implies that there is a unique abstract algebra homomorphism $\overline{\overline{h}} : T_{(\mathcal{F},\rho)}(S)/\theta \to A$ such that $\overline{h} = \overline{\overline{h}} \circ q_\theta$. For $s \in S$, this implies that $(\overline{\overline{h}} \circ q_\theta|_S)(s) = \overline{h}(s) = h(s)$. Hence we have found a factoring abstract algebra homomorphism $\overline{\overline{h}}$. It remains to show uniqueness. Suppose that $h_0 : T_{(\mathcal{F},\rho)}(S)/\theta \to A$ is an abstract algebra homomorphism such that $(h_0 \circ q_\theta|_S)(s) = h(s)$ for all $s \in S$. Then $h_0 \circ q_\theta : T_{(\mathcal{F},\rho)}(S) \to A$ is an abstract algebra homomorphism such that $(h_0 \circ q_\theta)(s) = (h_0 \circ q_\theta|_S)(s) = h(s) = \overline{h}(s)$ for $s \in S$. This implies that $h_0 \circ q_\theta = \overline{h}$ and this, in turn, shows that $h_0 = \overline{\overline{h}}$.

All in all, we have shown the following. Its main part is the existence of a object, but we have also included the construction of a concrete realisation of it.

Theorem 5.4 *Let $(\mathcal{F}, \rho)$ be a type, and take $\Sigma \subseteq T_{(\mathcal{F},\rho)}(S_{\aleph_0}) \times T_{(\mathcal{F},\rho)}(S_{\aleph_0})$. Let S be a non-empty set, and let θ be the smallest congruence relation on $T_{(\mathcal{F},\rho)}(S)$ that contains the pairs $(h'(t_1), h'(t_2))$ for all $(t_1, t_2) \in \Sigma$ and all abstract algebra homomorphisms $h' : T_{(\mathcal{F},\rho)}(S_{\aleph_0}) \to T_{(\mathcal{F},\rho)}(S)$. Then $T_{(\mathcal{F},\rho)}(S_{\aleph_0})/\theta$ is an abstract algebra of type $(\mathcal{F}, \rho)$ that satisfies Σ. Let $q_\theta : T_{(\mathcal{F},\rho)}(S) \to T_{(\mathcal{F},\rho)}(S)/\theta$ denote the quotient map, and let $q_\theta|_S$ denote its restriction to the subset S of $T_{(\mathcal{F},\rho)}(S)$.*

For every abstract algebra A of type $(\mathcal{F}, \rho)$ that satisfies Σ, and for every map $h : S \to A$, there is a unique abstract algebra homomorphism $\overline{h} : T_{(\mathcal{F},\rho)}(S)/\theta \to A$ such that $h = \overline{h} \circ q_\theta|_S$ for all $s \in S$:

$$\begin{array}{ccc} S & \xrightarrow{q_\theta|_S} & T_{(\mathcal{F},\rho)}(S)/\theta \\ & \searrow^{h} & \downarrow \overline{h} \\ & & A \end{array}$$

That is, $\mathrm{F}_{\mathsf{Set}}^{\mathsf{AbsAlg}_{(\mathcal{F},\rho);\Sigma}}[S]$ exists and is equal to $T_{(\mathcal{F},\rho)}(S)/\theta$.

Remark 5.5

(1) If $\mathsf{AbsAlg}_{(\mathcal{F},\rho);\Sigma}$ contains an object that has at least two elements, then $q_\theta|_S$ must be injective. Consequently, j is *not* injective if and only if S contains at least two elements and $\mathsf{AbsAlg}_{(\mathcal{F},\rho);\Sigma}$ consists only of the one-point algebra of type $(\mathcal{F}, \rho)$. For many equational classes of practical interest, $q_\theta|_S$ is, therefore, injective.

(2) For the choice $\Sigma = \emptyset$ one retrieves the fact from Theorem 3.10 that $\mathrm{F}_{\mathsf{Set}}^{\mathsf{AbsAlg}_{(\mathcal{F},\rho)}}[S]$ exists and equals $T_{(\mathcal{F},\rho)}(S)$. The injectivity of the accompanying

map j is not obtained in this fashion, but follows easily from the universal property since $T_{(\mathcal{F},\rho)}(S)$ is an abstract algebra of type $(\mathcal{F},\rho)$ that has at least two elements.

Theorem 5.4 is a classical result; see [16, Theorem 2.10], for example. It can also be found as (a part of) [7, Corollary 4.30], where it is actually proved—with a slightly different proof—that free objects over non-empty sets exist in varieties of abstract algebras. As in [7, p.9], we say that a *variety of abstract algebras* is a class of algebras of the same type that is closed under taking subalgebras, abstract algebra homomorphic images, and abstract algebra products. Some sources use both the terms 'variety' and 'equational class' for what we call an 'equational class'; see [16, p.81] and [13, p.152]. As we shall see in a moment, a theorem of Birkhoff's shows that there is no actual ambiguity in the terminology.

It is routine to verify that $\mathsf{AbsAlg}_{(\mathcal{F},\rho);\Sigma}$ is closed under the taking of abstract subalgebras and the formation of arbitrary abstract algebra products. Using the universal property of $T_{(\mathcal{F},\rho)}(S_{\aleph_0})$, one sees that it is also closed under the taking of abstract algebra homomorphic images. Every equational class (in our terminology) is, therefore, a variety (in our terminology). The converse is actually also true by Birkhoff's theorem; see [7, Theorem 4.41] and [16, Theorem 2.15], for example. Therefore, there is no ambiguity in terminology, and [7, Corollary 4.30] and Theorem 5.4 are actually equivalent.

6 Free Vector Lattices and Free Objects of Categories of Vector Lattice Algebras

It is clear from Proposition 4.5 that the category of unital vector lattice algebras is isomorphic to a category of abstract algebras that is an equational class. At the level of objects, the isomorphism keeps the underlying sets, but views them as different structures. At the level of morphisms, the maps between the underlying sets are kept, but are observed to have different pertinent properties. Theorem 5.4 therefore implies that free unital vector lattice algebras over non-empty sets exist. Let us spell out the details once more. The additional properties under (1), (2), and (3) follow from general principles.

Theorem 6.1 *Let S be a non-empty set. Then there exist a unital vector lattice algebra $F_{\mathsf{Set}}^{\mathsf{VLA}^1}[S]$ and a map $j : S \to F_{\mathsf{Set}}^{\mathsf{VLA}^1}[S]$ with the property that, for any map $\varphi : S \to A$ from S into a unital vector lattice algebra A^1, there exists a unique unital vector lattice algebra homomorphism $\overline{\varphi}$ such that the diagram*

$$\begin{array}{ccc} S & \xrightarrow{\;j\;} & F_{\mathsf{Set}}^{\mathsf{VLA}^1}[S] \\ & \searrow^{\varphi} & \downarrow{\scriptstyle\overline{\varphi}} \\ & & A^1 \end{array}$$

is commutative. Furthermore:

(1) *the pair* $(j, F_{\mathsf{Set}}^{\mathsf{VLA}^1}[S])$ *is unique up to a unique compatible isomorphism;*
(2) $F_{\mathsf{Set}}^{\mathsf{VLA}^1}[S]$ *equals its unital vector lattice subalgebra that is generated* $j(S)$*;*
(3) *the map* j *is injective.*

For precisely the same reason—the existence of a category isomorphism with an equational class—it is clear that free vector spaces over non-empty sets exist (which can be seen much easier, of course), as do free vector lattices, free vector lattice algebras, and free vector lattice algebras with a positive identity element.

Likewise, the combination of Lemma 4.2 and Theorem 5.4 shows that free lattices over non-empty sets exist. Adding the distributive laws to the identities in Lemma 4.2 shows that free distributive lattices over non-empty sets also exist.

Let us return to our original chain of categories

$$\mathsf{Set} \supset \mathsf{VS} \supset \mathsf{VL} \supset \mathsf{VLA} \supset \mathsf{VLA}^1 \supset \mathsf{VLA}^{1+}. \tag{6.1}$$

There are 15 instances of a category and a subcategory associated with this chain. For 5 of these, the ones where Set has a subcategory, we know that there are always free objects of the subcategory because of the general theorem for equational classes. How about the remaining 10?

For example, given a vector space V, do there exist a vector lattice $\mathrm{F}_{\mathsf{VS}}^{\mathsf{VL}}[V]$ and a linear map $j : V \to \mathrm{F}_{\mathsf{VS}}^{\mathsf{VL}}[V]$ with the property that, for every linear map $\varphi : E \to F$ from E into a vector lattice F, there exists a unique vector lattice homomorphism $\overline{\varphi} : \mathrm{F}_{\mathsf{VS}}^{\mathsf{VL}}[V] \to F$ such that the diagram

$$\begin{array}{ccc} V & \xrightarrow{\ j\ } & \mathrm{F}_{\mathsf{VS}}^{\mathsf{VLA}}[V] \\ & \searrow^{\varphi} & \downarrow{\scriptstyle\overline{\varphi}} \\ & & F \end{array}$$

is commutative? As another example, given a vector lattice E, do there exist a unital vector lattice algebra $\mathrm{F}_{\mathsf{VL}}^{\mathsf{VLA}^{1+}}[E]$ with a positive identity element and a vector lattice homomorphism $j : E \to \mathrm{F}_{\mathsf{VS}}^{\mathsf{VLA}^{1+}}[E]$ with the property that, for every vector lattice homomorphism $\varphi : E \to A^{1+}$ from E into a vector lattice algebra A^{1+} with a positive identity element, there exists a unique unital vector lattice algebra homomorphism $\overline{\varphi}$ such that the diagram

$$\begin{array}{ccc} E & \xrightarrow{\ j\ } & \mathrm{F}_{\mathsf{VS}}^{\mathsf{VLA}^{1+}}[V] \\ & \searrow^{\varphi} & \downarrow{\scriptstyle\overline{\varphi}} \\ & & A^{1+} \end{array}$$

is commutative?

As we shall see, the existence of all of these 'missing' 10 free objects can be derived from the existence of $\mathrm{F}_{\mathsf{Set}}^{\mathsf{VLA}^1}[S]$ for non-empty sets S. We shall, in fact, also use this basic existence result to derive the existence of $\mathrm{F}_{\mathsf{Set}}^{\mathsf{VS}}[S]$, $\mathrm{F}_{\mathsf{Set}}^{\mathsf{VL}}[S]$, $\mathrm{F}_{\mathsf{Set}}^{\mathsf{VLA}}[S]$, and $\mathrm{F}_{\mathsf{Set}}^{\mathsf{VLA}^{1+}}[S]$ once more, even though we had already observed this to be a consequence of the general result for equational classes. The methods that are used below to obtain 14 other existence results from a basic one for a free object with 'maximal' structure, can undoubtedly be formulated in general in terms of abstract algebras, their reducts (see [7, p.7]) and forgetful functors, inclusion of congruences relations and the general Second Isomorphism Theorem (see [7, Theorem 3.5]). We believe, however, that this would actually obscure the picture for the concrete cases we have in mind. Our approach, which is a very simple combination of passing to quotients and sub-objects, also leads to an overview of the (quite natural) relations between the various free objects of our interest; see Theorem 6.2, below. This overview would presumably be a little less obvious when using a more general abstract approach.

In order to solve the 14 remaining universal problems, we first make sure that free unital vector lattice algebras and free unital vector lattice algebras with positive identity elements over non-empty sets, over vector spaces, over vector lattices, and over vector lattice algebras all exist. These are 8 universal problems in all.

The first batch of 4 free objects consists of the free unital vector lattice algebras in this list. We already have $\mathrm{F}_{\mathsf{Set}}^{\mathsf{VLA}^1}[S]$ for a non-empty set. This will be our starting point to construct free unital vector lattice algebras over vector spaces. To this end, let V be a vector space. We let $\mathrm{Set}\,V$ be the underlying set of V and take a pair $(j, \mathrm{F}_{\mathsf{Set}}^{\mathsf{VLA}^1}[\mathrm{Set}\,V])$. Suppose that $\varphi : V \to A^1$ is a linear map from V into a unital vector lattice algebra A^1. There exists a unique unital vector lattice algebra homomorphism $\overline{\varphi} : \mathrm{F}_{\mathsf{Set}}^{\mathsf{VLA}^1}[\mathrm{Set}\,V] \to A^1$ such that $\overline{\varphi} \circ j = \varphi$. For $x, y \in V$, we have, since φ is actually linear, that

$$
\begin{aligned}
\overline{\varphi}(j(x+y) - j(x) - j(y)) &= \overline{\varphi}(j(x+y)) - \overline{\varphi}(j(x)) - \overline{\varphi}(j(y)) \\
&= \varphi(x+y) - \varphi(x) - \varphi(y) \\
&= 0.
\end{aligned}
$$

Likewise, one sees that $\overline{\varphi}(j(\lambda x) - \lambda j(x)) = 0$ for all $\lambda \in \mathbb{R}$ and $x \in V$. Hence $\overline{\varphi}$ vanishes on the bi-ideal I of $\mathrm{F}_{\mathsf{Set}}^{\mathsf{VLA}^1}[\mathrm{Set}\,V]$ that is generated by the elements $j(x+y) - j(x) - j(y)$ for $x, y \in V$ and the elements $j(\lambda x) - \lambda j(x)$ for $\lambda \in \mathbb{R}$ and $x \in V$. Consider the quotient $\mathrm{F}_{\mathsf{Set}}^{\mathsf{VLA}^1}[\mathrm{Set}\,V]/I$, which is a unital vector lattice algebra, and let $q_I : \mathrm{F}_{\mathsf{Set}}^{\mathsf{VLA}^1}[\mathrm{Set}\,V] \to \mathrm{F}_{\mathsf{Set}}^{\mathsf{VLA}^1}[\mathrm{Set}\,V]/I$ be the quotient map. There exists a unique unital vector lattice homomorphism $\overline{\overline{\varphi}} : \mathrm{F}_{\mathsf{Set}}^{\mathsf{VLA}^1}[\mathrm{Set}\,V]/I \to A^1$ such that $\overline{\varphi} = \overline{\overline{\varphi}} \circ q_I$. Hence we have a commutative diagram

$$
\begin{array}{ccccc}
V & \xrightarrow{\;j\;} & \mathrm{F}_{\mathsf{Set}}^{\mathsf{VLA}^1}[\mathrm{Set}\,V] & \xrightarrow{\;q_I\;} & \mathrm{F}_{\mathsf{Set}}^{\mathsf{VLA}^1}[\mathrm{Set}\,V]/I \\
& {\scriptstyle\varphi}\searrow & \downarrow{\scriptstyle\overline{\varphi}} & \swarrow{\scriptstyle\overline{\overline{\varphi}}} & \\
& & A & &
\end{array}
$$

Then $\overline{\overline{\varphi}} \circ (q_I \circ j) = \varphi$. Since $j(S)$ generates $\mathrm{F}_{\mathsf{Set}}^{\mathsf{VLA}^1}[\mathrm{Set}\, V]$ as a unital vector lattice algebra, $(q_I \circ j)(S)$ generates $\mathrm{F}_{\mathsf{Set}}^{\mathsf{VLA}^1}[\mathrm{Set}\, V]/I$ as a unital vector lattice algebra. This shows that the unital vector lattice homomorphism $\overline{\overline{\varphi}}$ is uniquely determined by the requirement that $\overline{\overline{\varphi}} \circ (q_I \circ j) = \varphi$. Since, furthermore, $q_I \circ j : V \to \mathrm{F}_{\mathsf{Set}}^{\mathsf{VLA}^1}[\mathrm{Set}\, V]/I$ is linear, we see that the pair $(q_I \circ j, \mathrm{F}_{\mathsf{Set}}^{\mathsf{VLA}^1}[\mathrm{Set}\, V]/I)$ solves the problem of finding a free unital vector lattice algebra $\mathrm{F}_{\mathsf{VS}}^{\mathsf{VLA}^1}[V]$ over the vector space V.

The same method yields a free unital vector lattice algebra $\mathrm{F}_{\mathsf{VL}}^{\mathsf{VLA}^1}[E]$ over a vector lattice E. One starts with a pair $(j, \mathrm{F}_{\mathsf{Set}}^{\mathsf{VLA}^1}[\mathrm{Set}\, E]$, and lets I be the bi-ideal in $\mathrm{F}_{\mathsf{Set}}^{\mathsf{VLA}^1}[\mathrm{Set}\, E]$ that is generated by the elements $j(x+y) - j(x) - j(y)$ for all $x, y \in E$, the elements $j(\lambda x) - \lambda j(x)$ for all $\lambda \in \mathbb{R}$ and $x \in E$, and now also the elements $j(x \vee y) - j(x) \vee j(y)$ for all $x, y \in E$. The pair $(q_I \circ j, \mathrm{F}_{\mathsf{Set}}^{\mathsf{VLA}^1}[\mathrm{Set}\, E]/I)$ then solves the problem of finding a free unital vector lattice algebra $\mathrm{F}_{\mathsf{VL}}^{\mathsf{VLA}^1}[E]$ over the vector lattice E.

In order to obtain a free unital vector lattice algebra over a vector lattice algebra A, one includes the previous three classes of elements into the generating set of the ideal I of $\mathrm{F}_{\mathsf{Set}}^{\mathsf{VLA}^1}[\mathrm{Set}\, A]$, and now also adds the elements $j(xy) - j(x)j(y)$ for all $x, y \in A$. The pair $(q_I \circ j, \mathrm{F}_{\mathsf{Set}}^{\mathsf{VLA}^1}[\mathrm{Set}\, A]/I))$ then solves the problem of finding a free unital vector lattice algebra $\mathrm{F}_{\mathsf{VLA}}^{\mathsf{VLA}^1}[A]$ over the vector lattice algebra A.

The second batch of 4 free objects from the list above consists of the free unital vector lattice algebras with positive identity elements over sets, over vector spaces, over vector lattices, and over unital vector lattice algebras.

Let S be a non-empty set. Take a pair $(j, \mathrm{F}_{\mathsf{Set}}^{\mathsf{VLA}^1}[S])$. Let I be the bi-ideal in $\mathrm{F}_{\mathsf{Set}}^{\mathsf{VLA}^1}[S]$ that is generated by $(|1| - 1)$, and let $q_I : \mathrm{F}_{\mathsf{Set}}^{\mathsf{VLA}^1}[S] \to \mathrm{F}_{\mathsf{Set}}^{\mathsf{VLA}^1}[S]/I$ be the quotient map. Suppose that $\varphi : S \to A^{1+}$ is a map from S into a unital vector lattice algebra with positive identity element A^{1+}. There exists a unique unital vector lattice algebra homomorphism $\overline{\varphi} : \mathrm{F}_{\mathsf{Set}}^{\mathsf{VLA}^1}[S]$ such that $\overline{\varphi} \circ j = \varphi$. Since the identity element in A^{1+} is positive, $\overline{\varphi}$ vanishes on I. Hence there exists a unique unital vector lattice algebra homomorphism $\overline{\overline{\varphi}} : \mathrm{F}_{\mathsf{Set}}^{\mathsf{VLA}^1}[S]/I \to A^{1+}$ such that $\overline{\varphi} = \overline{\overline{\varphi}} \circ q_I$. Then $\mathrm{F}_{\mathsf{Set}}^{\mathsf{VLA}^1}[S]/I$ is a unital vector lattice algebra with positive identity element, and the pair $(q_I \circ j, \mathrm{F}_{\mathsf{Set}}^{\mathsf{VLA}^1}[S]/I)$ solves the problem of finding a free unital vector lattice algebra with positive identity element $\mathrm{F}_{\mathsf{Set}}^{\mathsf{VLA}^{1+}}[S]$ over the non-empty set S.

Analogously to this, one can, for a vector space V, obtain $\mathrm{F}_{\mathsf{VS}}^{\mathsf{VLA}^{1+}}[V]$ as a quotient of $\mathrm{F}_{\mathsf{VS}}^{\mathsf{VLA}^1}[V]$ that we had already obtained. For a vector lattice E, $\mathrm{F}_{\mathsf{VL}}^{\mathsf{VLA}^{1+}}[E]$ is a quotient of $\mathrm{F}_{\mathsf{VS}}^{\mathsf{VLA}^1}[E]$ and, for a vector lattice algebra A, $\mathrm{F}_{\mathsf{VLA}}^{\mathsf{VLA}^{1+}}[A]$ is a quotient of $\mathrm{F}_{\mathsf{VLA}}^{\mathsf{VLA}^1}[A]$.

Alternatively, one can proceed as earlier, but now with the free unital vector lattice algebra with a positive identity element over a set as a starting point, instead of a free unital vector lattice algebra. For a vector space V, $\mathrm{F}_{\mathsf{VS}}^{\mathsf{VLA}^{1+}}[V]$ is then obtained as a quotient of $\mathrm{F}_{\mathsf{Set}}^{\mathsf{VLA}^{1+}}[\mathrm{Set}\, V]$; for a vector lattice E, $\mathrm{F}_{\mathsf{VL}}^{\mathsf{VLA}^{1+}}[E]$ is then obtained as a quotient of $\mathrm{F}_{\mathsf{Set}}^{\mathsf{VLA}^{1+}}[\mathrm{Set}\ E]$; and, for a vector lattice algebra A, $\mathrm{F}_{\mathsf{VLA}}^{\mathsf{VLA}^{1+}}[A]$ is then obtained as a quotient of $\mathrm{F}_{\mathsf{Set}}^{\mathsf{VLA}^{1+}}[\mathrm{Set}\ A]$.

We have now completed our first task of obtaining all 8 universal objects in the above list. What about the remaining 7? As it turns out, it is easy to locate these. For this, we use Lemma 2.4, which is concerned with the 'reversal of directions' in our chain of categories in Eq. (6.1). It enables us to locate the remaining 7 free objects.

Let S be a non-empty set. Take a pair $(j, \mathrm{F}_{\mathsf{Set}}^{\mathsf{VLA}^{1}}[S])$. Suppose that A is a vector lattice algebra, and that $\varphi : S \to A$ is a map. Lemma 2.4 shows that we may view A as a vector lattice subalgebra of a unital vector lattice algebra A^1. After doing this, there exists a unique unital vector lattice algebra homomorphism $\overline{\varphi} : \mathrm{F}_{\mathsf{Set}}^{\mathsf{VLA}^{1}}[S] \to A^1$ such that $\varphi = \overline{\varphi} \circ j$. Since $\overline{\varphi}$ maps $j(S)$ into the vector lattice subalgebra A of A^1, this is also the case for the vector lattice subalgebra of $\mathrm{F}_{\mathsf{Set}}^{\mathsf{VLA}^{1}}[S]$ that is generated by $j(S)$. Hence this vector lattice subalgebra solves the problem of finding a free vector lattice algebra $\mathrm{F}_{\mathsf{Set}}^{\mathsf{VLA}}[S]$ over the non-empty set S.

Let S be a non-empty set. Since Lemma 2.4 implies that every vector lattice can be viewed as a vector sublattice of a unital vector lattice algebra, a similar argument shows that $\mathrm{F}_{\mathsf{Set}}^{\mathsf{VL}}[S]$ is the vector sublattice of $\mathrm{F}_{\mathsf{Set}}^{\mathsf{VLA}^{1}}[S]$ that is generated by $j(S)$.

Let S be a non-empty set. Since Lemma 2.4 implies that every vector space can be viewed as a vector subspace of a unital vector lattice algebra, it is now clear that $\mathrm{F}_{\mathsf{Set}}^{\mathsf{VS}}[S]$ is the vector subspace of $\mathrm{F}_{\mathsf{Set}}^{\mathsf{VLA}^{1}}[S]$ that is generated by $j(S)$.

We have thus taken care of another 3 free objects. Before proceeding, we note that Lemma 2.4 shows that non-empty sets, vector spaces, vector lattices, and unital vector lattice algebras can all be found inside some unital vector lattice algebra with a positive identity element. A similar argument, therefore, shows that, for a non-empty set S, each of $\mathrm{F}_{\mathsf{Set}}^{\mathsf{VLA}}[S]$, $\mathrm{F}_{\mathsf{Set}}^{\mathsf{VL}}[S]$, and $\mathrm{F}_{\mathsf{Set}}^{\mathsf{VS}}[S]$ can also be found inside $\mathrm{F}_{\mathsf{Set}}^{\mathsf{VLA}^{1+}}[S]$.

Let V be a vector space. We have already shown that $\mathrm{F}_{\mathsf{VS}}^{\mathsf{VLA}^{1}}[V]$ exists. Using Lemma 2.4 again, it is then easily seen that $\mathrm{F}_{\mathsf{VS}}^{\mathsf{VLA}}[V]$ exists, and that it is the vector lattice subalgebra of $\mathrm{F}_{\mathsf{VS}}^{\mathsf{VLA}^{1}}[V]$ that is generated by $j(V)$. Likewise, $\mathrm{F}_{\mathsf{VS}}^{\mathsf{VL}}[V]$ is the vector sublattice of $\mathrm{F}_{\mathsf{VS}}^{\mathsf{VLA}^{1}}[V]$ that is generated by $j(V)$. We have thus found another 2 free objects. Before proceeding, we note that, for a vector space V, $\mathrm{F}_{\mathsf{VS}}^{\mathsf{VLA}}[V]$ and $\mathrm{F}_{\mathsf{VS}}^{\mathsf{VLA}^{1}}[V]$ can both also be found inside $\mathrm{F}_{\mathsf{VS}}^{\mathsf{VLA}^{1+}}[V]$ again.

Let E be a vector lattice. Using Lemma 2.4 again, we see that $\mathrm{F}_{\mathsf{VL}}^{\mathsf{VLA}}[E]$ exists and is equal to the vector lattice subalgebra of $\mathrm{F}_{\mathsf{VL}}^{\mathsf{VLA}^{1}}[E]$ that is generated by $j(E)$. We have now covered 14 free objects. Before proceeding, we note that $\mathrm{F}_{\mathsf{VL}}^{\mathsf{VLA}}[E]$ is also equal to the vector lattice subalgebra of $\mathrm{F}_{\mathsf{VL}}^{\mathsf{VLA}^{1+}}[E]$ that is generated by $j(E)$.

Finally, let A^1 be a unital vector lattice algebra. Let I be the bi-ideal in A^1 that is generated by $(|1| - 1)$. It is then obvious that $\mathrm{F}_{\mathsf{VLA}^1}^{\mathsf{VLA}^{1+}}[A^1]$ exists and is simply A^1/I. We can also write this as $\mathrm{F}_{\mathsf{VLA}^1}^{\mathsf{VLA}^{1}}[A^1]/I$, thus making clear that this is completely analogous to the way how, for example, $\mathrm{F}_{\mathsf{VL}}^{\mathsf{VLA}^{1+}}[E]$ can be obtained as a quotient of $\mathrm{F}_{\mathsf{VL}}^{\mathsf{VLA}^{1}}[E]$ for a vector lattice E.

In the above, we have not mentioned the accompanying maps j being injective or not. It is easy to see, using Remark 2.3 and Lemma 2.4, that these 15 maps are all injective.

We collect what we have found in the following theorem.

Theorem 6.2 *For a non-empty set S, a vector space V, a vector lattice E, a vector lattice algebra A, and a unital vector lattice algebra A^1, the 15 free objects below all exist. There are inclusions as indicated. The surjective unital vector lattice algebra homomorphisms as indicated are the quotient maps corresponding to dividing out the bi-ideal that is generated by* $(|1| - 1)$.

$$S \subset F^{\mathsf{VS}}_{\mathsf{Set}}[S] \subset F^{\mathsf{VL}}_{\mathsf{Set}}[S] \subset F^{\mathsf{VLA}}_{\mathsf{Set}}[S] \begin{array}{l} \subset F^{\mathsf{VLA}^1}_{\mathsf{Set}}[S] \\ \qquad \twoheaddownarrow \\ \subset F^{\mathsf{VLA}^{1+}}_{\mathsf{Set}}[S] \end{array}$$

$$V \subset F^{\mathsf{VL}}_{\mathsf{VS}}[V] \subset F^{\mathsf{VLA}}_{\mathsf{VS}}[V] \begin{array}{l} \subset F^{\mathsf{VLA}^1}_{\mathsf{VS}}[V] \\ \qquad \twoheaddownarrow \\ \subset F^{\mathsf{VLA}^{1+}}_{\mathsf{VS}}[V] \end{array}$$

$$E \subset F^{\mathsf{VLA}}_{\mathsf{VL}}[E] \begin{array}{l} \subset F^{\mathsf{VLA}^1}_{\mathsf{VL}}[E] \\ \qquad \twoheaddownarrow \\ \subset F^{\mathsf{VLA}^{1+}}_{\mathsf{VL}}[E] \end{array}$$

$$A \begin{array}{l} \subset F^{\mathsf{VLA}^1}_{\mathsf{VLA}}[A] \\ \qquad \twoheaddownarrow \\ \subset F^{\mathsf{VLA}^{1+}}_{\mathsf{VLA}}[A] \end{array}$$

$$\begin{array}{c} A^1 = F^{\mathsf{VLA}^1}_{\mathsf{VLA}^1}[A^1] \\ \twoheaddownarrow \\ F^{\mathsf{VLA}^{1+}}_{\mathsf{VLA}^1}[A^1] \end{array}$$

Remark 6.3 For a vector lattice algebra A, $\mathrm{F}^{\mathsf{VLA}^1}_{\mathsf{VLA}}[A]$ is what deserves to be called the unitisation of A.

Remark 6.4 One can also ask for free *commutative* vector lattice algebras, for free *commutative* unital vector lattice algebras, and for free *commutative* unital vector lattice algebras with a positive identity element. These can be obtained by taking the general free object and dividing out the bi-ideal that is generated by the elements $(j(x)j(y) - j(y)j(x))$ for all x, y in the starting object. Using Lemma 2.4, it is immediate that sets, vector spaces, and vector lattices still embed into these new free objects under the new map j, which is the composition of the quotient map and the original map j. Vector lattice algebras, however, embed precisely when they are commutative.

Remark 6.5 Remark 2.6 shows how compositions behave. For example, a free vector lattice algebra over a free vector lattice over a non-empty set S is a free vector lattice algebra over S.

Remark 6.6 There are 14 free objects in Theorem 6.2 that are vector lattices or vector lattice algebras. These correspond to the 14 occurrences of a category and a subcategory in the chain in Eq. (6.1) where the subcategory consists of lattices. For each of these 14 occurrences, one can define a new subcategory by considering only Archimedean objects of the original subcategory. One can then ask for a free object in that context, i.e., ask for an Archimedean object of the original subcategory that has the universal property for all morphisms from the initial object of the category into Archimedean objects of the original subcategory. Using [18, Theorem 60.2], it is easy to see that the Archimedean free object is obtained from the general one in Theorem 6.2 by taking its quotient with respect to the uniform closure of $\{0\}$.

The general free (lattice) object in Theorem 6.2 is sometimes already Archimedean because it can be realised as a lattice of real-valued functions. This is the case for the free vector lattice $\mathrm{F}^{\mathsf{VL}}_{\mathsf{Set}}[S]$ over a non-empty set S (see [10]) and for the free vector lattice $\mathrm{F}^{\mathsf{VL}}_{\mathsf{VS}}[V]$ over a vector space V (this can be inferred from [22, Theorem 3.1]). We do not know whether any other of the remaining 12 is already Archimedean or not.

7 Free Objects Over Lattices

The free objects obtained above, and the two methods of obtaining new ones that were used above (passing to quotients and sub-objects), can also be used in other context. As an example, consider the chain of categories

$$\mathsf{Set} \supset \mathsf{Lat} \supset \mathsf{VL} \supset \mathsf{VLA} \supset \mathsf{VLA}^1 \supset \mathsf{VLA}^{1+}.$$

We have already observed that the free lattice over a non-empty subset exists, as a consequence of Theorem 5.4. How about the other four existence problems that are not related to the original chain in Eq. (2.1)? We shall now show that the corresponding free objects all exist, as an easy consequence of our previous work. Suppose that L is a (not necessarily distributive) algebraic lattice. Take $\mathrm{F}^{\mathsf{VLA}^1}_{\mathsf{Set}}[\mathrm{Set}\, L]$, and let I be the bi-ideal in $\mathrm{F}^{\mathsf{VLA}^1}_{\mathsf{Set}}[\mathrm{Set}\, L]$ that is generated by the elements $(j(x \wedge y) - j(x) \wedge j(y))$ and $(j(x \vee y) - j(x) \vee j(y))$ for all $x, y \in L$. Then $\mathrm{F}^{\mathsf{VLA}^1}_{\mathsf{Set}}[\mathrm{Set}\, L]/I$ is the free unital vector lattice algebra $\mathrm{F}^{\mathsf{VLA}^1}_{\mathsf{Lat}}[L]$ over L. The quotient of $\mathrm{F}^{\mathsf{VLA}^1}_{\mathsf{Lat}}[L]$ modulo the bi-ideal that is generated by $(|1| - 1)$ is the free unital vector lattice algebra with a positive identity element $\mathrm{F}^{\mathsf{VLA}^{1+}}_{\mathsf{Lat}}[L]$ over L. Using Lemma 2.4, one sees that $\mathrm{F}^{\mathsf{VL}}_{\mathsf{Lat}}[L]$ and $\mathrm{F}^{\mathsf{VLA}}_{\mathsf{Lat}}[L]$ are the vector sublattice, resp. the vector lattice subalgebra, of $\mathrm{F}^{\mathsf{VLA}^1}_{\mathsf{Lat}}[S]$ (and also of $\mathrm{F}^{\mathsf{VLA}^{1+}}_{\mathsf{Lat}}[S]$) that is generated by $j(L)$.

It is not always the case that the maps $j : L \to \mathrm{F}^{\mathsf{VL}}_{\mathsf{Lat}}[L]$, $j : L \to \mathrm{F}^{\mathsf{VLA}}_{\mathsf{Lat}}[L]$, $j : L \to \mathrm{F}^{\mathsf{VLA}^1}_{\mathsf{Lat}}[L]$, or $j : L \to \mathrm{F}^{\mathsf{VLA}^{1+}}_{\mathsf{Lat}}[L]$ are injective. If one of these is injective, then L must be distributive. Conversely, suppose that L is distributive. Then L is isomorphic to a sublattice of the power set of some set X; see [7, Corollary 2.42] or [14, Theorem 119], for example. Passing to characteristic functions, we see that a distributive lattice L is isomorphic to a sublattice of the lattice of real-valued bounded functions on X. As earlier, since the real-valued bounded functions on X form a unital vector lattice algebra with a positive identity element, the existence of this single (non-trivial) embedding of a distributive lattice L already implies that all four maps j are injective. Thus we have shown the following.

Theorem 7.1 *Let L be a (partially ordered or algebraic) lattice. Then the 4 free objects in*

$$L \xrightarrow{j} F^{\mathsf{VL}}_{\mathsf{Lat}}[L] \subset F^{\mathsf{VLA}}_{\mathsf{Lat}}[V] \begin{array}{c} \subset \\ \subset \end{array} \begin{array}{c} F^{\mathsf{VLA}^1}_{\mathsf{Lat}}[L] \\ \not\Downarrow \\ F^{\mathsf{VLA}^{1+}}_{\mathsf{Lat}}[L] \end{array}$$

all exist. The map j is injective if and only if L is distributive.

Remark 7.2 The above argument linking the injectivity of j to the distributivity of L is taken from [3, proof of Proposition 3.1], where it was used in the context of free Banach lattices over a lattice. The bounded real-valued functions on X, when supplied with the supremum norm, also form a unital Banach lattice algebra with a positive identity element. The fact that the distributive lattice L embeds into its unit ball will have consequences for the injectivity of the maps j when considering free Banach lattice algebras over distributive lattices.

Remark 7.3 As in Remark 6.6, a free Archimedean object can be obtained by taking the quotient of a general free object in Theorem 7.1 with respect to the uniform closure of $\{0\}$. It can be inferred from [3, Theorem 2.1] that the free vector lattice $\mathrm{F}^{\mathsf{VL}}_{\mathsf{Lat}}[L]$ over a (not necessarily distributive) lattice L can be realised as a vector lattice of real-valued functions. Hence it is Archimedean. We do not know whether any of the other three free objects in Theorem 7.1 is already Archimedean or not.

8 Free f-algebras

We conclude with a discussion of free f-algebras over non-empty sets. We can neither prove nor disprove that they exist, but there are still a number of observations to be made.

We recall that a vector lattice algebra A is a member of a family of abstract algebras, each of which is supplied with a constant 0, a binary map $\oplus$, a unary map $\ominus$, a unary map m_λ for every $\lambda \in \mathbb{R}$, a binary map $\odot$, and binary maps $\owedge$ and $\ovee$.

We let $(\mathcal{T}, \rho)$ denote the obvious underlying type of such abstract algebras. Among all abstract algebra of this type $(\mathcal{T}, \rho)$, we can single out the vector lattice algebras as those in which a number of identities are satisfied: they form an equational class. We have seen that this implies that free vector lattice algebras over a non-empty set exist.

How is this with f-algebras? We recall that a vector lattice algebra is called an f-algebra if, for all $x, y \in A$ and $z \in A^+$, the fact that $x \wedge y = 0$ implies that $(xz) \wedge y = (zx) \wedge y = 0$. We can, therefore, also single out the f-algebras among all abstract algebras of type $(\mathcal{T}, \rho)$ by requiring that all the identities for vector lattice algebras be again satisfied, and requiring that this extra f-algebra implication be valid. Is this perhaps also an equational class? This would imply the existence of free f-algebras over non-empty sets. Likewise, if one can prove that the Archimedean f-algebras form an equational class, then this would establish the existence of free Archimedean f-algebras.

For the unital case, a similar setup can be given. One starts with abstract algebras, each of which is supplied with constants 0 and now also 1, a binary map $\oplus$, a unary map $\ominus$, a unary map m_λ for every $\lambda \in \mathbb{R}$, a binary map $\odot$, and binary maps $\bigcirc\!\!\!\!\wedge$ and $\bigcirc\!\!\!\!\vee$. There is an obvious underlying type again (slightly different from the previous one), and the unital vector lattice algebras are the abstract algebras of this type in which a (slightly different) number of identities are satisfied. They form an equational class and, therefore, free unital vector lattice algebras over non-empty sets exist.

How is this with unital f-algebras, which can be singled out as those unital vector lattice algebras where the f-algebra implication holds? Do they form an equational class? If so, then free unital f-algebras over non-empty sets exist. How about Archimedean unital f-algebras, unital f-algebras with a positive identity element, and Archimedean unital f-algebras with a positive element?

All in all, we have six classes of f-algebras that may or may not be equational classes. For three of them, we can show that they are not. The reason is that equational classes are closed under the taking of abstract homomorphic images, i.e., under the taking of vector lattice algebra homomorphic images. For the Archimedean f-algebras, the Archimedean unital f-algebras, and the Archimedean unital f-algebras with a positive identity, this is not the case as is demonstrated by the following example. It is based on [18, second part of Example 60.1], where it is used to show that a quotient of an Archimedean vector lattice need not be Archimedean. The particular context shows much more, however.

Example 8.1 Under pointwise algebra operations and ordering, the sequence space ℓ^∞ is a unital f-algebra with a positive identity element. Consider the order ideal I_u of ℓ^∞ that is generated by the element $u = (u_1, u_2, u_3 \ldots) := (1/1^2, 1/2^2, 1/3^2, \ldots)$. Since the elements of ℓ^∞ are bounded, I_u is also an algebra ideal. Hence it is a bi-ideal, so that the quotient ℓ^∞ / I_u is a vector lattice algebra again. This quotient is not Archimedean, however. To see this, we include the short argument from [18, Example 60.1]. Let $q : \ell^\infty \to \ell^\infty / I_u$ denote the quotient map. Set $e := (1, 1, 1, \ldots)$ and $v = (v_1, v_2, v_3, \ldots) := (1/1, 1/2, 1/3, \ldots)$. Then $q(v) \neq 0$. Let k be a positive integer. Then $v_n \leq e_n / k$ for all $n \geq k$. Hence there exists an element $x = (x_1, x_2, \dots) \in \ell^\infty$ such that $x_n = 0$ for all $n \geq k$ and $v \leq x + e/k$. Since $x \in I_u$, this implies that $q(v) \leq q(e)/k$. Since $q(v) > 0$, this shows that ℓ^∞/I_u is not Archimedean.

We therefore have a vector lattice algebra homomorphism (which is automatically unital by its surjectivity) $q : \ell^\infty \to \ell^\infty/I_u$, but the codomain is not even an Archimedean vector lattice, let alone an Archimedean f-algebra with or without additional properties.

Aside, although it is not relevant to our main issues, let us nevertheless note that the vector lattice algebra ℓ^∞/I_u is not just a vector lattice algebra, but even an f-algebra. To see this, suppose that $x, y \in \ell^\infty$ are such that $q(x) \wedge q(y) = 0$. We may suppose that $x, y \geq 0$. Take a positive element $q(z)$ of ℓ^∞/I_u, where $z = (z_1, z_2, \dots) \in \ell^\infty$. We may suppose that $z \geq 0$. Since $q(x \wedge y) = q(x) \wedge q(y) = 0$, we have $x \wedge y \in I_u$. Then the estimate

$$0 \leq ((xz) \wedge y)_n = (x_n z_n) \wedge y_n \leq (\|z\|_\infty + 1)(x_n \wedge y_n)$$

for all n shows that $(xz) \wedge y \in I_u$. Hence $(q(x)q(z)) \wedge q(y) = q((xz) \wedge y) = 0$, as required.

It seems worthwhile to note explicitly that losing the Archimedean property when passing to a homomorphic image is not only possible in the category of vector lattices, but also in the much smaller subcategory of unital f-algebras with a positive identity element.

Proposition 8.2 *There exist an Archimedean unital f-algebra A^{1+} with a positive identity element, a non-Archimedean f-algebra B (automatically unital with a positive identity element) and a surjective vector lattice algebra homomorphism $\varphi : A \to B$.*

Returning to the main line, let us remark that, for the remaining three classes of f-algebras we do not know whether they are equational classes or not. We do not have an example showing that they are not, but our attempts to 'equationalise' the validity of the f-algebra implication have failed. The latter does, of course, not show that it is impossible to do so.

A more indirect way to prove that they *are* equational classes would be to try to use Birkhoff's theorem (see [7, Theorem 4.41]), which shows that the equational classes are precisely the varieties of abstract algebras. We recall that a variety of abstract algebras is a class of abstract algebras, all of the same type, that is closed under taking abstract subalgebras, abstract product algebras, and abstract homomorphic images. It is clear that each of the classes of f-algebras, of unital f-algebras, and of unital f-algebras with a positive identity element are closed under taking (unital) vector lattice subalgebras and taking vector lattice algebraic products. It seems to be open, however, whether any of these three classes is closed under the taking of abstract algebra homomorphic images, i.e., under (unital) vector lattice algebra homomorphic images. Hence we have the following questions.

Questions 8.3 Let A be an f-algebra, let B be a vector lattice algebra, and let $\varphi : A \to B$ be a surjective vector lattice algebra homomorphism.

(1) Is B always an f-algebra? If so, then the f-algebras form an equational class and free f-algebras over non-empty sets exist.
(2) Is B always an f-algebra (automatically unital) when A is unital? If so, then the unital f-algebras form an equational class and free unital f-algebras over non-empty sets exist.
(3) Is B always an f-algebra (automatically unital with a positive identity element) when A is unital with a positive identity element? If so, then the unital f-algebras with a positive identity element form an equational class and free unital f-algebras with a positive identity element over non-empty sets exist.

We have no answers. Related to all three questions, we can only mention that it is known that the answers are affirmative, *provided* that we also know that both A and B are Archimedean; this follows from [11, Proposition 3.2]. Related to the second and third question, we can only mention that it is known that an Archimedean vector lattice algebra with an identity element is an f-algebra precisely when all squares are positive; see [21, Corollary 1]. This shows once more that the answers to the second and third questions are affirmative, *provided* that we also know that both A and B are Archimedean. Unfortunately, this is not what we need, and to the best of our knowledge these three issues are all open at the time of writing.

Even though the Archimedean f-algebras, the Archimedean unital f-algebras, and the Archimedean unital f-algebras with a positive identity are not an equational class, and the f-algebras, the unital f-algebras, or the unital f-algebras with a positive identity element might also not be, this does not preclude the possibility that one or more of these six classes still has free objects over non-empty sets. After all, the Archimedean vector lattices do not form an equational class because there exist quotients of such lattices that are no longer Archimedean, but free Archimedean vector lattices over non-empty sets still exist. The reason is simply that the general free vector lattice over a non-empty set, which has a model as a lattice of real-valued functions, just happens to be Archimedean. This fact does not appear to be accessible with methods from universal algebra or category theory alone; one really has to look at the internal structure of the free object once one knows that it exists as is done in [10], for example. It is conceivable that something similar may be the case for f-algebras, where free vector lattice algebras of various kinds over non-empty sets—the existence of which is guaranteed by the general result for equational classes—may happen to be objects of a much smaller subcategory of f-algebras, where they are then evidently also free objects over non-empty sets. The following questions are, therefore, natural to ask:

Questions 8.4 Let S be a non-empty set.

(1) Is $\mathrm{F}_{\mathsf{Set}}^{\mathsf{VLA}}[S]$ an f-algebra? Is it Archimedean?
(2) Is $\mathrm{F}_{\mathsf{Set}}^{\mathsf{VLA}^1}[S]$ an f-algebra? Is it Archimedean?
(3) Is $\mathrm{F}_{\mathsf{Set}}^{\mathsf{VLA}^{1+}}[S]$ an f-algebra? Is it Archimedean?

Of course, if $\mathrm{F}_{\mathsf{Set}}^{\mathsf{VLA}^1}[S]$ or $\mathrm{F}_{\mathsf{Set}}^{\mathsf{VLA}^{1+}}[S]$ is Archimedean, or an f-algebra, then the same is true for its vector lattice subalgebra $\mathrm{F}_{\mathsf{Set}}^{\mathsf{VLA}}[S]$.

Acknowledgments It is a pleasure to thank Karim Boulabiar, Ben de Pagter, Mitchell Taylor, and Vladimir Troitsky for helpful discussions.

References

1. J. Adámek, H. Herrlich, G.E. Strecker, Abstract and concrete categories: the joy of cats. Rep. Theory Appl. Categ. **17**, 1–507 (2006). Reprint of the 1990 original [Wiley, New York]
2. C.D. Aliprantis, E. Langford, Order completions of Archimedean Riesz spaces and l-groups. Algebra Universalis **19**(2), 151–159 (1984)
3. A. Avilés, J.D. Rodríguez Abellán, The free Banach lattice generated by a lattice. Positivity **23**(3), 581–597 (2019)
4. A. Avilés, J. Rodríguez, P. Tradacete, The free Banach lattice generated by a Banach space. J. Funct. Anal. **274**(10), 2955–2977 (2018)
5. A. Avilés, P. Tradacete, I. Villanueva, The free Banach lattices generated by ℓ_p and c_0. Rev. Mat. Complut. **32**(2), 353–364 (2019)
6. K.A. Baker, Free vector lattices. Canadian J. Math. **20**, 58–66 (1968)
7. C. Bergman, Universal algebra. Fundamentals and selected topics, in *Pure and Applied Mathematics (Boca Raton)*, vol. 301 (CRC Press, Boca Raton, 2012)
8. G.M. Bergman, *An Invitation to General Algebra and Universal Constructions*, 2nd edn. (Universitext, Springer, Cham, 2015)
9. G. Birkhoff, *Lattice Theory*, 3rd edn., in *American Mathematical Society Colloquium Publications*, vol. XXV (American Mathematical Society, Providence, 1967)
10. R.D. Bleier, Free vector lattices. Trans. Am. Math. Soc. **176**, 73–87 (1973)
11. K. Boulabiar, The range of lattice homomorphisms on f-algebras, in *Ordered Algebraic Structures*. Developments in Mathematics, vol. 7 (Kluwer Academic Publication, Dordrecht, 2002), pp. 179–188
12. B. de Pagter, A.W. Wickstead, Free and projective Banach lattices. Proc. R. Soc. Edinburgh Sect. A **145**(1), 105–143 (2015)
13. G. Grätzer, *Universal Algebra*, 2nd edn. (Springer, Heidelberg, 1979)
14. G. Grätzer, *Lattice Theory: Foundation* (Birkhäuser/Springer Basel, 2011)
15. H. Herrlich, G.E. Strecker, Category theory. An introduction, in *Sigma Series in Pure Mathematics*, 3rd edn., vol. 1 (Heldermann, Lemgo, 2007)
16. N. Jacobson, *Basic Algebra. II*, 2nd edn. (W.H. Freeman and Company, New York, 1989)
17. A.Y. Khelemskiĭ, Metric freeness and projectivity for classical and quantum normed modules. Mat. Sb. **204**(7), 127–158 (2013)
18. W.A.J. Luxemburg, A.C. Zaanen, *Riesz spaces*, vol. I (North-Holland Publishing Company/American Elsevier Publishing Company, Amsterdam/New York, 1971)
19. S. Mac Lane, *Categories for the Working Mathematician*, 2nd edn. Graduate Texts in Mathematics, vol. 5 (Springer, New York, 1998)
20. V.G. Pestov, Universal arrows to forgetful functors from categories of topological algebra. Bull. Aust. Math. Soc. **48**(2), 209–249 (1993)
21. S.A. Steinberg, On lattice-ordered rings in which the square of every element is positive. J. Aust. Math. Soc. Ser. A **22**(3), 362–370 (1976)
22. V.G. Troitsky, Simple constructions of FBL(A) and FBL[E]. Positivity **23**(5), 1173–1178 (2019)
23. A.W. Wickstead, Ordered Banach algebras and multi-norms: some open problems. Positivity **21**, 817–823 (2017)

On Disjointness, Bands and Projections in Partially Ordered Vector Spaces

Jochen Glück

Abstract Disjointness, bands, and band projections are a classical and essential part of the structure theory of vector lattices. If X is such a lattice, those notions seem—at first glance—intimately related to the lattice operations on X. The last 15 years, though, have seen an extension of all those concepts to a much larger class of ordered vector spaces.

In fact if X is an Archimedean ordered vector space with generating cone, or a member of the slightly larger class of pre-Riesz spaces, then the notions of disjointness, bands and band projections can be given proper meaning and give rise to a non-trivial structure theory.

The purpose of this note is twofold: (i) We show that, on any pre-Riesz space, the structure of the space of all band projections is remarkably close to what we have in the case of vector lattices. In particular, this space is a Boolean algebra. (ii) We give several criteria for a pre-Riesz space to already be a vector lattice. These criteria are coined in terms of disjointness and closely related concepts, and they mark how lattice-like the order structure of pre-Riesz spaces can get before the theory collapses to the vector lattice case.

Keywords Ordered vector space · Pre-Riesz space · Band · Projection · Disjointness

Mathematics Subject Classification (2010) 06F20, 46A40, 46B40

J. Glück (✉)
Fakultät für Informatik und Mathematik, Universität Passau, Passau, Germany
e-mail: jochen.glueck@uni-passau.de

E. Kikianty et al. (eds.), *Positivity and its Applications*, Trends in Mathematics,
https://doi.org/10.1007/978-3-030-70974-7_7

1 Introduction

1.1 Disjointness

Two elements x and y of a vector lattice X are called *disjoint* if $|x| \wedge |y| = 0$—a notion that is well-motivated by the case where X is one of the classical function spaces such as L^p. A straightforward generalisation to ordered vector spaces that are not lattices seems to be difficult at first glance, as there is no obvious replacement of the modulus of x and y. Van Gaans and Kalauch, though, observed more than a decade ago [30] that one can circumvent this obstacle by noting that any two elements x and y of a vector lattice X are disjoint if and only if $|x + y| = |x - y|$, and that this is in turn true if and only if the sets $\{x + y, -x - y\}$ and $\{x - y, y - x\}$ have the same set of upper bounds. The latter property clearly allows a generalisation to other ordered vector spaces, which gives rise to the following definition.

Let (X, X_+) be an *ordered vector space*, by which we mean that X is a real vector space and $X_+ \subseteq X$ is a non-empty subset of X which satisfies $X_+ \cap (-X_+) = \{0\}$ and $\alpha X_+ + \beta X_+ \subseteq X_+$ for all scalars $\alpha, \beta \in [0, \infty)$ (we call X_+ the *positive cone* in X_+). Two elements $x, y \in X$ are called *disjoint* if both sets $\{x + y, -x - y\}$ and $\{x - y, y - x\}$ have the same set of upper bounds in X. We use the notation $x \perp y$ to denote that x and y are disjoint. Note that $x \perp y$ if and only if $y \perp x$, and that $x \perp x$ if and only if $x = 0$.

If $x, y \in X$ are both *positive*—i.e., $x, y \in X_+$—then one can prove that $x \perp y$ if and only if the infimum of x and y in X exists and is equal to 0; see [7, Proposition 2.1].

1.2 Disjoint Complements and pre-Riesz Spaces

Let (X, X_+) be an ordered vector space and let $S \subseteq X$. The set

$$S^\perp := \{x \in X : x \perp s \text{ for all } s \in S\}$$

is called the *disjoint complement* of S. We note that $S_1^\perp \supseteq S_2^\perp$ whenever S_1 and S_2 are two subsets of X such that $S_1 \subseteq S_2$.

From the theory of vector lattices we would expect $S^\perp$ to always be a vector subspace of X—but it turns out that one can construct examples of ordered spaces where this is not true (see for instance [30, Example 4.3]). On the other hand though, such counterexamples are somewhat pathological: in fact, one can show that $S^\perp$ is always a vector subspace of X if the cone X_+ is *generating* in X (i.e., $X = X_+ - X_+$) and X is *Archimedean* (i.e., $nx \leq y$ for all $n \in \mathbb{N} := \{1, 2, \dots\}$ implies $x \leq 0$ whenever x, y are two vectors in X).

There is also the slightly more general class of *pre-Riesz spaces* that is relevant in this context: an ordered vector space (X, X_+) is called a pre-Riesz space if for every non-empty finite set $A \subseteq X$ and every $x \in X$ the following implication is true: if the set of upper bounds of $x + A$ is contained in the set of upper bounds of A, then $x \in X_+$. This concept was introduced by van Haandel in [33, Definition 1.1(viii)].

If (X, X_+) is a pre-Riesz space and $S \subseteq X$, then the disjoint complement $S^\perp$ is always a vector subspace of X; see [30, Corollary 2.2 and Section 3]. Moreover, we note that every pre-Riesz space has generating cone and that, conversely, every ordered vector space which has generating cone and is, in addition, Archimedean, is a pre-Riesz space [30, Theorem 3.3].

The theory of pre-Riesz spaces has undergone a considerable development over the last 15 years. References to papers that deal with bands and projection bands on pre-Riesz spaces are given at the beginning of Sects. 2 and 3. Further contributions to the theory of pre-Riesz spaces can be found in [10–12, 32] and, with a focus on operator theory, in [14, 15, 19, 20]. The present state of the art in the theory of pre-Riesz spaces is presented in the recent monograph [18].

1.3 Organisation of the Paper

In the rest of the introduction we recall a bit more terminology and a simple result about disjointness. In Sect. 2 we recall how a band is defined in a pre-Riesz space, and we show a few elementary results about the structure of the set of all bands. In Sect. 3 we discuss projection bands and band projections. We show, among other things, that the band projections on a pre-Riesz space constitute a Boolean algebra and that, under appropriate assumptions on the space, the intersection of arbitrarily many projection bands is again a projection band. In the final Sect. 4 we give various sufficient conditions for a pre-Riesz space to be a vector lattice; these conditions are related to several variations of the notion *disjointness*.

1.4 Setting the Stage

Throughout the rest of the paper, let (X, X_+) be a pre-Riesz space.

By an *operator* on X we always mean a linear map $X \to X$, and by a *projection* on X we always mean a linear projection $X \to X$.

We use standard terminology and notation from the theory of ordered vector spaces (which has, to some extent, already been employed above). In particular, we write $x \leq y$ (or $y \geq x$) for $x, y \in X$ if $y - x \in X_+$ and we note that the relation $\leq$ is a partial order on X which is compatible with the vector space structure. Elements of X_+ will be called *positive*. For $x, z \in X$ we denote the *order interval* between x and z by

$$[x, z] := \{y \in X : x \le y \le z\}.$$

A linear map $T : X \to X$ is said to be *positive*, which we denote by $T \ge 0$, if $TX_+ \subseteq X_+$, and for two linear maps $S, T : X \to X$ we write $S \le T$ if $T - S \ge 0$.

For each vector subspace $V \subseteq X$ we set $V_+ := V \cap X_+$, and we say that V has *generating cone* or that V is *directed* if $V = V_+ - V_+$. The following simple proposition is quite useful.

Proposition 1.1 *Let $V, W \subseteq X$ be vector subspaces of X with generating cone. If $V_+ \perp W_+$, then $V \perp W$.*

Here we use the notation $A \perp B$ for two subsets $A, B \subseteq X$ if $a \perp b$ for each $a \in A$ and each $b \in B$.

Proof of Proposition 1.1 We use that orthogonal complements in pre-Riesz spaces are always vector subspaces. Since $V_+ \subseteq (W_+)^\perp$ we conclude $V \subseteq (W_+)^\perp$. The latter inclusion is equivalent to $W_+ \subseteq V^\perp$, which in turn implies $W \subseteq V^\perp$. □

2 Bands

This section is in a sense prologue to our main results in Sects. 3 and 4. We briefly recall some basics about bands (Sect. 2.1), we show that the collection of all bands in X is a complete lattice with respect to set inclusion (Sect. 2.2) and we briefly discuss how the sum of two bands can be computed under certain assumptions (Sect. 2.3).

Bands in pre-Riesz spaces were first defined in [30, Section 5] and were further studied in [13, 16, 29, 31].

2.1 Basics

For $S \subseteq X$ we use the notation $S^{\perp\perp} := (S^\perp)^\perp$. Of course, we always have $S \subseteq S^{\perp\perp}$.

A subset $B \subseteq X$ is called a *band* if $B = B^{\perp\perp}$. For every set $S \subseteq X$ the disjoint complement $S^\perp$ is a band [30, Proposition 5.5(ii)]. As a consequence, for each $S \subseteq X$ the set $S^{\perp\perp}$ is the smallest band in X that contains S. Since (X, X_+) is a pre-Riesz space, every band in X is a vector subspace of X. Note that if B is a band in X and $0 \le x \le b$ for $x \in X$ and $b \in B$, then we also have $x \in B$.

In the classical theory of vector lattices, the concept of bands is of outstanding importance. For the convenience of the reader we recall a few examples of bands in vector lattices.

Examples 2.1

(a) Let (Ω, μ) be a σ-finite measure space, let $p \in [1, \infty]$ and let $X = L^p(\Omega, \mu)$ with the standard cone. If $A \subseteq \Omega$ is a measurable set, then

$$\begin{aligned} B &= \{f \in X : \text{ there is a representative of } f \text{ that vanishes a. e. on } A\} \\ &= \{f \in X : \text{ every representative of } f \text{ vanishes a. e. on } A\} \end{aligned}$$

is a band in X, and in fact all bands in X are of this form.

(b) Let $X = C([0, 1])$ be the space of continuous real-valued functions on $[0, 1]$ and let $0 \le a \le 1$. Then

$$B_a = \{f \in X : f \text{ vanishes on } [a, 1]\}$$

is a band in X (this example is further discussed in [27, Example 5 on p. 63]). A general description of bands in spaces of continuous functions over compact sets can, for instance, be found in [18, Proposition 1.3.13].

Interesting examples for bands in a non-lattice ordered pre-Riesz space can for instance be found in the space $\mathbb{R}^3$ ordered by the so-called *four ray cone*:

Examples 2.2 Let $X = \mathbb{R}^3$ and $X_+ := \{\sum_{k=1}^4 \alpha_k v_k : \alpha_1, \ldots, \alpha_4 \in [0, \infty)\}$, where

$$v_1 = \begin{pmatrix} 1 \\ 0 \\ 1 \end{pmatrix}, \quad v_2 = \begin{pmatrix} 0 \\ 1 \\ 1 \end{pmatrix}, \quad v_3 = \begin{pmatrix} -1 \\ 0 \\ 1 \end{pmatrix}, \quad v_4 = \begin{pmatrix} 0 \\ -1 \\ 1 \end{pmatrix}.$$

The cone X_+ is called the *four ray cone* in $\mathbb{R}^3$. All bands in X are computed in [18, Example 4.4.18]. Besides the two trivial bands $\{0\}$ and X there are four directed bands—namely the lines spanned by $v_1, \ldots, v_4$, respectively. Moreover, there exist two non-directed bands—namely the lines spanned by

$$\begin{pmatrix} 1 \\ 1 \\ 0 \end{pmatrix} \quad \text{and} \quad \begin{pmatrix} 1 \\ -1 \\ 0 \end{pmatrix},$$

respectively.

2.2 *The Lattice of All Bands*

The following proposition shows that the intersection of any collection of bands in X is again a band.

Proposition 2.3 *The intersection of arbitrarily many bands B_i in X (where the indices i are taken from a—possibly empty—index set I) is again a band, and it is given by*

$$\bigcap_{i \in I} B_i = \Big(\bigcup_{i \in I} B_i^{\perp}\Big)^{\perp}.$$

Proof It suffices to prove the formula. If $x \in \bigcap_{i \in I} B_i$, then x is disjoint to each set $B_i^{\perp}$, so x is also disjoint to the union $\bigcup_{i \in I} B_i^{\perp}$. Conversely, fix $i_0 \in I$. Then $B_{i_0}^{\perp} \subseteq \bigcup_{i \in I} B_i^{\perp}$ and hence, $B_{i_0} = B_{i_0}^{\perp\perp} \supseteq \big(\bigcup_{i \in I} B_i^{\perp}\big)^{\perp}$. □

It is an immediate consequence of this proposition that the set of all bands in X is a complete lattice with respect to set inclusion; let us state this explicitly in the following corollary.

Corollary 2.4 *Let* $\mathrm{Bands}(X)$ *denote the set of all bands in X, ordered by set inclusion. Then every subset $\mathcal{B}$ of* $\mathrm{Bands}(X)$ *has a supremum and an infimum in* $\mathrm{Bands}(X)$*, given by*

$$\inf \mathcal{B} = \cap \mathcal{B}$$

and

$$\sup \mathcal{B} = \cap \big\{ C \in \mathrm{Bands}(X) : C \supseteq B \text{ for all } B \in \mathcal{B} \big\};$$

in other words, $\mathrm{Bands}(X)$ *is a complete lattice.*

2.3 *The Sum of Two Bands*

Even in the case of vector lattices, the sum of two bands need not be a band, in general. Let us illustrate this by means of the following simple example.

Example 2.5 Let $X = C([-1, 1])$ denote the space of all continuous real-valued functions on the interval $[-1, 1]$ and endow this space with the standard cone. Then the sets

$$B = \{f \in X : f|_{[-1,0]} = 0\} \quad \text{and} \quad C = \{f \in X : f|_{[0,1]} = 0\}$$

are bands in X, but their sum $B + C = \{f \in X : f(0) = 0\}$ is not a band in X.

Another counterexample—in a non-lattice ordered pre-Riesz space—can be found in $\mathbb{R}^3$ endowed with the four ray cone from Example 2.2. In this space, all non-trivial bands are one-dimensional; hence, the sum of two distinct non-trivial bands cannot be a band in this space.

If, however, the sum of two bands B and C is a band, then we can compute it be means of the formula $B+C = (B^{\perp}\cap C^{\perp})^{\perp}$; this is part of the following proposition.

Proposition 2.6 *Let $B, C \subseteq X$ be bands.*

(a) *We have $B \cap C = (B^{\perp} + C^{\perp})^{\perp}$.*
(b) *We have $B + C = (B^{\perp} \cap C^{\perp})^{\perp}$ if (and only if) $B + C$ is a band.*
(c) *More generally than* (b), we always have

$$B + C \subseteq (B + C)^{\perp\perp} = (B^{\perp} \cap C^{\perp})^{\perp}.$$

Proof

(a) According to Proposition 2.3 we have $B \cap C = (B^{\perp} \cup C^{\perp})^{\perp}$, and the latter set clearly contains $(B^{\perp}+C^{\perp})^{\perp}$. On the other hand, if $x \in X$ is disjoint to $B^{\perp}\cup C^{\perp}$, then it is also disjoint to $B^{\perp} + C^{\perp}$ since the disjoint complement of $\{x\}$ is a vector subspace of X; this shows that we also have $(B^{\perp}\cup C^{\perp})^{\perp} \subseteq (B^{\perp}+C^{\perp})^{\perp}$.
(c) It follows from (a) that

$$B^{\perp} \cap C^{\perp} = (B^{\perp\perp} + C^{\perp\perp})^{\perp} = (B + C)^{\perp},$$

so $(B^{\perp} \cap C^{\perp})^{\perp} = (B + C)^{\perp\perp}$.
(b) This is an immediate consequence of (c).

□

The main point of the above proposition—and the reason for the title of this subsection—is assertion (b). Anyway, we chose to include assertion (a) in the same proposition in order to have an immediate comparison between (a) and (b).

We point out that the assumption of (b) that $B + C$ be a band is automatically satisfied of both B and C are projection bands; see Proposition 3.7 below. On the other hand, Example 2.5 shows that there are situations in which $B + C$ is not a band—and in this case the formula from Proposition 2.6 necessarily fails.

3 Band Projections

Band projections (and, accordingly, projection bands) in pre-Riesz spaces are a main subject of study in [7, 17]. In this section we further develop their theory.

3.1 Basics

If B is a band in X, then it intersects its orthogonal band $B^{\perp}$ only in 0. However, the sum of B and $B^{\perp}$ can be smaller than the entire space X, in general; this happens, for instance, in Example 2.5, where $C = B^{\perp}$.

We call a subset $B \subseteq X$ a *projection band* if B is a band and if, in addition, $X = B \oplus B^{\perp}$. It is not difficult to see that a band B is a projection band if and only if $B^{\perp}$ is a projection band. Every projection band B has generating cone according to [7, Proposition 2.5].

The notion of a projection band also gives rise to the following definition: a linear projection $P : X \to X$ is called a *band projection* if there exists a projection band $B \subseteq X$ such that P is the projection onto B along $B^{\perp}$. In other words, P is a band projection if and only if PX is a projection band and $\ker P$ equals the disjoint complement of PX.

The following proposition contains various characterisations of band projections.

Proposition 3.1 *For every linear projection $P : X \to X$ the following assertions are equivalent:*

(i) *P is a band projection.*
(ii) $\ker P = (PX)^{\perp}$.
(iii) $PX = (\ker P)^{\perp}$.
(iv) $PX \perp \ker P$.
(v) *Both projections P and $I - P$ are positive.*
(vi) *$I - P$ is a band projection.*

Proof "(i) $\Leftrightarrow$ (v)" This equivalence was proved in [7, Theorem 3.2].

"(i) $\Leftrightarrow$ (vi)" This equivalence follows from the fact that a band B is a projection band if and only if $B^{\perp}$ is a projection band (alternatively, it follows immediately from the equivalence of (i) and (v)).

"(i) $\Rightarrow$ (ii)" and "(i) $\Rightarrow$ (iii)" These implications follow immediately from the definition of a band projection.

"(ii) $\Rightarrow$ (iv)" and "(iii) $\Rightarrow$ (iv)" These implications are obvious.

"(iv) $\Rightarrow$ (v)" Let $x \in X_+$. Then Px and $(I - P)x$ are disjoint and sum up to x, so it follows from [7, Proposition 2.4(a)] that Px and $(I - P)x$ are positive, too. This shows (v). □

If P is a band projection in X, then both the range and the kernel of P are projection bands. In Corollary 3.17 below we will see that the converse implication is also true, which yields another characterisation of band projections.

We conclude this subsection with a few examples.

Examples 3.2

(a) If X is a Dedekind complete vector lattice, then every band in E is a projection band [27, Theorem II.2.10].
(b) Let (Ω, μ) be a σ-finite measure space and let $p \in [1, \infty]$. The bands in $L^p(\Omega, \mu)$ are described in Example 2.1(a). Since $L^p(\Omega, \mu)$ is Dedekind complete, it follows from (a) that each of these bands is actually a projection band.
(c) The bands B_a in $C([0, 1])$ from Example 2.1(b) are not projection bands unless $a = 0$ (see [27, Example 5 on p. 63]). More generally, it is not difficult to see

that, for a compact Hausdorff space K, there are no non-trivial projection bands in $C(K)$ if K is connected.

(d) Let X and Y be two pre-Riesz spaces and endow the product space $Z := X \times Y$ with the product order (i.e. $Z_+ = X_+ \times Y_+$). Then Z is a pre-Riesz space, too, and X and Y—which we identify with the subspaces $X \times \{0\}$ and $\{0\} \times Y$ of Z, respectively—are projection bands in Z. Indeed, we have $X^\perp = Y$, and vice versa.

On a related note, we will see in Theorem 4.7 below that every finite-dimensional pre-Riesz space can be written as the product of finitely many minimal projection bands.

(e) If X is a Banach lattice and we identify X with a subspace of its bi-dual space X'' by means of evaluation, then X is a band in X'' if and only if X is a projection band in X'' if and only if X is a so-called *KB-space*. This class of space includes all reflexive Banach lattices and all L^1-spaces (over arbitrary measure spaces). For details we refer for instance to [23, Section 2.4].

Example 3.2(e) can be extended to also include spaces that are not lattice-ordered. An ordered vector space (Y, Y_+) is called an *ordered Banach space* if Y carries a complete norm and Y_+ is closed. Note that the order in an ordered Banach space is always Archimedean. Hence, if Y_+ is, in addition, generating, then the ordered Banach space (Y, Y_+) is a pre-Riesz space.

Throughout the rest of the paper, we will tactily use some important concepts from the theory of ordered Banach spaces—such as *normality* of cones and the fact that the dual of an ordered Banach space with generating cone is again an ordered Banach space. For details about the theory of ordered Banach spaces we refer the reader for instance to the monograph [1], in particular to Section 2.5 there.

Example 3.3 Assume that the pre-Riesz space X is an ordered Banach space with normal cone. Then we can consider X as a subspace of the bi-dual space X'' by means of evaluation.

There are interesting examples where X is not a Banach lattice and not reflexive, but yet a projection band in X''. This is, for instance, the case if X is the pre-dual of a von Neumann algebra; see [25, Proposition 1.17.7] or [28, pp. 126–127].

Ordered Banach space that are projection bands in their bi-dual were employed in [8] to study the long-term behaviour of positive operator semigroups.

One can easily find examples where a pre-Riesz space X does not contain any projection bands except for $\{0\}$ and X itself. One situation of this type has already been discussed in Example 3.2(c) above. Here are two more examples.

Examples 3.4

(a) Let us endow $X = \mathbb{R}^3$ with the four ray cone X_+ from Example 2.2. Then every non-trivial band B in X_+ is one-dimensional, so there is no non-trivial projection band in X.

(b) Assume that X is a so-called *anti-lattice*, which means that any two vectors x, y in X have a supremum if and only if $x \geq y$ or $x \leq y$. Then there are, according to [18, Theorem 4.1.10(ii)], no non-trivial disjoint elements in X_+. Hence, there are no non-trivial projections bands in X.

We note that a classical example of an anti-lattice is the space of all self-adjoint bounded linear operators on a Hilbert space; this result goes back to Kadison [9, Theorem 6].

3.2 The Boolean Algebra of Band Projections

In this section we study the structure of the collection of all band projections on X. As in the vector lattice case, this collection turns out to be a Boolean algebra (Theorem 3.8).

We begin with the following proposition which shows that a band projection Q dominates a band projection P (in the sense of operators on the ordered vector space X) if and only if the range of Q contains the range of P:

Proposition 3.5 *For two band projections P and Q on X the following assertions are equivalent.*

(i) $PX \subseteq QX$.
(ii) $QP = P$.
(iii) $P \leq Q$.

Proof "(i) ⇔ (ii)" This can immediately be checked to be true for all projections on arbitrary vector spaces.

"(ii) ⇒ (iii)" We have $P = QP \leq Q \cdot I = Q$.

"(iii) ⇒ (ii)" We have $QP \leq I \cdot P = P = P^2 \leq QP$, so $QP = P$. □

Next we describe the interaction of two arbitrary band projections on X in a bit more detail; in particular, we prove that any two band projections commute.

Proposition 3.6 *For two band projections P and Q on X the following assertions hold:*

(a) *P leaves the range of Q invariant, and vice versa.*
(b) *P and Q commute.*
(c) *The mapping $PQ = QP$ is a band projection, too.*
(d) *We have $PQX = PX \cap QX$.*

Proof

(a) Let $0 \leq x \in QX$. For each $0 \leq z \in (QX)^\perp$ it follows from $0 \leq Px \leq x$ that $Px \perp z$ (see [7, Proposition 2.2]); since the positive cone in the projection band $(QX)^\perp$ is generating in $(QX)^\perp$, we conclude that $Px \perp (QX)^\perp$.

Now we also use that the positive cone in the projection band QX is generating in QX, which implies that $Px \perp (Qx)^{\perp}$ for each $x \in QX$. Hence, $Px \in (Qx)^{\perp\perp} = QX$ for each $x \in QX$, which shows that P leaves QX invariant. By interchanging the roles of P and Q we also obtain that Q leaves PX invariant.

(b) It follows from (a) that Q leaves both PX and $(I - P)X$ invariant. Thus,

$$PQP = QP \qquad \text{and} \qquad PQ(I - P) = 0.$$

The second equality is equivalent to $PQP = PQ$ which yields, in conjunction with the first equality, $QP = PQ$.

(c) Clearly, $0 \leq PQ \leq I \cdot I = I$, so it remains to show that PQ is a projection. Since P leaves QX invariant, we know that $QPQ = PQ$, so $(PQ)^2 = P(QPQ) = P(PQ) = PQ$.

(d) "$\supseteq$" For $x \in PX \cap QX$ we have $x = Px = PQx \in PQX$.
"$\subseteq$" If $x \in PQX$, then $x = PQx \in PX$ and $x = QPx \in QX$.

□

We point out that assertion (d) in the above proposition is in fact true for all commuting projection P and Q on an arbitrary vector space.

As a consequence of the fact that any two band projections commute we obtain the following proposition which shows, in particular, that the sum of two projection bands is a projection band (and a formula for such a sum can thus be found in Proposition 2.6(b) above).

Proposition 3.7 *Let P and Q be band projections on X. Then $P + Q - PQ$ is a band projection, too, and its range coincides with the set $PX + QX$. In particular, the sum of two projection bands is a projection band.*

Proof Since P and Q commute, a direct computation shows that $P + Q - PQ$ is a projection. Moreover,

$$P + Q - PQ = P + Q(I - P),$$

and the latter mapping is clearly positive and dominated by $P + I(I - P) = I$. Thus, $P + Q - PQ$ is a band projection.

Obviously, the range of $P + Q - PQ$ is contained in the vector space sum $PX + QX$. The converse inclusion follows from the formula

$$Px + Qy = (P + Q - PQ)(Px + Qy)$$

which holds for all $x, y \in X$. □

Now we can prove that the set of all band projections on X is a Boolean algebra. Recall (for instance from [27, Definition II.1.1]) that a *Boolean algebra* is a non-empty partially ordered set A with the following properties:

(a) For all $x, y \in A$ the infimum $x \wedge y$ and the supremum $x \vee y$ exist (i.e., A is a *lattice*).
(b) The lattice operations $\wedge$ and $\vee$ are *distributive*, i.e., we have

$$(x \vee y) \wedge z = (x \wedge z) \vee (y \wedge z)$$

for all $x, y, z \in A$ (this is equivalent to assuming that $(x \wedge y) \vee z = (x \vee z) \wedge (y \vee z)$ for all $x, y, z \in A$, see [3, Theorem 9 on p. 11]).
(c) There exists a smallest element 0 and a largest element 1 in A.
(d) A is *complemented*, i.e., for each $x \in A$ there exists a so-called *complement* $x^c \in A$ such that

$$x \wedge x^c = 0 \quad \text{and} \quad x \vee x^c = 1.$$

We note that, in a Boolean algebra A, the complement of each element is uniquely determined; this follows from [3, Theorem 10 on p. 12].

Theorem 3.8 *Let* $\mathrm{BandPr}(X)$ *denote the set of all band projections on X, ordered by the usual order of positive operators on X. Then* $\mathrm{BandPr}(X)$ *is a Boolean algebra with smallest element* 0 *and largest element I. The lattice operations* $\wedge$ *and* $\vee$ *on this Boolean algebra are given by*

$$P \wedge Q = PQ \quad \textit{and} \quad P \vee Q = P + Q - PQ$$

for all band projections P and Q, and the complement is given by

$$P^c = I - P$$

for each band projection P.

In seperable Hilbert spaces ordered by self-dual cones, a related result for the set of all self-adjoint band projections was shown in [24, Theorem II.1] (although the notion *band projection* was not used explicitly there).

In the proof we make use of the facts established in the propositions above; in particular we will frequently—and often tacitly—use that $P \leq Q$ for two band projections P and Q if and only if $PX \subseteq QX$.

Proof of Theorem 3.8 We first show that $\mathrm{BandPr}(X)$ is a lattice with respect to its given order, and that the lattice operators are given by the formulae in the theorem. Let $P, Q \in \mathrm{BandPr}(X)$.

It follows from Proposition 3.6(d) that PQ is a lower bound of P and Q. If $R \in \mathrm{BandPr}(X)$ is another lower bound of P and Q, then $RX \subseteq PX$ and $RX \subseteq QX$, so $RX \subseteq PQX$, again by Proposition 3.6(d); hence, $R \le PQ$. This proves that P and Q have infimum PQ in $\mathrm{BandPr}(X)$.

On the other hand, $P + Q - PQ$ is an upper bound of P and Q according to Proposition 3.7. If $R \in \mathrm{BandPr}(X)$ is another upper bound of PX and QX, then $RX \supseteq PX \cup QX$, hence $RX \supseteq PX + QX$ and thus, it follows again from Proposition 3.7 that $R \ge P + Q - PQ$. This proves that P and Q have supremum $P + Q - PQ$ in $\mathrm{BandPr}(X)$.

In particular, $\mathrm{BandPr}(X)$ is a lattice. The fact that it is even a distributive lattice, i.e., that the distributive law

$$(P \vee Q) \wedge R = (P \wedge R) \vee (Q \wedge R)$$

is satisfied for all band projections P, Q, R, can now be checked by a straightforward computation that uses the formulae for $\wedge$ and $\vee$ established above.

Clearly, $\mathrm{BandPr}(X)$ has the smallest element 0 and the largest element I, and for every band projection P, the projection $Q := I - P$ satisfies $P \wedge Q = PQ = 0$ and $P \vee Q = P + Q - PQ = P + Q = I$; hence, $I - P$ is the complement of any $P \in \mathrm{BandPr}(X)$ and $\mathrm{BandPr}(X)$ is indeed a Boolean algebra. □

Corollary 3.9 *Let* $\mathrm{PrBands}(X)$ *denote the set of all projection bands in X, ordered via set inclusion. The mapping*

$$\varphi : \mathrm{BandPr}(X) \to \mathrm{PrBands}(X),$$
$$P \mapsto PX$$

is an order isomorphism between the partially ordered sets $\mathrm{BandPr}(X)$ *and* $\mathrm{PrBands}(X)$. *In particular,* $\mathrm{PrBands}(X)$ *is a Boolean algebra with infimum and supremum given by*

$$B \wedge C = B \cap C \quad \text{and} \quad B \vee C = B + C$$

for all projections bands B, C in X, and with the complement operation given by

$$B^c = B^\perp$$

for each projection band B in X.

Proof The mapping φ is surjective by definition of the notions "projection band" and "band projection", and it is injective since every band projection P is uniquely determined by its range PX. It follows from Proposition 3.5 that φ and its inverse map φ^{-1} are monotone. Thus, $\mathrm{PrBands}(X)$ is indeed a Boolean algebra and φ is an isomorphism between the boolean algebras $\mathrm{BandPr}(X)$ and $\mathrm{PrBands}(X)$.

The formulae for the lattice operations on $\mathrm{PrBands}(X)$ now follow from Propositions 3.6 and 3.7, and the formula for the complement follows from the fact that $(I - P)X = \ker P = (PX)^\perp$ for each band projection P. □

3.3 The Intersection of Arbitrarily Many Projection Bands

According to Proposition 3.6, the intersection of finitely many projection bands is again a projection band. In general, this is no longer true for infinitely many projections bands (not even in the case of Banach lattices) as the following simple example shows:

Example 3.10 Consider the compact space $K = [-1, 0] \cup \{\frac{1}{n} : n \in \mathbb{N}\}$ and the Banach lattice $C(K)$ of continuous real-valued functions on K.

For each $n \in \mathbb{N}$ the set $B_n := \{f \in C(K) : f(x) = 0 \text{ for all } x \geq \frac{1}{n}\}$ is a projection band in $C(K)$. However, the intersection

$$\begin{aligned}\bigcap_{n\in\mathbb{N}} B_n &= \{f \in C(K) : f(x) = 0 \text{ for all } x > 0\} \\ &= \{f \in C(K) : f(x) = 0 \text{ for all } x \geq 0\}\end{aligned}$$

is not a projection band in $C(K)$.

However, in a Dedekind complete vector lattice every band is a projection band and hence, the intersection of arbitrarily many projection bands is still a projection band.

Motivated by this we show in this subsection that the intersection of arbitrarily many projection bands in a Dedekind complete pre-Riesz space is again a projection band. Here, we call the pre-Riesz space X *Dedekind complete* if the supremum $\sup A$ exists in X for every non-empty upwards directed set $A \subseteq X$ that is bounded above.

Assume for a moment that X is Dekekind complete. If (x_j) and (y_j) are decreasing nets in X (with the same index set) that are bounded below, then the net (x_j) has an infimum x (we write $x_j \downarrow x$ for this), the net (y_i) has an infimum y, and it is not difficult to show that the sum $(x_j + y_j)$ has infimum $x + y$; similarly, for $\lambda \in [0, \infty)$ the net (λx_j) has infimum λx.

Theorem 3.11 *Assume that X is Dedekind complete and let (P_j) be a net of band projections on X such that $P_j \leq P_i$ (equivalently: $P_j X \subseteq P_i X$) whenever $j \geq i$. Then there exists a band projection P_0 on X with the following two properties:*

(a) *We have $P_j x \downarrow P_0 x$ for each $x \in X_+$.*
(b) *$P_0 X = \bigcap_j P_j X$.*

Proof First we define a mapping $P_0 : X_+ \to X_+$ by means of $P_0 x = \inf_j P_j x$ for each $x \in X_+$. By the remarks we made before the theorem, P_0 is linear in the sense that $P_0(\alpha x + \beta y) = \alpha P_0 x + \beta P_0 y$ for all $x, y \in X_+$ and all $\alpha, \beta \in [0, \infty)$. As X_+ is generating in X, we can extend P_0 to a (uniquely determined) linear map—that we again denote by P_0—from X to X. For each $x \in X_+$ we have $0 \leq P_0 x \leq x$.

Let us show next that P_0 is a projection; to this end, it suffices to consider $x \in X_+$ and show that $P_0^2 x = P_0 x$. For each index j we have $0 \leq P_0 x \leq P_j x$, so $P_0 x \in$

P_jX and hence, $P_j(P_0x) = P_0x$. This shows that $P_0(P_0x) = P_0x$, so $P_0^2 = P_0$. Consequently, P_0 is a band projection that has property (a). Let us now show (b).

"$\subseteq$" Let $x \in P_0X$. Then we can write x as $x = y - z$, where y and z are positive vectors in P_0X. For each index j, we then have $0 \le y = P_0y \le P_jy$; hence, $y \in P_jX$, and likewise for z. Thus, $x = y - z \in P_jX$ for each j.

"$\supseteq$" Let $x \in \bigcap_j P_jX$. We decompose x as $x = y - z$ for $y, z \in X_+$. For each j we have $x = P_jx = P_jy - P_jz$, so

$$x \le P_jy \qquad \text{and} \qquad -x \le P_jz.$$

Consequently, $x \le P_0y$ and $-x \le P_0z$, so $x \in [-P_0z, P_0y]$, which proves that $x \in P_0X$. □

In order to derive from Theorem 3.11 that the intersection of an arbitrary—maybe non-directed—collection of projection bands is still a projection band, we need the following lemma (which is true on every pre-Riesz space, be it Dedekind complete or not).

Lemma 3.12 *Let $P_1, \dots, P_n$ be band projections on X and let $x \in X_+$.*

(a) *If $z \in X$ and $z \le P_1x, \dots, z \le P_nx$, then also $z \le P_1 \cdots P_nx$.*
(b) *We have $P_1 \cdots P_nx = \inf\{P_1x, \dots, P_nx\}$.*

Proof

(a) We first note that, if $z \le Px$ for a band projection P, then $(I - P)z \le 0$. Now we prove the assertion by induction over n. For $n = 1$ the assertion is obvious, so assume that it has already been proved for some $n \in \mathbb{N}$. If P_{n+1} is another band projection such that $z \le P_{n+1}x$, then

$$z = (I - P_{n+1})z + P_{n+1}z \le P_{n+1}z \le P_{n+1}P_1 \cdots P_nx = P_1 \cdots P_{n+1}x.$$

(b) Clearly, $P_1 \cdots P_nx$ is a lower bound of $\{P_1x, \dots, P_nx\}$, and according to (a) it is also the greatest lower bound of this set.

□

In the proof of the following corollary we only need assertion (a) of the lemma. We included assertion (b) in the lemma anyway since we think it is interesting in its own right.

Corollary 3.13 *Assume that X is Dedekind complete. Then the intersection of arbitrarily many projection bands in X is again a projection band. More precisely, if $\mathcal{P}$ is a set of band projections on X, then there exists a (unique) band projection P_0 on X with range $\bigcap_{P \in \mathcal{P}} PX$; if $\mathcal{P}$ is non-empty, then we have*

$$P_0x = \inf\{Px : P \in \mathcal{P}\}$$

for each $x \in X_+$

Proof We may assume that $\mathscr{P}$ is non-empty. Let $\hat{\mathscr{P}}$ denote the set of all finite products of elements from $\mathscr{P}$. Then it is easy to see that $\bigcap_{P \in \mathscr{P}} PX = \bigcap_{P \in \hat{\mathscr{P}}} PX$. Moreover, $\hat{\mathscr{P}}$ is directed by the converse of the usual order $\leq$ on linear operators (since $\hat{\mathscr{P}}$ is closed with respect to taking finite products). Thus, $(P)_{P \in \hat{\mathscr{P}}}$ is a decreasing net of band projections, so Theorem 3.11 shows the existence of a band projection P_0 on X such that $P_0 X = \bigcap_{P \in \hat{\mathscr{P}}} PX$.

It remains to prove the formula for $P_0 x$, so let $x \in X_+$. By Theorem 3.11(a) we have

$$P_0 x = \inf\{Px : P \in \hat{\mathscr{P}}\}.$$

Clearly, $P_0 x$ is a lower bound of $\{Px : P \in \mathscr{P}\}$, so let $z \in X$ be another lower bound of this set. Then z is, according to Lemma 3.12(a), also a lower bound of $\{Px : P \in \hat{\mathscr{P}}\}$, and hence $z \leq P_0 x$. □

If X is Dedekind complete, then it follows from Corollary 3.13 that, for every set $S \subseteq X$, there exists a smallest projection band that contains S. This projection band can, however, be much larger than the band generated by S, as the following example shows:

Example 3.14 Let $X = \mathbb{R}^3$, let X_+ be the four ray cone from Example 2.2 and let v_1 by the vector introduced in that example. The span of $\{v_1\}$ is a band, but according to Example 3.4(a) there are no non-trival projection bands in X.

Hence, the band generated by $\{v_1\}$ equals $\operatorname{span}\{v_1\}$, while the projection band generated by $\{v_1\}$ equals X.

3.4 *Another Characterisation of Band Projections*

The following propositions shows that if two projections bands B and C have trivial intersection, than we automatically have $B \perp C$.

Proposition 3.15 *For two band projections P and Q on X the following assertions are equivalent.*

(i) $PQ = 0$
(ii) $PX \cap QX = \{0\}$.
(iii) $PX \perp QX$.

Proof "(i) ⇔ (ii)" This equivalence follows from Proposition 3.6(d) (and is thus true for arbitrary commuting projections on every vector space).

"(i) ⇒ (iii)" According to Proposition 1.1 it suffices to show that $PX_+ \perp QX_+$, so let $x \in PX_+$ and $y \in QX_+$. In order to show that $x \perp y$ it is necessary and sufficient to prove that x and y have infimum 0 in X. Obviously, 0 is a lower bound of x and y, so let b be another lower bound of those vectors. We then have

$$Pb \leq Py = PQy = 0 \quad \text{and} \quad Qb \leq Qx = QPx = 0,$$

so $(P+Q)b \leq 0$. On the other hand, we know from Proposition 3.7 that $P+Q$ is a band projection (since $PQ = 0$), so $I-(P+Q)$ is positive. Hence,

$$(I-(P+Q))b \leq (1-(P+Q))x = x - Px - Qx = -Qx = 0.$$

Consequently, $b = (I-(P+Q))b + (P+Q)b \leq 0$. This proves that x and y indeed have infimum 0.

"(iii) $\Rightarrow$ (ii)" For each $x \in PX \cap QX$ we have $x \perp x$, so $x = 0$. □

We remark that the implication "(iii) $\Rightarrow$ (ii)" in Proposition 3.15 remains true if PX and QX are replaced with arbitrary bands (over even arbitrary vector subspaces) in X. However, the converse implication fails for general bands, even if they are assumed to be directed. We illustrated this, again, in the space $\mathbb{R}^3$ endowed with the four ray cone.

Example 3.16 Let $X = \mathbb{R}^3$, let X_+ denote the four ray cone from Example 2.2, and let $v_1, \ldots, v_4$ denote the four vectors defined in the same example.

Then $B_1 := \operatorname{span}\{v_1\}$ and $B_2 := \operatorname{span}\{v_2\}$ are bands in X that intersect only in 0. However, we do not have $B_1 \perp B_2$ since v_1 is not disjoint to v_2. To see this, consider the vector

$$w = \begin{pmatrix} 1 \\ 1 \\ 0 \end{pmatrix}.$$

Then $v_1 - w = v_4 \in X_+$ and $v_2 - w = v_3 \in X_+$. Hence, w is a lower bound of both v_1 and v_2. On the other hand, w is not an element of the negative cone $-X_+$. Thus, 0 is not the greatest lower bound of v_1 and v_2, so $v_1 \not\perp v_2$.

As a consequence of Proposition 3.15 we obtain another characterisation of band projections.

Corollary 3.17 *For every linear projection $P : X \to X$ the following assertions are equivalent:*

(i) *P is a band projection.*
(ii) *Both PX and* $\ker P$ *are projection bands.*

Proof "(i) $\Rightarrow$ (ii)" If P is a band projection, then PX is a projection band by definition, and hence $\ker P = (PX)^\perp$ is also a projection band.

"(ii) $\Rightarrow$ (i)" If PX and $\ker P$ are projection bands, than there exist band projections $Q_1, Q_2 : X \to X$ such that $Q_1X = PX$ and $Q_2X = \ker P$. Since $Q_1X \cap Q_2X = \{0\}$, it follows from Proposition 3.15 that $Q_1X \perp Q_2X$, i.e., $PX \perp \ker P$. According to Proposition 3.1 this implies that P is a band projection. □

We note that the implication "(ii) $\Rightarrow$ (i)" in Corollary 3.17 does not remain true, in general, if we replace "projection bands" in (ii) with "bands". More precisely, we have the following situation:

Remarks 3.18

(a) There exists a (*weakly pervasive*, see Definition 4.9) pre-Riesz space X and two bands B and C in X such that $X = B \oplus C$, but $C \neq B^{\perp}$. A concrete example of this situation can be found in [17, Example 19]; it is, however, important to observe that one of the bands is not directed in this example.
(b) If X is weakly pervasive and $X = B \oplus C$ for two directed bands—or, more generally, two directed ideals—B and C, then it is shown in [17, Theorem 18] that B and C are projections bands and $B = C^{\perp}$.
(c) If X is even *pervasive* (see Definition 4.9), then the implication mentioned in (b) remains true even if B and C are only ideals in X (which are not a priori assumed to be directed); this is shown in [17, Theorem 17].
(d) Now, let X be a general pre-Riesz space and let $X = B \oplus C$ for two directed bands—or, more generally, directed ideals—B and C. It seems to be open whether this implies $C = B^{\perp}$.

4 Characterisations of Vector Lattices

In this section we give various criteria for a pre-Riesz space to actually be a vector lattice. All these criteria are in some way related to disjointness. We note that, in the important special case where X is finite dimensional and Archimedean, several sufficient criteria for X to be a vector lattice are known. It suffices, for instance, if X has the Riesz decomposition property (see for instance [1, Corollary 2.48]) or if X is pervasive [17, Theorem 39]. In Corollary 4.14 below we give a simultaneous generalisation of those two results.

4.1 Criteria in Terms of One-Dimensional Projection Bands

In this subsection we prove that a finite-dimensional pre-Riesz space is automatically a vector lattice if there exist sufficiently many projection bands in it. We begin with the following proposition about linear independence.

Proposition 4.1 *Let $m \in \mathbb{N}$ and let $x_1, \dots, x_m \in X \setminus \{0\}$ be pairwise disjoint. Then the tuple $(x_1, \dots, x_m)$ is linearly independent.*

Proof For $m = 1$ the assertions is obvious, and we next show it for $m = 2$. So let $\alpha_1 x_1 + \alpha_2 x_2 = 0$ for real numbers α_1, α_2. Since the sum of the disjoint vectors $\alpha_1 x_1$ and $\alpha_2 x_2$ is both positive and negative, it follows from [7, Proposition 2.4(a)] that both vectors $\alpha_1 x_1$ and $\alpha_2 x_2$ are both positive and negative, and thus 0. This implies that $\alpha_1 = \alpha_2 = 0$ since $x_1, x_2 \neq 0$ by assumption.

Now assume that the assertion has been proved for a fixed integer $m \geq 2$ and let $\alpha_1, \dots, \alpha_{m+1} \in \mathbb{R}$ such that $\sum_{k=1}^{m+1} \alpha_k x_k = 0$. Since the vectors $\sum_{k=1}^{m} \alpha_k x_k$ and x_{m+1} are disjoint and linearly dependent, it follows from the case $m = 2$ considered above that $\sum_{k=1}^{m} \alpha_k x_k = 0$. Using that the assertion has already been proved for the number m, we conclude that $\alpha_1 = \dots = \alpha_m = 0$. Finally, we observe that $\alpha_{m+1} x_{m+1} = 0$, so $\alpha_{m+1} = 0$ since $x_{m+1} \neq 0$. □

Theorem 4.2 *Assume that $n = \dim X < \infty$. If there exist (at least) n distinct band projections of rank 1 on X, then X is an Archimedean vector lattice, i.e., X is linearly order isomorphic to $\mathbb{R}^n$ with the standard cone.*

Proof We may assume that $n \neq 0$. Let $P_1, \dots, P_n$ denote n distinct band projections of rank 1. Then we have $P_k P_j X = \{0\}$ for $j \neq k$. Indeed, if we assumed $\dim(P_k P_j X) = 1$, then $P_k P_j X = P_k X = P_j X$—which would imply $P_k = P_j$ since band projections are uniquely determined by their range.

Since each projection band is spanned by its positive elements, each space $P_k X$ is spanned by a vector $x_k > 0$. According to Proposition 3.15 the vectors $x_1, \dots, x_n$ are pairwise disjoint. Hence, they are linearly independent by Proposition 4.1. Since $\dim X = n$, this implies that the vectors $x_1, \dots, x_n$ span X.

It follows from Proposition 3.15 that $P_k P_j = 0$ whenever $j \neq k$, so

$$(P_1 + \dots + P_n) x_k = x_k \qquad \text{for each } k \in \{1, \dots, n\};$$

hence, $P_1 + \dots + P_n = I$, and we conclude that the linear mapping

$$J : Y := P_1 X \times \dots \times P_n X \ni (z_1, \dots, z_n) \to z_1 + \dots + z_n \in X$$

is a bijection. Each $P_k X$ is an ordered space with respect to the order inherited from X, and as such it is isomorphic to $\mathbb{R}$ with the cone $[0, \infty)$. If we endow Y with the product order, then Y is isomorphic to $\mathbb{R}^n$ with the standard cone, and the mapping J is an order isomorphism between Y and X, which proves the assertion. □

There is a certain conceptual similarity between the above proof and the approach taken in [17, Section 6] to prove [17, Theorem 39]: the authors of [17] prove that every finite dimensional Archimedean pervasive pre-Riesz space is actually a vector lattice by considering atoms in such a space and by showing that if atoms $a_1, \dots, a_m$ in a pervasive pre-Riesz space are pairwise linearly independent, then the entire system $(a_1, \dots, a_m)$ is linearly independent. Our usage of Proposition 4.1 and of rank-1 band projections in the above proof is somewhat reminiscent of this approach (as it is easy to see that the range of a rank-1 band projection is always spanned by an atom).

As a simple consequence of Theorem 4.2 we obtain the following numerical bound on the number of rank-1 band projections in X:

Corollary 4.3 *Assume that $n = \dim X < \infty$. Then there exists at most n distinct band projections of rank 1 on X.*

Proof If there were strictly more than n distinct band projections of rank 1, then Theorem 4.2 would imply that X is isomorphic to $\mathbb{R}^n$ with the standard cone—but on this space there exist precisely n distinct rank-1 band projections, so we arrive at a contradiction. □

4.2 Criteria in Terms of the Number of Projection Bands

If the space X is finite-dimensional and has closed cone, then it follows from [18, Theorem 4.4.26] that there exist only finitely many bands in X. In particular, the Boolean algebra $\mathrm{PrBands}(X)$ is finite, so we conclude that the number of projection bands in X is a power 2^m of 2. In the following we are going to prove a bit more: we will not assume X_+ to be closed a priori, we will show that we always have $m \le \dim X$, and that equality holds if and only if X is an Archimedean vector lattice.

Let us start with the following slightly more sophisticated version of Proposition 4.1.

Proposition 4.4 *Let $A_1, \dots, A_m \subseteq X$ be subsets of X such that $A_i \perp A_j$ whenever $i \neq j$. For each $j \in \{1, \dots, m\}$, let $(x_{j,1}, \dots, x_{j,n_j})$ be a linearly independent system of vectors in A_j. Then the entire system*

$$(x_{1,1}, \dots, x_{1,n_1}, \dots, x_{m,1}, \dots, x_{m,n_m})$$

is linearly independent.

Proof First we note that, for any two distinct indices $i, j \in \{1, \dots, m\}$, we have $\mathrm{span}(A_i) \perp A_j$ and hence $\mathrm{span}(A_i) \perp \mathrm{span}(A_j)$ since X is a pre-Riesz space. Now assume that

$$\sum_{j=1}^{m} \sum_{k=1}^{n_j} \lambda_{j,k} x_{j,k} = 0$$

for scalars $\lambda_{j,k} \in \mathbb{R}$. We define vectors $y_j = \sum_{k=1}^{n_j} \lambda_{j,k} x_{j,k} \in \mathrm{span}(A_j)$ for $j \in \{1, \dots, m\}$. Thus, the vectors $y_1, \dots, y_m$ are pairwise disjoint. Since

$$y_1 + \dots + y_m = 0,$$

it follows from Proposition 4.1 that one of these vectors is 0, and inductively we then derive that actually all vectors $y_1, \dots, y_m$ are 0.

Now, fix $j \in \{1, \dots, m\}$. Since $0 = y_j = \sum_{k=1}^{n_j} \lambda_{j,k} x_{j,k}$, we conclude from the linear independence of the system $(x_{j,1}, \dots, x_{j,n_j})$ that $\lambda_{j,1} = \dots = \lambda_{j,n_j} = 0$. This proves the assertion. □

A second ingredient that we need is the following simple observation about band projections.

Lemma 4.5 *Let $P_1, \dots, P_m : X \to X$ be band projections and assume that $P_i P_j = 0$ whenever $i \neq j$. Then*

$$P_1 + \cdots + P_m$$

is also a band projection.

Proof The assumptions clearly imply that $P_1 + \cdots + P_m$ is a positive projection. Next, we show by induction over m that

$$I - (P_1 + \cdots + P_m) = (I - P_1)(I - P_2) \cdots (I - P_m).$$

For $m = 1$ this is obvious, so assume that it has been proved for some fixed $m \in \mathbb{N}$ and consider now one more band projection P_{m+1} such that $P_j P_{m+1} = 0$ for all $j \in \{1, \dots, m\}$. Then

$$\begin{aligned}(I - P_1)(I - P_2) \cdots (I - P_m)(I - P_{m+1}) &= \big(I - (P_1 + \cdots + P_m)\big)(I - P_{m+1}) \\ &= I - (P_1 + \cdots + P_{m+1}),\end{aligned}$$

as claimed. We thus conclude that $I - (P_1 + \cdots + P_m)$ is positive, too. □

Now we can prove the first main result of this subsection.

Theorem 4.6 *If $n = \dim X < \infty$, the following assertions hold:*

(a) *The number of band projections on X is equal to 2^m for some $m \in \{0, \dots, n\}$.*
(b) *We have $m = n$ if and only if X is an Archimedean vector lattice.*

It is hardly surprising that the proof of Theorem 4.6 below is strongly related to the Boolean algebra structure of the set of all projection bands. However, we cannot rely on this Boolean structure alone since we want to relate the number m to the dimension of X—i.e., we need to take the linear structure of the underlying space into account.

Proof of Theorem 4.6 We may assume throughout the proof that $n \neq 0$.

(a) *Step 1* Within this proof, let us call a projection band B minimal if it is non-zero and if it does not contain any non-zero projection band except itself. Since X is finite dimensional, every non-zero projection band contains a minimal projection band. Let $\mathfrak{M}$ denote the set of all minimal projection bands in X. If $B, C \in \mathfrak{M}$ are two distinct projection bands, then $B \cap C = \{0\}$; indeed, $B \cap C$ is a projection band that is contained in both B and C. Hence, if it were non-zero, we would have $B \cap C = B$ and $B \cap C = C$, so $B = C$.

Consequently, $B \perp C$ for any two distinct $B, C \in \mathfrak{M}$ by Proposition 3.15. It thus follows from Proposition 4.1 that there exist at most n distinct minimal projection bands in X; we enumerate them as $B_1, \ldots, B_m$ (where $1 \le m \le n$), and we denote the corresponding band projections by $P_1, \ldots, P_m$.

Since $P_i P_j = 0$ whenever $i \neq j$, it follows from Lemma 4.5 that $P_1 + \cdots + P_m$ is a band projection. Actually, this band projection coincides with I, since otherwise the range of the complementary band projection $I - (P_1 + \cdots + P_m)$ would contain one of the minimal projection bands $B_1, \ldots, B_m$, which is a contradiction. Hence,

$$P_1 + \cdots + P_m = I.$$

Consequently, $B_1 + \cdots + B_m = X$.

Step 2 Next we note that, for each projection band C in X and each $k \in \{1, \ldots, m\}$ we have either $B_j \subseteq C$ or $B_j \cap C = \{0\}$; this is a consequence of the minimality of B_j. Hence, for every band projection P on X and every $j \in \{1, \ldots, m\}$ we have either $PP_j = P_j$ or $PP_j = 0$.

Thus, for every band projection P on X we have

$$P = \sum_{j \in I_P} P_j,$$

where $I_P := \{j \in \{1, \ldots, m\} : P_j P \neq 0\}$. Conversely, we note that the sum $P_I := \sum_{j \in I} P_j$ is, for any $I \subseteq \{1, \ldots, m\}$, a band projection (according to Lemma 4.5), and the set I is uniquely determined by this sum (since it is the set of all k such that $P_k P_I \neq 0$). This proves that there exist exactly 2^m band projections on X, and we have already observed above that $m \le n$. We have thus proved (a)

(b) Assume now that $m = n$.

For every $j \in \{1, \ldots, m\}$ we now choose a basis $(x_{j,1}, \ldots, x_{j,n_j})$ of the space B_j. It follows from Proposition 4.4 that the system

$$(x_{1,1}, \ldots, x_{1,n_1}, \ldots, x_{m,1}, \ldots, x_{m,n_m})$$

is linearly independent. Hence, $n_1 + \cdots + n_m \le n$. As $m = n$, it follows that none of the numbers n_j can be larger than 1, so each of the n projection bands $B_1, \ldots, B_m = B_n$ is one-dimensional. Theorem 4.2 thus shows that X is an Archimedean vector lattice.

Conversely, if X is an Archimedean vector lattice, then it is isomorphic to $\mathbb{R}^n$ with the standard cone, so there exist indeed 2^n band projections on X, so $m = n$.

□

Step 1 in the proof of Theorem 4.6(a) also provides us with another interesting insight into the structure of finite-dimensional pre-Riesz spaces. The facts that

$P_i P_j = 0$ for any two distinct $i, j \in \{1, \dots, m\}$ and that $P_1 + \cdots + P_m = I$ imply that the mapping

$$X \to B_1 \times \cdots \times B_m,$$
$$x \mapsto (P_1 x, \dots, P_m x)$$

is an isomorphism of ordered vector spaces, where $B_1 \times \cdots \times B_m$ is endowed with the product order. Moreover, it is not difficult to see that every projection band in a pre-Riesz space is itself a pre-Riesz space; hence, each B_j is a pre-Riesz space.

We also observe that none of the pre-Riesz spaces B_j contains a non-trivial projection band. Indeed, if $Q : B_j \to B_j$ is a band projection, then $QP_j : X \to X$ is a band projection with the same range as Q; by the minimality if B_j this implies that this range is either $\{0\}$ or B_j. We thus have the following structure result, which is the second main result of this subsection.

Theorem 4.7 *Assume that* $1 \leq \dim X < \infty$. *Then there exists a number* $m \in \{1, \dots, \dim X\}$ *such that* X *is isomorpic (as an ordered vector space) to the product of* m *non-zero pre-Riesz spaces none of which contains a non-trivial projection band.*

For Hilbert spaces ordered by self-dual cones a related structure result (even in infinite dimensions) can be found in [24, Corollary II.12]. On finite-dimensional ordered vector spaces a related result, formulated as a decomposition theorem for cones, is given in [5, Theorem 9.3].

4.3 Criteria in the Class of Weakly Pervasive Spaces

Assume for a moment that X is finite dimensional with closed cone. In [17, Theorem 39] it was shown that if X is pervasive (see Definition 4.9 below), then X is in fact a vector lattice. The same is true if X is assumed to have the Riesz decomposition property instead if being pervasive (see for instance [1, Corollary 2.49]). These observations suggest to study the following two questions:

(a) Since the Riesz decomposition property and the property of beging pervasive are logically independent for general pre-Riesz spaces (see [10, Example 13] and [22, Example 23]), it is natural to seek for a simultaneous generalisations of the two above mentioned results.

(b) The fact that the Riesz decomposition property implies that X is a vector lattice is actually not only true in finite-dimensional spaces (with closed cone), but for instance also for the more general case of reflexive ordered Banach spaces with generating cone [1, Corollary 2.48]. This suggests that searching for sufficient criteria for infinite dimensional spaces to be a vector lattice is a worthwhile endeavour.

In this subsection we pursue both goals outlined above. As before, we assume that X is a general pre-Riesz space.

Two vectors $x, y \in X_+$ are called *D-disjoint* if $[0, x] \cap [0, y] = \{0\}$. For a more detailed discussion of this notion and of its origin, we refer to [18, Section 4.1.3] and [21, Defintion 8 and Proposition 9].

Every two disjoint elements in X are clearly D-disjoint, but the converse implication is not true, in general; this can, for instance, again be seen by considering the four ray cone in $\mathbb{R}^3$:

Example 4.8 Let $X = \mathbb{R}^3$ and let X_+ denote the four ray cone from Example 2.2; let v_1 and v_2 denote the vectors given in the same example. According to Example 3.16 the vectors v_1 and v_2 are not disjoint. However, both elements v_1 and v_2 are so-called *atoms* in X (see [17, Definition 27 and Proposition 28] or Sect. 4.4 below), so it follows that v_1 and v_2 are D-disjoint.

Hence, disjointness of two vectors $x, y \in X_+$ is, in general, a much stronger property than D-disjointness. There are, however, spaces in which both notions coincide; this gives rise to part (a) of the following definition.

Definition 4.9

(a) The pre-Riesz space X is called *weakly pervasive* if any two D-disjoint vectors in X_+ are disjoint.
(b) The pre-Riesz space X is called *pervasive* if for every $b \in X$ such that $b \not\leq 0$ there exists $x \in X_+ \setminus \{0\}$ such that every positive upper bound of b is also an upper bound of x.

The concept of a weakly pervasive pre-Riesz space was coined in [10, Definition 8 and Lemma 9]. The usual definition of a pervasive pre-Riesz space in the literature is somewhat different and employs the Riesz completion of X (see [18, Definition 2.8.1]). However, this definition is equivalent to the one given above according to [10, Theorem 7].

If one uses that two vectors $x, y \in X_+$ are disjoint if and only if they have infimum 0, it is easy to show that every pervasive pre-Riesz space is also weakly pervasive. Moreover, every vector lattice is pervasive and hence weakly pervasive.

We also note that every pre-Riesz space with the Riesz decomposition property is weakly pervasive [10, Proposition 11]; hence, weakly pervasive spaces are a simultaneous generalisation of pervasive pre-Riesz spaces and pre-Riesz spaces with the Riesz decomposition property.

Let us give a simple criterion in order to check that several function spaces are pervasive.

Proposition 4.10 *Let $\Omega \subseteq \mathbb{R}^d$ be open, let $\Omega \subseteq L \subseteq \overline{\Omega}$ and let X be a directed vector subspace of $C(L)$ (where $C(L)$ denotes the space of all real-valued continuous functions on L). Then X is a pre-Riesz space; if, in addition, X contains all test functions on Ω, then X is pervasive.*

Proof As $C(L)$ is Archimedean, so is X, and since X_+ is generating in X by assumption, it follows that X is a pre-Riesz space.

Now assume that X contains all test functions on Ω. Let $b \in X$ and $b \not\leq 0$. Since $C(L)$ is a vector lattice, we can take the positive part b^+ in $C(L)$. This is a non-zero positive continuous function on L, so there exists a positive non-zero test function x on Ω such that $x \leq b^+$. We note that $x \in X$ by assumption. Now, if $u \in X_+$ is an upper bound of b in X, then it is also an upper bound of b^+ in $C(L)$. Hence, $u \geq x$. □

As a consequence of the above proposition we obtain, for instance, the following examples of pervasive spaces.

Examples 4.11

(a) Let $\emptyset \neq \Omega \subseteq \mathbb{R}^d$ be open and bounded and let $k \in \mathbb{N}_0$. Then the space $C_b^k(\overline{\Omega})$ of functions that are k-times continuously differentiable on Ω and whose partial derivatives up to order k all have a continuous extension to $\overline{\Omega}$ is pervasive.
(b) Let $\emptyset \neq \Omega \subseteq \mathbb{R}^d$ be open and bounded with Lipschitz boundary, let $p \in [1, \infty]$ and $k \in \mathbb{N}$ such that $kp > d$. Then the positive cone in the Sobolev space $W^{k,p}(\Omega)$ is closed (with respect to the usual Sobolev norm), and it is also generating (see [2, Examples 2.3(c) and (d)]); hence, $W^{k,p}(\Omega)$ is a pre-Riesz space. Moreover, $W^{k,p}(\Omega)$ is also pervasive since it embeds into $C(\Omega)$ and since it contains all test functions on Ω.

In [10, Example 13] one can find an example of a pre-Riesz space that is not pervasive, but has the Riesz decomposition property and is thus weakly pervasive.

We now prove the main result of this subsection; it gives a sufficient criterion for a weakly pervasive pre-Riesz space to already be a vector lattice.

Theorem 4.12 *Assume that every non-empty totally ordered subset of X that is bounded from above has a supremum. If X is weakly pervasive, then X is a lattice.*

Proof If suffices to show that any two positive elements in X have an infimum, so let $x, y \in X_+$. It follows from Zorn's lemma and from the assumption on X that the set $[0, x] \cap [0, y]$ has a maximal element a. Let us show that $[0, x-a] \cap [0, y-a] = \{0\}$: if z is an element of this set, then $0 \leq z \leq x-a$ and $0 \leq z \leq y-a$, so $0 \leq z+a \leq x$ and $0 \leq z+a \leq y$. Hence, $z+a$ is an element of $[0, x] \cap [0, y]$ that dominates a; it thus follows from the maximality of a that $z = 0$.

As X is weakly pervasive, this implies that the positive vectors $x - a$ and $y - a$ are disjoint, i.e. they have infimum 0. Consequently, x and y also have an infimum (namely a). □

In the context of ordered Banach spaces, the following proposition gives sufficient criteria for the first assumption of Theorem 4.12 to be satisfied.

Proposition 4.13 *Assume that the pre-Riesz space X is an ordered Banach space. Consider the following assertions:*

(i) *The cone X_+ is normal and the space X is reflexive.*
(ii) *The cone X_+ is normal and X is a projection band in its bi-dual (compare Example 3.3).*

(iii) *Every order interval in X is weakly compact.*
(iv) *X is the dual space of an ordered Banach space Y such that Y has generating cone.*
(v) *The norm is additive on* X_+ *(i.e.,* $\|x + y\| = \|x\| + \|y\|$ *for all* $x, y \in X_+$*).*
(vi) *Every increasing norm bounded net in* X_+ *is norm convergent.*
(vii) *The cone* X_+ *is normal and Every increasing net in* X_+ *that is bounded from above is norm convergent.*
(viii) *Every non-empty upwards directed set in X that is bounded above has a supremum (i.e., in the terminology of Sect. 3.3, X is Dedekind complete).*
(ix) *Every non-empty totally ordered set in X that is bounded above has a supremum.*

Then the following implications hold:

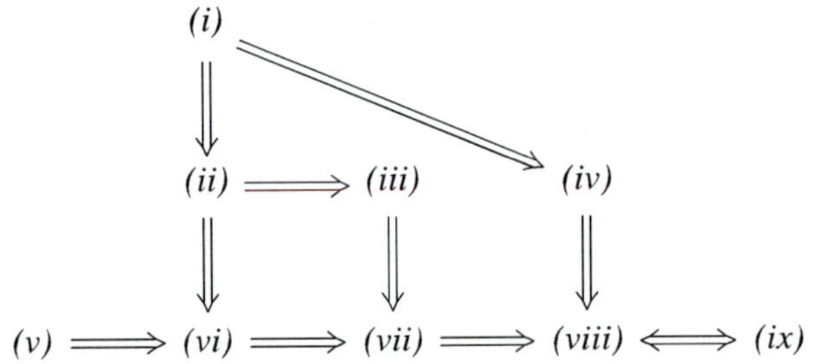

Proof "(i) ⇒ (ii)" This is obvious.

"(i) ⇒ (iv)" This is obvious.

"(ii) ⇒ (iii)" This was proved in [8, Proposition 2.6].

"(ii) ⇒ (vi)" The proof of this implication has already been sketched in [8, Remark 6.2]; we give a few more details here:

Let $P : X'' \to X''$ be the band projection with range X and let (x_j) be an increasing and norm-bounded net in X_+. Then (x_j) converges to a vector $x'' \in X''_+$ with respect to the weak*-topology. We have $Px'' \leq x''$. On the other hand, $Px'' \geq Px_j = x_j$ for each index j, which implies that $Px'' \geq x''$. We have thus shown that $Px'' = x''$, i.e., $x := x''$ is an element of X.

The increasing net (x_j) converges weakly to x, so it follows from [26, Theorem V.4.3] that (x_j) actually converges in norm to x.

"(iii) ⇒ (vii)" Assertion (iii) implies that every order interval in X is bounded; hence, the cone X_+ is normal. Now, let (x_j) is an increasing net in X_+ that is bounded above. Then (x_j) is contained in an order interval. Hence, (x_j) is weakly convergent and therefore also norm convergent according to [26, Theorem V.4.3].

"(iv) ⇒ (viii)" Let $A \subseteq X$ be a non-empty upwards directed set that is bounded above. Then the increasing net $(a)_{a \in A}$ is weak*-convergent to an element $x \in X$, and one readily checks that x is the supremum of A.

"(v) ⇒ (vi)" Let $(x_j) \subseteq X_+$ be an increasing norm bounded net. We show that this net is Cauchy and thus norm convergent. To this end, set $\alpha := \sup_j \|x_j\| \in [0, \infty)$ and let $\varepsilon > 0$. Choose j_0 such that $\|x_{j_0}\| \geq \alpha - \varepsilon$. For all indices $j \geq j_0$ we then obtain

$$\alpha \geq \|x_j\| = \|x_{j_0}\| + \|x_j - x_{j_0}\| \geq \alpha - \varepsilon + \|x_j - x_{j_0}\|$$

so $\|x_j - x_{j_0}\| \leq \varepsilon$. This proves that (x_j) is indeed Cauchy.

"(vi) ⇒ (vii)" As every increasing norm-bounded sequence in X_+ is norm convergent, it follows that X_+ is normal; see for instance [1, Theorem 2.45]. Hence, every increasing net in X_+ which is bounded above is also norm bounded and thus norm convergent according to (vi).

"(vii) ⇒ (viii)" Let $D \subseteq X$ be a non-empty upwards directed set which is bounded above by a vector $u \in X$. Choose $b \in X_+$ such that $b + D$ intersects the positive cone X_+. Then $\tilde{D} := X_+ \cap (b + D)$ is an upwards directed set, too, and $\tilde{D}$ is bounded above by $u + b$. Hence, the increasing net $(x)_{x \in \tilde{D}}$ converges to a vector $y \in X$. Clearly, y is the supremum of $\tilde{D}$, and thus it is also the supremum of $b + D$ (here we used again that $b + D$ is directed). Therefore, D has the supremum $y - b$.

"(viii) ⇒ (ix)" This is obvious.

"(ix) ⇒ (viii)" A general result in the theory of ordered sets says that, if every non-empty totally ordered subset of a partially ordered set Z has a supremum in Z, then every non-empty upwards directed subset of Z has a supremum in Z, too; see for instance [6, Proposition 1.5.9]. We can apply this to the partially set

$$Z := X \cup \{\infty\},$$

where we define ∞ as an object that is larger than each element of X:

It follows from (ix) that every non-empty totally ordered subset of $X \cup \{\infty\}$ has a supremum in $X \cup \{\infty\}$. So if $D \subseteq X$ is non-empty, directed and bounded above by an element $u \in X$, then we first conclude that D has a supremum s in $X \cup \{\infty\}$; since D is bounded above by u, it follows that $s \leq u$, so in particular, $s \in X$. Now one can immediately check that s is also the supremum of D within X. □

As a consequence of Theorem 4.12 we observe that if the pre-Riesz space X is a weakly pervasive ordered Banach space and satisfies at least one of the assertions (i)–(ix) in Proposition 4.13, then X is actually a vector lattice. Since every finite-dimensional Banach space is reflexive and every closed cone in such a space is normal, we obtain in particular the following corollary.

Corollary 4.14 *Let X be finite dimensional and assume that X_+ is closed. If X is weakly pervasive, then X is a vector lattice (and thus isomorphic to $\mathbb{R}^n$ with the standard order).*

We note once again that, for the special case where X is pervasive, Corollary 4.14 has been recently proved in [17, Theorem 39]. Let us remark a few further consequences of Theorem 4.12 in conjuction with Proposition 4.13.

Remarks 4.15

(a) Recall that several examples of ordered Banach spaces that are pervasive (and hence weakly pervasive) are listed in Examples 4.11. Theorem 4.12 and Proposition 4.13 show that such examples have to satisfy many restrictions if we do not want to end up in the category of vector lattices.
(b) If the pre-Riesz space X is an ordered Banach space with normal cone, then the dual space X' also has generating cone and is thus a pre-Riesz space. Proposition 4.13 shows that every non-empty totally ordered set in X' that is bounded above has a supremum. Hence, if X' is weakly pervasive, it follows from Theorem 4.12 that X' is in fact a vector lattice, and thus we conclude in turn that X has the Riesz decomposition property [1, Theorem 2.47].

    Hence, the dual space X' of an ordered Banach space X with normal (and generating) cone cannot be weakly pervasive unless X itself has the Riesz decomposition property. This suggests that the property "weakly pervasive" is not particularly well-behaved with respect to duality (at least not in the category of ordered Banach spaces).

4.4 Criteria in Terms of Other Concepts of Disjointness

Recall that weakly pervasive spaces are precisely those pre-Riesz spaces in which any two D-disjoint elements of the positive cone are automatically disjoint. In this context is interesting to observe that, in general pre-Riesz spaces, there exists an intermediate concept between disjointness and D-disjointness; this is the content of the following proposition.

Proposition 4.16 *For all $x, y \in X_+$ we have the following implications:*

$$\begin{aligned} &x \text{ and } y \text{ are disjoint} \\ &\qquad \Rightarrow \quad [-x, x] \cap [-y, y] = \{0\} \\ &\qquad\qquad \Rightarrow \quad x \text{ and } y \text{ are D-disjoint.} \end{aligned}$$

Proof Assume first that x and y are disjoint. If $f \in [-x, x] \cap [-y, y]$, then f is a lower bound of x and y, so $f \leq 0$. Moreover, $-f$ is also a lower bound of x and y, so $-f \leq 0$. Hence, $f = 0$. The second implication is obvious. □

We will see in Example 4.18 below that none of the two implications in Proposition 4.16 can be reversed in general pre-Riesz spaces. Before we give this example, we need a small auxiliary result.

We recall from [17, Definition 27] that an element $a \in X_+ \setminus \{0\}$ is called an *atom* in X if every vector $x \in [0, a]$ is a multiple of a; equivalently, the order interval $[0, a]$ equals the line segment $\{\lambda a : \lambda \in [0, 1]\}$.

Lemma 4.17 *Let a be an atom in X. Then the order interval $[-a, a]$ equals the line segment $\{\lambda a : \lambda \in [-1, 1]\}$.*

Proof Each $x \in [-a, a]$ can be written as

$$x = \frac{a + x}{2} - \frac{a - x}{2},$$

where both $\frac{a+x}{2}$ and $\frac{a-x}{2}$ are elements of $[0, a]$; hence, we have $[-a, a] = [0, a] - [0, a]$ (for this, we did not use that a is an atom). Since $[0, a]$ is the line segment $\{\lambda a : \lambda \in [0, 1]\}$, this implies the assertion. □

Example 4.18 Let $X = \mathbb{R}^3$, let X_+ denote the four ray cone from Example 2.2, and let $v_1, \dots, v_4 \in X$ be the vectors from that example.

(a) Let $w = v_1 + v_2$ and $\tilde{w} = v_3 + v_4$. Then w and $\tilde{w}$ are D-disjoint, but the set $[-w, w] \cap [-\tilde{w}, \tilde{w}]$ is non-zero since it contains the vector $(1, -1, 0)^T$.

(b) The order interval $[-v_1, v_1]$ is precisely the line segment $\{\lambda v_1 : \lambda \in [-1, 1]\}$; this follows from Lemma 4.17 since v_1 is an atom in X (which in turn follows from [17, Proposition 28]). Similarly, the order interval $[-v_2, v_2]$ is the line segment $\{\lambda v_2 : \lambda \in [-1, 1]\}$.

We thus conclude that $[-v_1, v_1] \cap [-v_2, v_2] = \{0\}$. Yet, we have seen in Example 3.16 that v_1 and v_2 are not disjoint.

We now consider pre-Riesz spaces in which, for all $x, y \in X_+$, the property $[-x, x] \cap [-y, y] = \{0\}$ implies that $x \perp y$. This property of a pre-Riesz space is (at least formally) weaker than being weakly pervasive. In finite dimensions, though, this property still suffices to conclude that a pre-Riesz space with closed cone is a vector lattice; we prove this in the following theorem.

Theorem 4.19 *Let X be finite dimensional and assume that X_+ is closed. Suppose that all vectors $x, y \in X_+$ that satisfy $[-x, x] \cap [-y, y] = \{0\}$ are disjoint. Then X is a vector lattice.*

For the proof we need the notion of an *extreme ray*. Let $a \in X_+ \setminus \{0\}$. If the half ray $\{\lambda a : \lambda \in [0, \infty)\}$ is a face of X_+, then we call this half ray an extreme ray of X_+. We note that $\{\lambda a : \lambda \in [0, \infty)\}$ is an extreme ray of X_+ if and only if a is an atom in X ([17, Proposition 28]). If X is finite dimensional and non-zero and X_+ is closed, then X_+ is the convex hull of its extreme rays.

Proof of Theorem 4.19 Set $n := \dim X$; we may assume that $n \geq 1$. Let E denote the set of all extreme rays of X_+ and for each $R \in E$, choose a non-zero vector $x_R \in R$. Then the set $\{x_R : R \in E\}$ spans X, so E has at least n elements.

On the other hand, each point x_R is an atom in X, so for any two distinct rays $R, S \in E$ we have $[-x_R, x_R] \cap [-x_S, x_S] = \{0\}$ according to Lemma 4.17. Thus, it follows from the assumption that $x_S \perp x_R$ for any two distinct rays $R, S \in E$. Hence, we conclude from Proposition 4.1 that the family of vectors $(x_R)_{R \in E}$ is linearly independent. Hence, E has exactly n elements. This proves that the positive cone X_+ is generated by exactly n extreme rays, so X is a vector lattice. □

Acknowledgments It is my pleasure to thank Anke Kalauch, Helena Malinowski and Onno van Gaans for various suggestions and discussions which helped me to considerably improve the paper.

I am also indebted to Andreas Blass for an answer on MathOverflow [4] that was very helpful for the proof of the implication "(ix) ⇒ (viii)" of Proposition 4.13.

This paper was originally motivated by the question how many bands and projection bands can exist in a finite dimensional pre-Riesz space—a question I first became aware of during a plenary talk of Anke Kalauch at the *Positivity X* conference that took place in July 2019 in Pretoria, South Africa. I am indebted to the organisers of the conference for financially supporting my participation.

References

1. C.D. Aliprantis, R. Tourky, *Cones and Duality*. Graduate Studies in Mathematics, vol. 84. (American Mathematical Society, Providence, 2007)
2. W. Arendt, R. Nittka, Equivalent complete norms and positivity. Arch. Math. (Basel) **92**(5), 414–427 (2009)
3. G. Birkhoff, *Lattice Theory*. American Mathematical Society Colloquium Publications, vol. XXV, 3rd edn. (American Mathematical Society, Providence, 1967)
4. A. Blass, Suprema of directed sets. MathOverflow (2020). https://mathoverflow.net/q/378005 Accessed 3 Dec 2020
5. M.N. Bleicher, H. Schneider, The decomposition of cones in modules over ordered rings. J. Algebra **1**, 233–258 (1964)
6. P.M. Cohn, *Universal Algebra*. Revised Edition. Mathematics and Its Applications, vol. 6. (D. Reidel Publishing Company, Dordrecht, 1981)
7. J. Glück, A short note on band projections in partially ordered vector spaces. Indag. Math. New Ser. **30**(1), 1–6 (2019)
8. J. Glück, M.P.H. Wolff, Long-term analysis of positive operator semigroups via asymptotic domination. Positivity **23**(5), 1113–1146 (2019)
9. R.V. Kadison, Order properties of bounded self-adjoint operators. Proc. Am. Math. Soc. **2**, 505–510 (1951)
10. A. Kalauch, H. Malinowski, Pervasive and weakly pervasive pre-Riesz spaces. Indag. Math. (N.S.) **30**(2), 375–385 (2019)
11. A. Kalauch, O. van Gaans, Directed ideals in partially ordered vector spaces. Indag. Math. (N.S.) **25**(2), 296–304 (2014)
12. A. Kalauch, B. Lemmens, O. van Gaans, Riesz completions, functional representations, and anti-lattices. Positivity **18**(1), 201–218 (2014)
13. A. Kalauch, B. Lemmens, O. van Gaans, Bands in partially ordered vector spaces with order unit. Positivity **19**(3), 489–511 (2015)
14. A. Kalauch, B. Lemmens, O. van Gaans, Inverses of disjointness preserving operators in finite dimensional pre-Riesz spaces. Quaest. Math. **42**(4), 423–430 (2019)
15. A. Kalauch, H. Malinowski, Order continuous operators on pre-Riesz spaces and embeddings. Z. Anal. Anwend. **38**(4), 375–395 (2019)
16. A. Kalauch, H. Malinowski, Vector lattice covers of ideals and bands in pre-Riesz spaces. Quaest. Math. **42**(7), 919–937 (2019

17. A. Kalauch, H. Malinowski, Projection bands and atoms in pervasive pre-Riesz spaces. Positivity **25**(1), 177–203 (2021). https://doi.org/10.1007/s11117-020-00757-7
18. A. Kalauch, O. van Gaans, *Pre-Riesz Spaces*. De Gruyter Expositions in Mathematics, vol. 66 (De Gruyter, Berlin, 2019)
19. A. Kalauch, O. van Gaans, F. Zhang, Disjointness preserving C_0-semigroups and local operators on ordered Banach spaces. Indag. Math. New Ser. **29**(2), 535–547 (2018)
20. A. Kalauch, O. van Gaans, F. Zhang, Disjointness preserving operators on normed pre-Riesz spaces: extensions and inverses. Positivity **24**(2), 481–504 (2020)
21. V. Katsikis, I.A. Polyrakis, Positive bases in ordered subspaces with the Riesz decomposition property. Stud. Math. **174**(3), 233–253 (2006)
22. H. Malinowski, Order closed ideals in pre-Riesz spaces and their relationship to bands. Positivity **22**(4), 1039–1063 (2018)
23. P. Meyer-Nieberg, *Banach Lattices*. Universitext (Springer, Berlin, 1991)
24. R.C. Penney, Self-dual cones in Hilbert space. J. Funct. Anal. **21**, 305–315 (1976)
25. S. Sakai, *C^*-Algebras and W^*-Algebras* (Springer, New York-Heidelberg, 1971). Ergebnisse der Mathematik und ihrer Grenzgebiete, Band 60
26. H.H. Schaefer, M.P. Wolff, *Topological Vector Spaces*. Graduate Texts in Mathematics, vol. 3, 2nd edn. (Springer, New York, 1999)
27. H.H. Schaefer, *Banach Lattices and Positive Operators* (Springer, New York-Heidelberg, 1974). Die Grundlehren der mathematischen Wissenschaften, Band 215
28. M. Takesaki, *Theory of Operator Algebras. I* (Springer, New York-Heidelberg, 1979)
29. O. van Gaans, A. Kalauch, Bands in pervasive pre-Riesz spaces. Oper. Matrices **2**(2), 177–191 (2008)
30. O. van Gaans, A. Kalauch, Disjointness in partially ordered vector spaces. Positivity **10**(3), 573–589 (2006)
31. O. van Gaans, A. Kalauch, Ideals and bands in pre-Riesz spaces. Positivity **12**(4), 591–611 (2008)
32. O. van Gaans, A. Kalauch, Tensor products of Archimedean partially ordered vector spaces. Positivity **14**(4), 705–714 (2010)
33. M. van Haandel, *Completions in Riesz Space Theory*. Ph.D. Thesis, University of Nijmegen, 1993

101 Years of Vector Lattice Theory: A Vector Lattice-Valued Daniell Integral

Jacobus J. Grobler

Abstract We show that the paper in which P.J. Daniell introduced his well-known integral, used modern Riesz space techniques to derive the properties of the integral and to prove a fundamental decomposition result for the integral. The latter result was proved a decade later by F. Riesz and was considered to be the origin of Riesz space theory. After a survey of Daniell's paper, we generalize P.E. Protter's version of the L^p-valued $(0 \le p \le \infty)$ Daniell integral to a vector lattice-valued Daniell integral, following closely Daniell's original method. A.C.M van Rooij and W.B. van Zuijlen also introduced integrals for functions with values in a partially ordered vector space a more general setting than the one we use.

Keywords Vector-valued Daniell integral · Vector lattice · Riesz space

Mathematics Subject Classification (2010) Primary 28C05, 46A40, 46G12; Secondary 28B05, 28B15

1 Introduction

The aim of this paper is to define a lattice vector-valued integral. Such an integral is needed in the study of stochastic processes in Riesz spaces (see [5–7]). Since it is needed in a wide variety of applications, we need a general integral, and the Daniell integral seems to be ideally suited to fulfill this role. This was first realized by P.E. Protter [10], who used the Daniell integral to integrate L^p-valued functions, $0 \le p \le \infty$. Since we do not use metric spaces, we have to generalize Protter's integral. This led us to a study of the original paper of P.J. Daniell [2], in which any reference to a measure or an integrator is avoided. The methods used in his paper are very general and easily generalized. In [5, 6] the vector-valued integral used

J. J. Grobler (✉)
Unit for Business Mathematics and Informatics, North-West University, Potchefstroom, South Africa

E. Kikianty et al. (eds.), *Positivity and its Applications*, Trends in Mathematics,
https://doi.org/10.1007/978-3-030-70974-7_8

was a version of the Bochner integral and in [7] a version of the Dobrakov integral. Both these integrals are designed for Banach space-valued functions and they are sometimes not so easy to apply in the theory of Riesz space-valued functions. We note that A.C.M. van Rooij and W.B. van Zuijlen also introduced an even more general integral in [15] for functions with values in an ordered vector space.

Daniell's original paper was surprisingly modern in its use of Riesz space techniques, which, according to the literature, only originated a decade later. We therefore include a historic first part in the paper, surveying the fundamental methods of Riesz space theory already present in Daniell's paper and come to the conclusion that Daniell's paper might have been the first paper to exhibit the fundamentals of the subject, contrary to what is widely accepted at present. Hence the title of our paper expressing the view that Riesz space theory originated a decade before 1928, the date usually regarded as its date of origin.

Our paper will consist of two parts: the first part gives a historic perspective of Daniell's work and the second part is devoted to the lattice-valued Daniell integral.

We expect the reader to be informed on general Riesz space theory and will supply references to relevant notions and literature as we proceed.

Part 1. A Note on the History

2 Riesz Spaces or Daniell Spaces?

The consentient wisdom about the history of the theory of vector lattices is that it was founded by F. Riesz, H. Freudenthal and L.V. Kantorovitch in the years around 1935. It is believed that the theory was initiated by a short note delivered to the International Mathematical Congress at Bologna in 1928 by F. Riesz [11], the results of which were generalized in a 1940 paper in the Annals of Mathematics (see [8, 11, 12] and the well known textbooks by W.A.J. Luxemburg and A.C. Zaanen [9], H.H. Schaefer [14], C.D. Aliprantis and O. Burkinshaw [1]).

This belief was expressed as follows by W.A.J. Luxemburg [8] and echoed by authors of most textbooks:

"In the more recent development of real analysis a more sophisticated version of an ordered structure defined by cones of "positive" elements emerged. It has its origin in Jordan's celebrated theorem characterizing the real functions that can be written as the difference of two monotone increasing (decreasing) functions. It expresses that the linear space of functions of bounded variation is generated by the cone of its non-negative elements, the increasing functions."

"It was F. Riesz however, who *for the first time* at the 1928 I.C.M. Congress in Bologna [11] pointed out this significant order theoretic aspect of Jordan's theorem. In this address, F. Riesz showed that such decomposition results hold more generally for certain spaces of continuous linear functionals in terms of positive linear functionals. To be more precise, F. Riesz showed then, *for the first time*, that the linear spaces of all continuous linear functionals defined on linear spaces of continuous real functions are generated by the cones of the positive linear functionals" (my italics).

However, already in 1918, P.J. Daniell had published his remarkable and influential paper "A general form of integral" [2].

In the paper he defined an integral S (S for Stieltjes) as a σ-order continuous order bounded linear functional on a certain Riesz space of real functions. He then showed that it can be decomposed as the difference or two positive linear functionals, both of which are again σ-order continuous. His proof of the decomposition does not rely on σ-order continuity, but only on the order boundedness of the functional S. We present a short survey of his paper that reads like a modern paper on Riesz space theory (see also [3]).

Dr. Percy John Daniell, Rice Institute (1913) Rice University: https://hdl.handle.net/1911/68457

Percy John Daniell (9 January 1889–25 May 1946) was a pure and applied mathematician, born in Valparaiso, Chile. He returned to England as a child in 1895, attended King Edward's School, Birmingham and proceeded to Trinity College, Cambridge. He lectured at the University of Liverpool and was subsequently appointed to the new Rice Institute in Houston, Texas. He was an Invited Speaker of the ICM in 1920 at Strasbourg.

3 Daniell's Paper

In this review of the paper we use Daniell's original definitions and notation.

Let P be as set of elements p. Let T_0 be an initial class of numerically valued functions $f(p)$, closed with respect to multiplication by a constant, addition, logical addition and logical multiplication. Assume furthermore that the functions of T_0 are limited, that is, corresponding to any $f(p)$, a finite number $K(f)$ can be found such that $|f(p)| < K(f)$ for all p.

Thus, T_0 is a Riesz subspace of the principal ideal generated in the Dedekind complete Riesz space $\mathbb{R}^P$ by the function that is equal to 1. Any Riesz subspace of a

Dedekind complete Riesz space with order unit can be represented as a space fitting Daniell's description.

The following relations hold:[1]

$$
\begin{aligned}
f_1 \vee f_2 + f_1 \wedge f_2 &= f_1 + f_2, \\
(-f_1) \vee (-f_2) &= -(f_1 \wedge f_2), \\
|f| &= f \vee (-f), \\
(f_1 - f_2)^+ &= f_1 - f_1 \wedge f_2.
\end{aligned}
$$

One also has the following decomposition property (now known as the Riesz decomposition property):

$$0 \le \phi \le f_1 + f_2 \implies \exists\, 0 \le \phi_1 \le f_1, 0 \le \phi_2 \le f_2,\ \phi = \phi_1 + \phi_2.$$

Remarks 1

1. We note that, working with real-valued functions, a well known algebraic proof can be given, but Daniell used only order properties and his proof is the same as the one found for instance in H.H. Schaefer's monograph [14].
2. A very important aspect of Daniell's first paper was that he developed the theory of integration without any reference to measure theory. P.E. Protter's comment on this is: "While measuring sets is a pleasurable pursuit, integrating functions sure is what measure theory ultimately is all about".
3. Daniell's idea was to define an elementary integral right away and then expand the integral.

In modern terminology, Daniell's definition of the integrals can be stated as follows:

Definition 3.1

1. An I-integral defined on T_0 is a *positive σ-order continuous linear functional defined on T_0*.
2. An S-integral defined on T_0 is an *order bounded σ-order continuous linear functional defined on T_0*.

Daniell remarked that every I-integral is an S-integral, by showing that

$$-I(|f|) \le I(f) \le I(|f|).$$

[1] Daniell claimed that the symbols $\vee$ and $\wedge$ were first used in this context by himself and by W.H. Young [16].

Remarks

1. Most texts on measure and integration theory, if they bother to mention the Daniell integral, defines only the I-integral (see [10, 13, 17]) and call it the Daniell integral.
2. It is clear that the S-integral is a generalization of the Stieltjes integral.

The important fact that we want to emphasize in this part on the history of Riesz spaces, is that, similar to Jordan's decomposition of Stieltjes integrals, Daniell shows that an S-integral is expressible as the difference of two I-integrals.

This is how he did it. For the S-integral (σ-order continuous order bounded linear functional) S defined on T_0, he defined three associated I-integrals S^+, S^- and $|S|$ as follows: for $0 \leq f \in T_0$,

$$\begin{aligned} S^+(f) &:= \sup\{S(\phi) \,:\, \phi \in T_0,\ 0 \leq \phi \leq f\}, \\ S^-(f) &:= S^+(f) - S(f), \\ |S| &:= S^+ + S^-. \end{aligned}$$

In order to prove that S^+ and S^- are I-integrals, he used the Riesz decomposition property as well as the Dedekind completeness of $\mathbb{R}$. Also, for the proof of the existence and linearity of the decomposition only the fact that S is order bounded was used.

He called the I-integral $|S| := S^+ + S^-$ the *modular integral associated with S* and showed that for $0 \leq f \in T_0$,

$$|S|(f) = \sup\{S(\phi) \,:\, \phi \in T_0,\ |\phi| \leq f\},$$

and

$$|S(f)| \leq |S|(|f|) \text{ for all } f \in \mathbb{T}_0.$$

The result above is the main result of F. Riesz presented in 1928 in a talk given at the ICM in Bologna. Riesz considered in his address linear functionals on a space $C(P)$ of continuous functions defined on an interval $P \subset \mathbb{R}$, a framework much less general than that of Daniell. He used, for $0 \leq f \in C(P)$, the formula

$$S^+(f) := \sup\{\sum_{j=1}^{n} S(f_n) \,:\, 0 \leq f_j \in C(P),\ \sum_{j=1}^{n} f_j = f\},$$

which we know is equivalent to Daniell's formula and the names of Riesz and Kantorovich are linked to this representation.

Daniell's proof did not use the fact that the elements of T_0 are functions and can be copied verbally for a general Riesz space.

It is hard to explain why Daniell's decomposition of an order bounded linear functional was overlooked, because in his 1928 address at Bologna, Riesz did refer to Daniell's paper as follows[2]:

"The advantage of this method—the methods always have their advantages—is that it is of a very general scope, since it is applicable to the analysis of linear operations involving functions defined on abstract sets as studied by MR. DANIELL in generalizing the notion of the integral" (my translation).

But, he never mentioned that Daniell also proved the decomposition for order bounded linear functionals that he was about to prove.

An explanation may be that most authors mentioning Daniell's integral in textbooks, are only interested in extending the positive integral I without mentioning the integral S. This is also true of Riesz, who introduced in his lectures on integration theory the Lebesgue integral using Daniell's method (see [13]). The extension process does not use the results of the third and fourth (uncapped) sections of Daniell's paper in which he proved the decomposition and the formula for the modular integral.

The next step is to extend the I-integral to a larger space than T_0. This process is well known and we will return to this matter in Part II where we will follow this method to extend the lattice vector-valued integral, and to derive further properties of the integral (the Monotone convergence theorem and Lebesgue's dominated convergence theorem).

Following the extension of the positive integral I, the S-integral is extended in the obvious way by writing $S = S^+ - S^-$ with S^+ and S^- two I-integrals. Hence the S-integral can be extended from T_0 to a wider class of functions, by extending the I-integrals S^+, S^-.

Part 2. The Abstract Vector Lattice-Valued Daniell Integral

P.E. Protter [10] was perhaps the first to realize that the Daniell integral is a powerful tool to define and study stochastic integrals.

Our setting will be that of a Dedekind complete Riesz space $\mathfrak{E}$ that has a weak order unit E. We assume that $\mathfrak{E}$ is separated by its order continuous dual $\mathfrak{E}_{00}^{\sim}$. The ideal generated by E in $\mathfrak{E}$ will be denoted by $\mathfrak{E}_E$. $\mathfrak{E}^s$ and $\mathfrak{E}^u$ will denote respectively the sup-completion and the universal completion of $\mathfrak{E}$. Everything about Riesz spaces the reader need to know can be found in the standard texts [1, 7, 9, 14, 18]. In [4], the reader will find the definition and properties of the sup-completion.

Consider an arbitrary set T and the vector space $\mathfrak{E}^T$ of all functions defined on T with values in the Riesz space $\mathfrak{E}$. By defining the operations point wise, $\mathfrak{E}^T$ is a Dedekind complete Riesz space with weak order unit (E_t), where $E_t = E$ for all $t \in T$.

[2]"L'avantage que présente cette méthode—les méthodes ont toujours leur avantage—est qu'elle est d'une portée toute générale, jusqu'á étre applicable á l'analyse des opérations linéaires portant sur des fonctions définies dans des ensembles *abstraits* comme les a étudiées M. DANIELL en généralisant la notion d'intégrale."

4 The Vector-Valued I-Integral

Let $\mathbb{L}$ be a Riesz subspace of $\mathfrak{E}^T$ that is contained in the ideal generated in $\mathfrak{E}^T$ by the weak order unit (E_t).

Definition 4.1 A positive linear operator $I : \mathbb{L} \to \mathfrak{E}$ is called a *vector-valued I-integral* whenever, for every sequence (X_n) in $\mathbb{L}$ that satisfies $X_n(t) \downarrow 0$ for every $t \in T$, it follows that $I(X_n) \downarrow 0$.

We now extend the I-integral from $\mathbb{L}$ to $\mathbb{L}^\uparrow$, where $\mathbb{L}^\uparrow$ is defined by

$$\mathbb{L}^\uparrow := \{X \in (\mathfrak{E}^s)^T \ : \ \exists X_n \in \mathbb{L},\ n \in \mathbb{N}, X_n \uparrow X\}.$$

Since I is positive, we get

$$I(X_1) \leq I(X_2) \leq \cdots,$$

and the supremum exists again in $\mathfrak{E}^s$.

If $X \in \mathbb{L}^\uparrow$, define

$$I(X) = \sup\{I(X_n) \ : \ X_n \in \mathbb{L}, X_n \uparrow X\}.$$

Then I is well-defined with values in $\mathfrak{E}^s$.

The proof that I is well-defined by Daniell can be reproduced, however, here is a shorter proof:

Let $X_1 \leq X_2 \leq \ldots \in \mathbb{L}$, $Y_1 \leq Y_2 \leq \ldots \in \mathbb{L}$, and $\sup X_n \geq \sup Y_n$. Let $X = \sup X_n$ and $Y = \sup Y_n$. Then, as $m \to \infty$, we have $X_m \wedge Y_n \uparrow_m X \wedge Y_n = Y_n$. Hence,

$$I(Y_n) = \lim_m I(X_m \wedge Y_n) \leq \lim I(X_m).$$

This holds for every n and so letting $n \to \infty$, we get $\lim I(Y_n) \leq \lim I(X_n)$. If, therefore, $X = Y$, then $I(X) = I(Y)$. □

The integral on $\mathbb{L}^\uparrow$ has the following properties:

Proposition 4.2

1. $X_1 \leq X_2 \leq \ldots \uparrow X$, $X_n \in \mathbb{L}^\uparrow$, *then* $X \in \mathbb{L}^\uparrow$ *and* $I(X_n) \uparrow I(X)$.
2. $0 \leq I(X)$ *for all* $X \in \mathbb{L}^\uparrow$.
3. *If* $X, Y \in \mathbb{L}^\uparrow$, $X \leq Y$, *then* $I(X) \leq I(Y)$.
4. *If* $X \in \mathbb{L}^\uparrow$ *and* $0 \leq c < \infty$, *then* $cX \in \mathbb{L}^\uparrow$ *and* $I(cX) = cI(X)$.
5. *If* $X, Y \in \mathbb{L}^\uparrow$, *then* $X + Y$, $X \vee Y$, $X \wedge Y \in \mathbb{L}^\uparrow$ *and*

$$I(X + Y) = I(X \vee Y) + I(X \wedge Y) = I(X) + I(Y).$$

Proof

1. Daniell's proof for functionals can be copied. Here is a shorter version: Let $0 \leq X_n \in \mathbb{L}^{\uparrow}$ and let $X_n \uparrow X$. Suppose that for every n we have that $0 \leq X_{nm} \uparrow_m X_n$ with $X_{nm} \in \mathbb{L}^{\uparrow}$. Put $Y_m := X_{1m} \vee X_{2m} \vee \ldots \vee X_{mm} \in \mathbb{L}^{\uparrow}$; then $Y_m \uparrow$ and
$$X_{nm} \leq Y_m \leq X_m \qquad \text{for all } n \leq m; \tag{4.1}$$
consequently,
$$I(X_{nm}) \leq I(Y_m) \leq I(X_m) \qquad \text{for all } n \leq m. \tag{4.2}$$
Let $m \to \infty$ in (4.1) to yield the result that $X_n \leq \lim_m Y_m \leq \lim_m X_m = X$; then, let $n \to \infty$ to see that $Y_m \uparrow_m X$. Hence, $X \in \mathbb{L}^{\uparrow}$ and $I(Y_m) \uparrow I(X)$; now let $m \to \infty$ in (4.2) and use the preceding to conclude that $I(X_n) \leq \lim_m I(Y_m) = I(X) \leq \lim_m I(X_m)$. If we let $n \to \infty$ in this inequality we find that $I(X_n) \uparrow I(X)$. This proves the desired result.

The proofs of the other properties are simple and Daniell's proofs for the case of functionals can be taken almost without change. □

Comment This method of proof was later used for extending positive order continuous (σ-order continuous) linear operators from subspaces that are order dense (σ-order dense) in a space to the whole space. To ensure that the extended operator's range is in the given larger space, it is additionally assumed that the subspace is majorizing (see for instance a result of A.I. Veksler, [1, Theorem 4.12]).

The element $X \in \mathbb{L}^{\uparrow}$ is called *summable* if $I(X) \in \mathfrak{E}$.

5 Semi-Integrals

We define the set $\mathbb{L}^{\uparrow\Downarrow}$ as follows:
$$\mathbb{L}^{\uparrow\Downarrow} = \{X \in (\mathfrak{E}^s)^T \,:\, X = \inf X_\alpha, X_\alpha \in \mathbb{L}^{\uparrow}, X_\alpha \geq X\}.$$

Note that, since we have for $X, Y \in \mathbb{L}^{\uparrow}$ that $X \wedge Y \in \mathbb{L}^{\uparrow}$, we have that X_α is downwards directed and therefore $X_\alpha \downarrow X$ for $X \in \mathbb{L}^{\uparrow\Downarrow}$.

Definition 5.1 For $X \in \mathbb{L}^{\uparrow\Downarrow}$ we define
$$\dot{I}(X) := \inf_\alpha \{I(X_\alpha) \,:\, X_\alpha \in \mathbb{L}^{\uparrow}, X_\alpha \geq X\}.$$

$\dot{I}(X)$ is called the *upper semi-integral* of X.

In Protter's terminology it is simply called the *upper integral*.

Proposition 5.2 *The upper semi-integral is positive homogeneous, sub-additive and monotone.*

Proof For $c \geq 0$ and $Z_\alpha \geq cX$, $Z_\alpha \in \mathbb{L}^\uparrow$, we have, since $c^{-1}Z_\alpha \geq X$ and $c^{-1}Z_\alpha \in \mathbb{L}^\uparrow$, that

$$\dot{I}(cX) = \inf\{I(Z_\alpha) \,:\, Z_\alpha \geq cX\} = c \inf_\alpha I(c^{-1}Z_\alpha) \geq c\dot{I}(X).$$

Similarly, for $c \geq 0$ and $Z_\alpha \geq X$, $Z_\alpha \in \mathbb{L}^\uparrow$, we have, since $cZ_\alpha \geq cX$ and $cZ_\alpha \in \mathbb{L}^\uparrow$, that

$$\dot{I}(X) = \inf\{I(Z_\alpha) \,:\, Z_\alpha \geq X\} = c^{-1} \inf_\alpha I(cZ_\alpha) \geq c^{-1}\dot{I}(cX).$$

This proves the positive homogeneity.

If $X_1 \leq X_\alpha \in \mathbb{L}^\uparrow$, $X_2 \leq X_\beta \in \mathbb{L}^\uparrow$, then $X_\alpha + X_\beta \geq X_1 + X_2$ and so

$$\dot{I}(X_1 + X_2) \leq I(X_\alpha + X_\beta) = I(X_\alpha) + I(X_\beta).$$

Taking the infimum over all such X_α and X_β we obtain the required sub-additivity.

Finally, if $X \leq Y$, then $\{Z \geq Y\} \subset \{Z \geq X\}$. Thus $\dot{I}(X) \leq \dot{I}(Y)$. □

Definition 5.3 For $X \in \mathfrak{E}^u$ satisfying $-X \in \mathbb{L}^{\uparrow\Downarrow}$, the *lower semi-integral of* X is defined as $\underset{\cdot}{I}(X) := -\dot{I}(-X)$.

Again, in Protter's terminology, it is called the *lower integral* of X. We note that $-X \in \mathbb{L}^{\uparrow\Downarrow}$ is equivalent to $X \in \mathbb{L}^{\downarrow\Uparrow}$, i.e., if

$$\mathbb{L}^\downarrow := \{X \in (\mathfrak{E}^s)^T \,:\, \exists X_n \in \mathbb{L}, X_n \downarrow X\},$$

then,

$$\mathbb{L}^{\downarrow\Uparrow} := \{X \in (\mathfrak{E}^s)^T \,:\, X = \sup X_\alpha, X_\alpha \in \mathbb{L}^\downarrow, X_\alpha \leq X\}.$$

Here we use $\mathfrak{E}^u$, because the invertible elements in $\mathfrak{E}^s$ are in $\mathfrak{E}^u$. The element $-X$ does not necessarily exist for $X \in (\mathfrak{E}^s)^T$.

Proposition 5.4

1. *For* $X, -X \in \mathbb{L}^{\uparrow\Downarrow}$ *we have* $\underset{\cdot}{I}(X) \leq \dot{I}(X)$.
2. *For* $X \in \mathbb{L}^{\uparrow\Downarrow}$, $\dot{I}(X \vee Y) + \dot{I}(X \wedge Y) \leq \dot{I}(X) + \dot{I}(Y)$.

Proof

1. $0 = \dot{I}(0) = \dot{I}(X - X) \leq \dot{I}(X) + \dot{I}(-X) = \dot{I}(X) - \underset{\cdot}{I}(X)$.
2. If $X_\alpha, Y_\beta \in \mathbb{L}^\uparrow$ with $X_\alpha \geq X$, $Y_\beta \geq Y$, we have

$$X_\alpha \vee Y_\beta \geq X \vee Y, \; X_\alpha \wedge Y_\beta \geq X \wedge Y.$$

Hence,

$$\dot{I}(X \vee Y) + \dot{I}(X \wedge Y) \leq I(X_\alpha \vee Y_\beta) + I(X_\alpha \wedge Y_\beta) = I(X_\alpha) + I(Y_\beta).$$

Taking the infimum over α and β the result follows.

□

Corollary 5.5 *If* $X, -X \in \mathbb{L}^{\uparrow\Downarrow}$

$$\dot{I}(|X|) - \underset{\cdot}{I}(|X|) \leq \dot{I}(X) - \underset{\cdot}{I}(X).$$

Proof $|X| = X \vee (-X),\ -|X| = X \wedge (-X)$. Hence,

$$\dot{I}(|X|) - \underset{\cdot}{I}(|X|) = \dot{I}(|X|) + \dot{I}(-|X|) \leq \dot{I}(X) + \dot{I}(-X) = \dot{I}(X) - \underset{\cdot}{I}(X).$$

It follows that $\underset{\cdot}{I}(X) \leq \dot{I}(X)$ and that $\dot{I}(|X|) - \underset{\cdot}{I}(|X|) \leq \dot{I}(X) - \underset{\cdot}{I}(X)$. □

Remark The proofs of Proposition 5.4 and Corollary 5.5 are copies of Daniell's proofs for the corresponding results for the real-valued case. They are reproduced to stress the point that Daniell used pure Riesz space techniques.

6 Summability

Definition 6.1 For $X, -X \in \mathbb{L}^{\uparrow\Downarrow}$ we call X *summable* if $\dot{I}(X) = \underset{\cdot}{I}(X) \in \mathfrak{E}$ and we define

$$I(X) := \dot{I}(X) = \underset{\cdot}{I}(X).$$

Theorem 6.2 *The set $\mathcal{L}$ of all summable vector valued functions in $\mathfrak{E}^T$ is a Riesz space and the integral I defined on $\mathcal{L}$ is a positive linear operator mapping $\mathcal{L}$ into $\mathfrak{E}$. Moreover the formula $|I(X)| \leq I(|X|)$ holds.*

Proof Since $\dot{I}$ is positive, $I = \dot{I}$ is positive.

We next show that if c is a constant and X is summable, then cX is summable and $I(cX) = cI(X)$. If c is positive,

$$\dot{I}(cX) = c\dot{I}(X) = cI(X) \qquad \text{(by Theorem 5.2)}$$
$$-\underset{\cdot}{I}(cX) = \dot{I}(-cX) = c\dot{I}(-X) = -c\underset{\cdot}{I}(X) = -cI(X).$$

Hence,

$$\dot{I}(cX) = cI(X) = \underset{\cdot}{I}(cX).$$

If c is negative,

$$\dot{I}(cX) = \dot{I}((-c)(-X)) = -c\dot{I}(-X) = c\underset{\cdot}{I}(X) = cI(X) \qquad \text{(by Theorem 5.2)}$$
$$-\underset{\cdot}{I}(cX) = \dot{I}(-cX) = -c\dot{I}(X) = -cI(X).$$

Hence,

$$\dot{I}(cX) = cI(X) = \underset{\cdot}{I}(cX).$$

Let X_1 and X_2 be summable. Then,

$$\dot{I}(X_1 + X_2) \leq \dot{I}(X_1) + dI(X_2) = I(X_1) + I(X_2) \text{ by subadditivity}$$

and

$$-\underset{\cdot}{I}(X_1 + X_2) = \dot{I}(-X_1 - X_2) \leq \dot{I}(-X_1) + \dot{I}(-X_2) = -I(X_1) - I(X_2),$$

i.e.,

$$\underset{\cdot}{I}(X_1 + X_2) \geq I(X_1) + I(X_2).$$

So,

$$I(X_1) + I(X_2) \leq \underset{\cdot}{I}(X_1 + X_2) \leq \dot{I}(X_1 + X_2) \leq I(X_1) + I(X_2).$$

Hence, I is additive and so $\mathcal{L}$ is a real vector space.

To see that $\mathcal{L}$ is a Riesz subspace, we need only proof that if X is summable, then $|X|$ is also summable. Let X be summable. It then follows from Corrollary 5.4 that

$$\dot{I}(|X|) - \underset{\cdot}{I}(|X|) \leq \dot{I}(X) - \underset{\cdot}{I}(X) = 0.$$

Hence, $\dot{I}(|X|) \leq \underset{\cdot}{I}(|X|) \leq \dot{I}(|X|)$ and so equality holds. Hence, $|X|$ is summable. Moreover, it follows from $-|X| \leq X \leq |X|$ that

$$-I(|X|) = \dot{I}(-|X|) \leq \dot{I}(X) = I(X) \leq \dot{I}(|X|) = I(|X|).$$

Hence the formula $|I(X)| \leq I(|X|)$ holds. □

From our knowledge of Riesz spaces, we know now that if X_1 and X_2 are summable so are $X_1 \vee X_2$ and $X_1 \wedge X_2$. This was not known by Daniell, who supplied an elegant proof of both these facts (see [2, section 7(5)]).

The monotone convergence theorem holds. For its proof, we use the fact that $\mathfrak{E}_{00}^{\sim}$ separates the points of $\mathfrak{E}$.

Theorem 6.3 *If $X_1 \leq X_2 \leq \cdots \uparrow X$ is an increasing sequence of summable functions and if* $\sup I(X_n) \in \mathfrak{E}$, *then X is summable and*

$$I(X) = \sup I(X_n).$$

Proof Let $X, -X \in \mathbb{L}^{\uparrow\Downarrow}$. We have to show that $\dot{I}(X) = \underset{\cdot}{I}(X)$. Now, $-X \leq -X_n$ and so, by Theorem 5.2, we have that $\dot{I}(-X) \leq \dot{I}(-X_n) = I(-X_n)$. This implies by definition that $\underset{\cdot}{I}(X) \geq I(X_n)$ and since this holds for all n, we have that

$$\underset{\cdot}{I}(X) \geq \sup I(X_n). \tag{6.1}$$

We complete the proof by showing that $\sup I(X_n) \geq \dot{I}(X)$.

Since X_n is summable for each n, we have for the downwards directed nets

$$\Gamma_1 := \{Y \in \mathbb{L}^{\uparrow} : Y \geq X_1\},$$
$$\Gamma_n := \{Y \in \mathbb{L}^{\uparrow} : Y \geq (X_n - X_{n-1})\},\ n \geq 2,$$

that $I(Y) \downarrow_{Y \in \Gamma_1} I(X_1)$, and for $n \geq 2$, $I(Y) \downarrow_{Y \in \Gamma_n} I(X_n - X_{n-1})$.

Let $\epsilon > 0$ and $0 \leq \phi \in \mathfrak{E}^{\sim}_{00}$ be arbitrary.

Then, taking $X_0 = 0$, we have for all n that

$$\phi I(Y) \downarrow_{Y \in \Gamma_n} \phi I(X_n - X_{n-1}),$$

and so there exists for each n an element $Y_n \geq (X_n - X_{n-1})$ satisfying

$$\phi(I(Y_1)) < \phi(I(X_1)) + \epsilon/2$$
$$\phi(I(Y_2)) < \phi(I(X_2 - X_1)) + \epsilon/4,$$
$$\vdots$$
$$\phi(I(Y_n)) < \phi(I(X_n - X_{n-1})) + \epsilon/2^n.$$

For $n \geq 2$, we have that $Y_n \geq X_n - X_{n-1} \geq 0$. Let

$$Z_n := Y_1 + Y_2 + \cdots + Y_n.$$

Then $Z_n \in \mathbb{L}^{\uparrow}$ and $Z_1 \leq Z_2 \leq \cdots$. By Theorem 4.2, $\sup Z_n \in \mathbb{L}^{\uparrow}$ and $I(\sup Z_n) = \sup I(Z_n)$. Noting that $Z_n \geq X_n$ we have that $\sup Z_n \geq \sup X_n = X$ and therefore

$$\dot{I}(X) \leq I(\sup Z_n) = \sup I(Z_n).$$

But,

$$\begin{aligned}\phi(I(Z_n)) &= \phi(I(Y_1) + \phi I(Y_2) + \cdots + \phi I(Y_n)) \\ &< \phi(I(X_1)) + \phi(I(X_2 - X_1)) + \cdots + \phi(I(X_n - X_{n-1})) \\ &\quad + \epsilon(\tfrac{1}{2} + \tfrac{1}{4} + \cdots + \tfrac{1}{2^n}) \\ &< \phi(I(X_n)) + \epsilon.\end{aligned}$$

Thus, since this holds for any $\epsilon > 0$,

$$\phi(\dot{I}(X)) \leq \sup \phi(I(Z_n)) \leq \sup \phi(I(X_n)),$$

which by the order continuity of ϕ yields

$$\phi(\dot{I}(X)) \leq \phi(\sup I(X_n)).$$

But this holds for any $0 \leq \phi \in \mathfrak{E}_{00}^{\sim}$ and since, by assumption, $\mathfrak{E}_{00}^{\sim}$ separates the points of $\mathfrak{E}$, we have

$$\dot{I}(X) \leq \sup I(X_n).$$

This completes the proof. □

Having proved the monotone convergence theorem, the Lebesgue dominated convergence theorem follows easily. We formulate the theorem using order convergence as the mode of convergence, i.e., (X_n) converges to X if

$$\liminf X_n = \limsup X_n = X.$$

Daniell used the monotone convergence theorem and his proof is standard. We modify his proof somewhat in order to avoid ϵ-arguments.

Theorem 6.4 *Let $X_1, X_2, \ldots$ be a sequence of summable functions and let $\lim X_n = X$. If there exists a summable function Z such that $|X_n| \leq Z$ for all n, then X is summable,* $\lim I(X_n)$ *exists and*

$$I(X) = \lim_{n\to\infty} I(X_n).$$

Proof Let

$$P_n := \sup\{X_n, X_{n+1,\ldots}\} := \sup_{m\geq n}(X_n \vee X_{n+1} \vee \cdots \vee X_m) = \sup_{m\geq n} P_{n,m}.$$

For every $m \geq n$ we have that $P_{n,m}$ is summable by Theorem 6.2. Moreover, $P_{n,m} \uparrow_m P_n$, and since $|X_n| \leq Z$ for all n, $P_{n,m} \leq Z$ with Z summable. Therefore, $I(P_{n,m}) \leq I(Z)$ and so by Theorem 6.3, P_n is summable.

Now, $P_n \downarrow X$ since $X = \limsup X_n$ (by definition of the order limit). But $X \geq -Z$ and so we have $P_{n,m} \geq -Z$, i.e., $-P_{n,m} \leq Z$. So also $-P_n \leq -P_{n,m} \leq Z$. Since P_n is decreasing,

$$-P_n \leq -P_{n+1} \leq \cdots \uparrow -X$$

and again by Theorem 6.3, $-X$ is summable and $I(-X) = \sup I(-P_n)$. Therefore, by Theorem 6.2 X is summable and $I(X) = \inf I(P_n)$.

Let

$$Q_n := \inf\{X_n, X_{n+1,\ldots}\} := \inf_{m \geq n} (X_n \wedge X_{n+1} \wedge \cdots \wedge X_m) = \inf_{m \geq n} Q_{n,m}.$$

Then $Q_{n,m} \downarrow_m Q_n$ and as before we proof that Q_n is summable. Now $Q_n \uparrow X$ and $Q_n \leq Z$ with Z summable. Again X is summable and $I(Q_n) \uparrow I(X)$.

Since $Q_n := \inf_{m \geq n} X_m$, we have $Q_n \leq X_n$ and similarly, $P_n \geq X_n$. Hence, since I is positive, $I(Q_n) \leq I(X_n)$ and we get

$$I(X) = \lim I(Q_n) = \liminf I(Q_n) \leq \liminf I(X_n).$$

Also, $I(X_n) \leq I(P_n)$ and we get

$$I(X) = \lim I(P_n) = \limsup I(P_n) \geq \limsup I(X_n).$$

Stringing this together, we get

$$I(X) \leq \liminf I(X_n) \leq \limsup I(X_n) \leq I(X).$$

This gives equality and it shows that the sequence $(I(X_n))$ is order convergent with limit $I(X)$. □

7 *S*-Integrals

Following the exposition given by Daniell for real valued functions, one can define also a vector lattice-valued S integral on $\mathbb{L}$ as being a σ-order continuous order bounded linear operator from $\mathbb{L}$ into $\mathfrak{E}$. In the remainder of this section we consider such an S-integral, and denote it by I (so I will not necessarily be positive).

Since the range of the operator I is Dedekind complete, the operators I^+, I^- and $|I|$ exist and are also σ-order continuous. They are therefore I-integrals and for all $X \in \mathbb{L}$ we have

$$I = I^+ - I^-, \quad |I| = I^+ + I^-.$$

If we extend these positive integrals as in the preceding paragraph, we again get for the extended integrals that

$$I(X) = I^{+}(X) - I^{-}(X) \text{ and } |I|(X) = I^{+}(X) + I^{-}(X)$$

for all $X \in \mathbb{L}^{\uparrow}$.

Proposition 7.1 *For the upper extensions of* I^{+}, I^{-} *and* $|I|$, *which we denote by* $\dot{I}^{+}$, $\dot{I}^{-}$ *and* $|\dot{I}|$, *we have for all* $X \in \mathfrak{E}^{T}$ *that*

$$\dot{I}^{+}(X) + \dot{I}^{-}(X) = |\dot{I}|(X).$$

Proof The set of all $X_{\alpha} \geq X,\ X_{\alpha} \in \mathbb{L}^{\uparrow}$ is a downwards directed net and since $|I|,\ I^{+}$ and I^{-} are positive operators, the nets $(|I|(X_{\alpha})),\ (I^{+}(X_{\alpha}))$ and $(I^{-}(X_{\alpha}))$ are also downwards directed nets with infima (or limits) $(|\dot{I}|(X_{\alpha}),\ \dot{I}^{+}(X_{\alpha})$ and $\dot{I}^{-}(X_{\alpha})$ respectively. Hence,

$$\begin{aligned} |\dot{I}|(X) &= \inf |I|(X_{\alpha}) = \lim |I|(X_{\alpha}) \\ &= \lim I^{+}(X_{\alpha}) + \lim I^{-}(X_{\alpha}) = \dot{I}^{+}(X) + \dot{I}^{-}(X). \end{aligned}$$

□

Proposition 7.2 *If* X *is* $|I|$*-summable, then it is also* I^{+}*- and* I^{-}*-summable and*

$$|I|(X) = I^{+}(X) + I^{-}(X).$$

Proof For X $|I|$-summable, we have

$$\dot{I}^{+}(X) + \dot{I}^{-}(X) = |\dot{I}|(X) = |I|(X),$$

$$\dot{I}^{+}(-X) + \dot{I}^{-}(-X) = |\dot{I}|(-X) = -|I|(X).$$

So,

$$\underset{\cdot}{I}^{+}(X) + \underset{\cdot}{I}^{-}(X) = |\dot{I}|(X) = |I|(X).$$

Subtracting, we get

$$(\dot{I}^{+}(X) - \underset{\cdot}{I}^{+}(X)) + (\dot{I}^{-}(X) - \underset{\cdot}{I}^{-}(X)) = 0$$

and since both terms in the sum are positive, they are both zero. Therefore X is I^{+}-summable and I^{-}-summable and

$$|I|(X) = \dot{I}^{+}(X) + \dot{I}^{-}(X) = I^{+}(X) + I^{-}(X).$$

□

Definition 7.3 X is said to be *I-summable* if it is $|I|$-summable. If X is I-summable, we define

$$I(X) := I^+(X) - I^-(X).$$

If X is I-summable it is also I^+-summable and I^--summable by Proposition 7.2.

The theorems already obtained for the positive integral also holds for the S-integral I; one applies the theorem to I^+ and I^- in order to derive it for I. Thus, Corollary 5.5 becomes:

Proposition 7.4 *If X is I-summable, so is $|X|$ and*

$$|I(X)| \leq |I|(|X|).$$

Proof If X is I-summable, it is by definition $|I|$-summable and also I^+-summable and I^--summable. But by the result for positive integrals we have that $|X|$ is also $|I|$-summable, I^+-summable and I^--summable, and

$$|I^+(X)| \leq I^+(|X|), \quad |I^-(X)| \leq I^-(|X|).$$

It follows also that

$$|I(X)| \leq |I^+(X)| + |I^-(X)| \leq I^+(|X|) + I^-(|X|) = |I|(|X|).$$

□

The monotone convergence theorem, Theorem 6.3, becomes:

Theorem 7.5 *If $X_1 \leq X_2 \leq \cdots$ are I-summable and if* $\lim |I|(X_n)$ *exists, then* $\lim X_n = X$ *is I-summable and*

$$I(X) = \lim I(X_n).$$

Proof With these assumptions, X is $|I|$-summable by the result for positive integrals. But then, by definition X is I-summable and moreover

$$|I(X) - I(X_n)| = |I(X - X_n)| \leq |I|(|X - X_n|) = |I|(X - X_n) \to 0.$$

Hence,

$$I(X) = \lim I(X_n).$$

□

A very important result proved by Daniell is that the initial space $\mathbb{L}$ is, in a sense, dense in $\mathcal{L}$. This result is often used to define the Daniell extended integral. The convergence theorems are then easy to derive. It has the advantage of using only the upper integral. This is the approach chosen by P.E. Protter to define the integral with values in L^p, $0 \leq p \leq \infty$ and works it well if one has a metric on the range space. In our general case such a metric is not assumed and we prove the general result.

We recall, see [9, p. 84] that the Riesz space $\mathfrak{E}$ has the *diagonal property* (with respect to order convergence) whenever, given any double sequence $(X_{n,k})_{n,k\in\mathbb{N}}$ in $\mathfrak{E}$, any sequence (X_n) in $\mathfrak{E}$ and any X_0 in $\mathfrak{E}$ such that $X_{n,k} \to X_n$ for all n as $k \to \infty$, and $X_n \to X_0$, there exists for every n and appropriate $k(n)$ such that $X_{n,k(n)} \to X_0$ as $n \to \infty$.

Theorem 7.6 *A necessary and sufficient condition for an element $X \in \mathfrak{E}^T$ to be I-summable is that there exists a net (X_α) of elements in $\mathbb{L}$ such that* $\lim_\alpha |\dot{I}|(|X - X_\alpha|) = 0$. *Also in this case*

$$I(X) = \lim I(X_\alpha).$$

The convergence is $|\sigma|(\mathfrak{E}, \mathfrak{E}_{00}^{\sim})$-convergence. The condition is sufficient for order convergence and necessary for order convergence if $\mathfrak{E}$ is super-Dedekind complete and has the diagonal property.

Proof *The condition is sufficient for order convergence:* Suppose that X satisfies the condition and let $X_\alpha \in \mathbb{L}$ be such that $|\dot{I}|(|X - X_\alpha|) \to 0$ in order. Then

$$X = X - X_\alpha + X_\alpha \leq X_\alpha + |X - X_\alpha|.$$

Hence,

$$|\dot{I}|(X) \leq |I|(X_\alpha) + |\dot{I}|(|X - X_\alpha|).$$

Substitute X by $-X$ in this to get

$$|\dot{I}|(-X) \leq |I|(-X_\alpha) + |\dot{I}|(|X - X_\alpha|),$$

i.e.,

$$-|\dot{I}|(-X) \geq -|I|(-X_\alpha) - |\dot{I}|(|X - X_\alpha|),$$

i.e.,

$$|\underset{.}{I}|(X) \geq |I|(X_\alpha) - |\dot{I}|(|X - X_\alpha|).$$

This gives us

$$|\dot{I}|(X) - |\underset{.}{I}|(X) \leq 2|\dot{I}|(|X - X_\alpha|) \to 0. \tag{7.1}$$

It follows that $|\dot{I}|(X) = |\underset{.}{I}|(X)$ and so X is $|I|$-summable and by definition I-summable. Moreover,

$$|I(X) - I(X_\alpha)| = |I(X - X_\alpha)| \leq |I|(|X - X_\alpha|) \to 0. \tag{7.2}$$

The condition is sufficient for $|\sigma|(\mathfrak{E}, \mathfrak{E}_{00}^{\sim})$ *convergence:* Suppose that there exists a net $X_\alpha \in \mathbb{L}$ such that $|\dot{I}|(|X - X_\alpha|) \to 0$ in $|\sigma|(\mathfrak{E}, \mathfrak{E}_{00}^{\sim})$. Then, from (7.1) we get for any $\phi \in \mathfrak{E}_{00}^{\sim}$, that

$$|\phi|(|\dot{I}|(X) - |\underset{.}{I}|(X)) \leq 2|\phi|(|\dot{I}|(|X - X_\alpha|)) \to 0. \tag{7.3}$$

It follows that $|\phi|(|\dot{I}|(X)) = |\phi|(|\underset{.}{I}|(X))$ for all ϕ. Therefore, $|\dot{I}|(X) = |\underset{.}{I}|(X)$ and so X is $|I|$-summable.

From (7.2) we get, since $\phi \in \mathfrak{E}_{00}^{\sim}$, that

$$|\phi|(|I(X) - I(X_\alpha)|) = |\phi|(|I(X - X_\alpha)|) \leq |\phi|(|I|(|X - X_\alpha|)) \to 0, \tag{7.4}$$

showing that, in the topology $|\sigma|(\mathfrak{E}, \mathfrak{E}_{00}^{\sim})$,

$$I(X) = \lim I(X_\alpha).$$

The condition is necessary for $|\sigma|(\mathfrak{E}, \mathfrak{E}_{00}^{\sim})$ *convergence:* Let X be summable and let (X_α) be the decreasing net of all $X_\alpha \in \mathbb{L}^\uparrow$ satisfying $X_\alpha \geq X$. We then know that $|I|(X_\alpha) \downarrow |I|(X) = |\dot{I}|(X)$. Hence,

$$|\dot{I}|(X_\alpha - X) = |\dot{I}|(|X_\alpha - X|) \downarrow 0.$$

Let $\epsilon > 0$ and $0 \leq \phi \in \mathfrak{E}_{00}^{\sim}$ be fixed. Then there exists some α_0 such that

$$\phi|\dot{I}|(|X_{\alpha_0} - X|) < \epsilon/2.$$

But $X_{\alpha_0} \in \mathbb{L}^\uparrow$ and so there exists an element $X_{\alpha_0,n} \in \mathbb{L}$ such that $\phi|I|(X_{\alpha_0} - X_{\alpha_0,n}) < \epsilon/2$. Hence,

$$\phi|I|(|X - X_{\alpha_0,n}|) \leq \phi|I|(|X - X_{\alpha_0}|) + \phi|I|(|X_{\alpha_0} - X_{\alpha_0,n}|) < \epsilon.$$

It follows that every neighbourhood of $|I|(X)$ in the topology $|\sigma|(\mathfrak{E}, \mathfrak{E}_{00}^{\sim})$ contains an element of $|I|(\mathbb{L})$.

We have shown that $\mathbb{L}$ is dense in the function space with topology generated by the semi-norms $p(X) := \phi(|I|(|X|))$.

To prove the necessity with order convergence replacing topological convergence, one has to prove that the following holds:

$$|I|(X_{\alpha,n}) \uparrow_n |I|(X_\alpha) \downarrow_\alpha |I|(X) \quad \text{i.e.,} \quad |I|(X_{\alpha,n}) \xrightarrow{o} |I|(X_\alpha) \xrightarrow{o} |I|(X),$$

implies that there exists for each α some $\alpha(n)$ such that $|I|(|X_{\alpha,\alpha(n)}|) \xrightarrow{o} |I|(X)$.

If the space $\mathfrak{E}$ is super Dedekind complete, we can replace the net (X_α) by a sequence (X_n). Then the condition becomes:

$$|I|(X_{m,n}) \xrightarrow{o} |I|(X_m) \xrightarrow{o} |I|(X)$$

implies that for each m there exists an $m(n)$ such that $|I|(X_{m,m(n)}) \xrightarrow{o} |I|(X)$.

The latter condition holds if and only if the Riesz space $\mathfrak{E}$ has the diagonal property for order convergence. □

It is important to note that the spaces L^p, $0 \leq p \leq \infty$, are super-Dedekind complete and have the diagonal property (see [9, Exercise 71.9]) and so our integral generalizes the Daniell integral defined by P.E. Protter.

8 Conclusion

We now have a very general vector lattice-valued integral for which Lebesgue's theorem holds. It is ideally suited to apply to the abstract theory of stochastic processes in vector lattices. In this theory we often have an order continuous σ-additive vector measure and one can define a primitive Daniell integral on the set of simple processes using this measure. The stochastic integral is then obtained as the extension of this integral. A case in point is when the measure is obtained as the spectral measure of a stopping time. The Daniell integral then yields a stopped process. This is the topic of a forthcoming paper.

Acknowledgments Financial support of the NRF is gratefully acknowledged (Grant No CSUR 160429163509). This paper is a version of a plenary talk delivered at Pretoria at the conference *Positivity X, 8–12 July 2019.*

References

1. C.D. Aliprantis, O. Burkinshaw, *Positive Operators* (Academic Press, New York, 1985)
2. P.J. Daniell, A general form of integral, in *Annals of Mathematics, Second Series*, vol. 19(4) (1918), pp. 279–294. https://www.jstor.org/stable/1967495
3. P.J. Daniell, Further properties of the general integral, in *Annals of Mathematics, Second Series*, vol. 21(3) (1920), pp. 203–220. https://www.jstor.org/stable/2007258
4. K. Donner, *Extension of Positive Operators and Korovkin Theorems*. Lecture Notes in Mathematics, vol. 904 (Springer, Berlin, 1982)
5. J.J. Grobler, C.C.A. Labuschagne, The Itô integral for Brownian motion in vector lattices: Part 1. J. Math. Anal. Appl. **423**. 797–819 (2015). https://doi.org/10.1016/j.jmaa.2014.08.013
6. J.J. Grobler, C.C.A. Labuschagne, The Itô integral for Brownian motion in vector lattices: Part 2. J. Math. Anal. Appl. **423**, 820–833 (2015). https://doi.org/10.1016/j.jmaa.2014.09.063

7. J.J. Grobler, C.C.A. Labuschagne, The Itô integral for martingales in vector lattices. J. Math. Anal. Appl. **450**, 1245–1274 (2017). https://doi.org/10.1016/j.jmaa.2017.01.081
8. W.A.J. Luxemburg, Ordered algebraic structures in Analysis, in *Ordered Algebraic Structures*, ed. by J. Martinez. Mathematics and Its Applications, vol. 55 (Springer, Dordrecht, 1989)
9. W.A.J. Luxemburg, A.C. Zaanen, *Riesz Spaces 1* (North-Holland Publishing, Amsterdam, 1971)
10. P.E. Protter, *Stochastic Integration and Differential Equations* (Springer, Berlin, 1990)
11. F. Riesz, Sur la décomposition des opérations fonctionnelles linéaires. Atti Congr. Internaz. Mat. Bologna **1928**(3), 143–148 (1928)
12. F. Riesz, Sur quelques notions fondamentals dans la theorie générale des opérations linéaires. Ann. Math. **41**, 174–206 (1940)
13. F. Riesz, B. Sz.-Nagy, *Functional Analysis*. Translated from *Lecons D'analyse Fonctionelle* (Blackie and Son Limited, London, 1955)
14. H.H. Schaefer, *Banach Lattices and Positive Operators* (Springer, Berlin, 1974)
15. A.C.M. van Rooij, W.B. van Zuijlen, Integrals for functions with values in a partially ordered vector space. Positivity **20**, 877–916 (2016)
16. W.H. Young, On integration with respect to a function of bounded variation. Proc. London Math. Soc. **13**, 109–150 (1914)
17. A.C. Zaanen, *Integration* (North-Holland, Amsterdam, 1967)
18. A.C. Zaanen, *Riesz spaces II* (North-Holland, Amsterdam, 1982)

Ergodicity in Riesz Spaces

Jonathan Homann, Wen-Chi Kuo, and Bruce A. Watson

Dedicated to the memory of our friend, collaborator and colleague Professor Coenraad Labuschagne, with whom our work on stochastic processes in Riesz spaces began.

Abstract The ergodic theorems of Hopf, Wiener and Birkhoff were extended to the context of Riesz spaces with a weak order unit and conditional expectation operator by Kuo, Labuschagne and Watson in [*Ergodic Theory and the Strong Law of Large Numbers on Riesz Spaces.* J Math Anal Appl **325** (2007), 422–437]. However, the precise concept of what constitutes ergodicity in Riesz spaces was not considered. In this short paper we fill in this omission and give some explanations of the choices made. In addition, we consider the interplay between mixing and ergodicity in the Riesz space setting.

Keywords Ergodicity · Riesz spaces · Weak mixing · Conditional expectation operators

Mathematics Subject Classification (2010) Primary 46A40; 47A35; Secondary 37A25; 60F05

1 Introduction

Ergodic theory seeks to study the long-term behaviour of a dynamical system. One typically works in a probability space, $(\Omega, \mathcal{A}, \mu)$, with a transformation $\tau : \Omega \to \Omega$. Here the quadruple $(\Omega, \mathcal{A}, \mu, \tau)$ is called a measure preserving system

This work was completed with the support of our TEX-pert.

J. Homann · W.-C. Kuo · B. A. Watson (✉)
School of Mathematics, University of the Witwatersrand, Johannesburg, South Africa
e-mail: wen.kuo@wits.ac.za

E. Kikianty et al. (eds.), *Positivity and its Applications*, Trends in Mathematics,
https://doi.org/10.1007/978-3-030-70974-7_9

if $\mu\left(\tau^{-1}(A)\right) = \mu(A)$ for each $A \in \mathcal{A}$. A set $A \in \mathcal{A}$ is called τ-invariant if $\tau^{-1}(A) \subset A$ and the measure preserving system $(\Omega, \mathcal{A}, \mu, \tau)$ is said to be ergodic if $\mu(A) \in \{0, 1\}$ for each τ-invariant A, see Petersen [13]. An equivalent definition of ergodicity is that $f \circ \tau = f$ for measurable f only if f is a.e. constant, see [14, page 15]. It is this definition that leads naturally to a conditional version and, from there, to a Riesz space version.

In [8], Kuo, Labuschagne, and Watson generalised the ergodic theorems of Birkhoff, Wiener, Hopf, and Garsia to the setting of Riesz spaces. They also generalised the Kolmogorov 0-1 law to Riesz spaces and were thus able to give a strong law of large numbers for conditionally independent sequences in Riesz spaces. However, they omitted to consider what constituted ergodicity in a Riesz space. In this paper we fill this gap and, using the works of Kuo, Labuschagne and Watson in [6, 8], are able to relate ergodicity to weak mixing in the Riesz space setting. Further to this, we apply these concepts to sequences with conditionally independent shifts and show the ergodicity of such processes.

This work continues the studies of mixing processes [6, 10], and of stochastic processes [2, 4, 7, 8, 11, 15], in the Riesz space setting.

2 Preliminaries

The reader is referred to Aliprantis and Border [1, pages 263–300], Fremlin [3, Chapter 35, pages 219–274], Meyer-Nieberg [12], and Zaanen [19] for background in Riesz spaces.

We recall from [7] the definition of a conditional expectation operator on a Riesz space.

Definition 2.1 Let E be a Riesz space with weak order unit. A positive order continuous projection $T: E \to E$, with range, $R(T)$, a Dedekind complete Riesz subspace of E, is called a conditional expectation operator on E if Te is a weak order unit of E for each weak order unit e of E.

The Riesz space analogue of a measure preserving system is introduced in the following definition.

Definition 2.2 If E is a Dedekind complete Riesz space with weak order unit, say, e, T is a conditional expectation operator on E with $Te = e$ and S is an order continuous Riesz homomorphism on E with $Se = e$ and, further, $TSPe = TPe$, for all band projections P on E, then (E, T, S, e), is called a conditional expectation preserving system.

Due to Freudenthal's Spectral Theorem [19, Theorem 33.2], the condition $TSPe = TPe$ for each band projection P on E in the above definition is equivalent to $TSf = Tf$ for all $f \in E$.

Let (E, T, S, e) be a conditional expectation preserving system. For $f \in E$ and $n \in \mathbb{N}$ we denote

$$S_n f := \frac{1}{n}\sum_{k=0}^{n-1} S^k f, \tag{2.1a}$$

$$L_S f := \lim_{n\to\infty} S_n f, \tag{2.1b}$$

where the above limit is the order limit, if it exits. We say that $f \in E$ is S-invariant if $Sf = f$. The set of all S-invariant $f \in E$ will be denoted $\mathcal{I}_S := \{f \in E \mid Sf = f\}$. The set of $f \in E$ for which $L_S f$ exists will be denoted $\mathcal{E}_S$ and thus $L_S : \mathcal{E}_S \to E$.

Lemma 2.3 *Let (E, T, S, e) be a conditional expectation preserving system and define $\mathcal{I}_S$ and $\mathcal{E}_S$ as before, then $\mathcal{I}_S \subset \mathcal{E}_S$ and $L_S f = f$, for all $f \in \mathcal{I}_S$, so $\mathcal{I}_S \subset R(L_S)$.*

Proof If $f \in \mathcal{I}_S$, then $S^k f = f$, for all $k \in \mathbb{N}_0$. Thus $S_n f = \frac{1}{n}\sum_{k=0}^{n-1} f = f$, for all $n \in \mathbb{N}$, giving $S_n f \to f$, in order, as $n \to \infty$. Hence $f \in \mathcal{E}_S$ and $L_S f = f$, thus $\mathcal{I}_S \subset \mathcal{E}_S$. □

We recall Birkhoff's bounded ergodic theorem from [8, Theorem 3.7].

Theorem 2.4 (Birkhoff's (Bounded) Ergodic Theorem) *Let (E, T, S, e) be a conditional expectation preserving system. For $f \in E$, the sequence $(S_n f)_{n\in\mathbb{N}}$ is order bounded in E if and only if $f \in \mathcal{E}_S$. For each $f \in \mathcal{E}_S$ we have $L_S f = SL_S f$ and $TL_S f = Tf$. If $E = \mathcal{E}_S$ then L_S is a conditional expectation operator on E with $L_S e = e$.*

Note 2.5 *If we restict our attention to E_e, then $(S_n f)_{n\in\mathbb{N}}$ is order bounded in E_e for each $f \in E_e$, so $E_e \subset \mathcal{E}_S$. Furthermore, $L_S f \in E_e$, giving that $L_S|_{E_e}$ is a conditional expectation on E_e.*

E is said to be universally complete with respect to T (T-universally complete), if, for each increasing net (f_α) in E_+ with (Tf_α) order bounded in E_u, we have that (f_α) is order convergent in E.

We recall Birkhoff's ergodic theorem for a T-universally complete Riesz space from [8, Theorem 3.9].

Theorem 2.6 (Birkhoff's (Complete) Ergodic Theorem) *Let (E, T, S, e) be a conditional expectation preserving system with E T-universally complete then $E = \mathcal{E}_S$ and hence $L_S = SL_S$. In addition, $TL_S = T$ and L_S is a conditional expectation operator on E.*

From [6, Lemma 2.1] we have.

Lemma 2.7 *Let $(f_n)_{n\in\mathbb{N}_0}$ be a sequence in E, a Dedekind complete Riesz space. If* $\sum_{k=0}^{\infty}|f_k|$ *is order convergent in E then* $\sum_{k=0}^{\infty} f_k$ *is order convergent in E.*

From the above lemma and [11, Lemma 2.1] we have the following.

Theorem 2.8 *Let E be a Dedekind complete Riesz space and $(f_n)_{n\in\mathbb{N}_0}$ a sequence in E with $f_n \to 0$, in order, as $n \to \infty$, then $\frac{1}{n}\sum_{k=0}^{n-1}|f_k| \to 0$, in order, as $n \to \infty$.*

Corollary 2.9 *Let E be a Dedekind complete Riesz space and $(f_n)_{n\in\mathbb{N}_0}$ a sequence in E with order limit f, then $\frac{1}{n}\sum_{k=0}^{n-1} f_k \to f$, in order, as $n \to \infty$.*

3 Ergodicity on Riesz Spaces

Consider the Riesz space $E = L^1(\Omega, \mathcal{A}, \mu)$, where μ is a probability measure and T is the expectation operator $T = \mathbb{E}[\cdot]\mathbf{1}$, where $\mathbf{1}$ is the equivalence class, in $L^1(\Omega, \mathcal{A}, \mu)$, of function with value 1 almost everywhere. For τ a measure preserving transformation on Ω setting $Sf = f \circ \tau$ we have that $(E, T, S, \mathbf{1})$ is a conditional expectation preserving system. The definition of ergodicity as $f \circ \tau = f$ for measurable f only if f is almost everywhere constant, can now be written as $Sf = f$ only if $f \in R(T)$. This leads naturally to the following definition in Riesz spaces, for conditional expectation preserving systems.

Definition 3.1 (Ergodicity) The conditional expectation preserving system (E, T, S, e) is said to be ergodic if $L_S f \in R(T)$ for all $f \in \mathcal{I}_S$.

As will be seen later, this definition preserves the original philosophy of an ergodic process as one in which the time mean (the Cesàro mean) is equal to the spatial mean (conditional expectation) in the limit.

We note that E_e is a $R(T)$-module, see [10], an hence, when dealing with conditioned systems, the role of the scalars ($\mathbb{R}$ or $\{ke \mid k \in \mathbb{R}\}$) is replaced by the ring $R(T)$.

Theorem 3.2 *The conditional expectation preserving system (E, T, S, e) is ergodic if and only $L_S f = Tf$ for all $f \in \mathcal{I}_S$.*

Proof Let (E, T, S, e) be ergodic and $f \in \mathcal{I}_S$, then, by Theorem 2.4, $L_S f$ exists and $TL_S f = Tf$. As (E, T, S, e) is ergodic $L_S f \in R(T)$, so there exists $g \in E$ such that $L_S f = Tg$. As T is a projection $TTg = Tg$, so $Tf = TL_S f = TTg = Tg = L_S f$.

Conversely, if $f \in \mathcal{I}_S$ then, as $\mathcal{I}_S \subset \mathcal{E}_S$, $L_S f$ exists and if we assume $L_S f = Tf$ then $L_S f \in R(T)$. Thus (E, T, S, e) is ergodic. □

For (E, T, S, e) a conditional expectation preserving system we denote by $\mathcal{B}$ the band projections on E, by $\mathcal{P} := \{P \in \mathcal{B} \mid TPe = PTe = Pe\}$ the band projections on E which commute with T and by $\mathcal{A} := \{P \in \mathcal{B} \mid SPe = Pe\}$ the set of S invariant band projections on E. Since S is a positive operator with $Se = e$ and P is positive and below the identity, we have that $SPe, Pe \in E_e$, for all $P \in \mathcal{B}$. Hence, L_S can always be applied to SPe and Pe. We now show that $\mathcal{A}$ characterises L_S.

Theorem 3.3 *For (E, T, S, e) a conditional expectation preserving system we have that $\mathcal{P} \subset \mathcal{A} = \mathcal{F}$ where $\mathcal{F} := \{P \in \mathcal{B} \mid Pe = L_S Pe\}$ is the set of band projections on E which commute with L_S.*

Proof For the purpose of this theorem we can assume that $E = E_e$ as the sets $\mathcal{A}, \mathcal{B}, \mathcal{P}$ do not change when E is replaced by E_e. Further, by Theorem 2.4, L_S restricted to E_e is everywhere defined and a conditional expectation on E_e with $SL_S = L_S$.

If $P \in \mathcal{A}$ then $Pe \in \mathcal{I}_S$, so, by Lemma 2.3, $L_S Pe = Pe$, hence $P \in \mathcal{F}$. Conversely, if $P \in \mathcal{F}$ then $L_S Pe = PL_S e = Pe$, so $SPe = SL_S Pe = L_S Pe = Pe$, giving $P \in \mathcal{A}$. Hence, $\mathcal{A} = \mathcal{F}$.

From Theorem 2.4, $L_S T = T = TL_S$ giving that if $P \in \mathcal{P}$ then $TPe = Pe$ so $L_S Pe = L_S TPe = TPe = Pe$ and $P \in \mathcal{F}$. □

The following corollary to Theorem 3.3 is critical when applying the concepts of ergodicity in Riesz spaces, further to this, it shows that ergodicity is equivalent to the time and spatial means coinciding in the limit.

Corollary 3.4 *The conditional expectation preserving system (E, T, S, e) is ergodic if and only if $T = L_S$, where $E = \mathcal{E}_S$.*

Proof Theorem 2.4 gives that $SL_S = L_S$, hence $R(L_S) \subset \mathcal{I}_S$. However, by Lemma 2.3 $\mathcal{I}_S \subset R(L_S)$. Hence $\mathcal{I}_S = R(L_S)$.

If (E, T, S, e) is ergodic, then $Tf = L_S f = f$, for all $f \in \mathcal{I}_S$, so $f \in R(T)$, giving $R(L_S) = \mathcal{I}_S \subset R(T)$. Hence, $R(L_S) = R(T)$, but $L_S T = T = TL_S$ so, from [18], $L_S = T$.

Conversely, suppose that $T = L_S$, then by Theorem 3.2, (E, T, S, e) is ergodic. □

An operator, say A, on a Riesz space is said to be strictly positive if A is a positive operator ($Af \geq 0$ for each $f \geq 0$) and $Af = 0$, $f \geq 0$, if and only if $f = 0$. We recall that $E_e = \{f \in E \mid |f| \leq ke \text{ for some } k \in \mathbb{R}_+\}$, the subspace of E of e bounded elements of E, is an f-algebra, see [2, 17, 20]. The f-algebra structure on E_e is generated by setting $Pe \cdot Qe = PQe$ for all band projections P and Q on E (here, $\cdot$ represents the f-algebra multiplication on E_e). The linear extension of this multiplication and use of order limits extends this multiplication to E_e. We refer the reader to Azouzi and Trabelsi [2], Venter and van Eldik [17] and Zaanen [20, Chapter 20] for further background on f-algebras.

If T is a conditional expectation operator on E with $Te = e$, then T is also a conditional expectation operator on E_e, since, if $f \in E_e$, then $|f| \leq ke$, for some $k \in \mathbb{R}^+$, giving $|Tf| \leq T|f| \leq Tke = ke$. Further T is an averaging operator on E_e, that is

$$T(f \cdot g) = f \cdot Tg$$

for $f, g \in E_e$ with $f \in R(T)$, see [7].

The following theorem will be seen to be fundamental in linking the concepts of ergodicity and conditional weak mixing in Riesz spaces.

Theorem 3.5 *A conditional expectation preserving system (E, T, S, e), with T being strictly positive, is ergodic if and only if*

$$\frac{1}{n}\sum_{k=0}^{n-1} T\left(S^k f \cdot g\right) \to Tf \cdot Tg, \tag{3.1}$$

in order, as $n \to \infty$, for $f, g \in E_e$.

Proof For $f, g \in E_e$, we note that

$$\frac{1}{n}\sum_{k=0}^{n-1} T\left(\left(S^k f\right) \cdot g\right) = T\left(\left(\frac{1}{n}\sum_{k=0}^{n-1} S^k f\right) \cdot g\right) = T\left(S_n f \cdot g\right),$$

and in E_e, $S_n f \to L_S f$, in order, as $n \to \infty$. Thus (3.1) is equivalent to

$$T(L_S f \cdot g) = Tf \cdot Tg. \tag{3.2}$$

Suppose that (E, T, S, e) is ergodic. As $L_S f \in R(T)$ and T is an averaging operator $T(L_S f \cdot g) = L_S f \cdot Tg$. Again from the ergodicity of (E, T, S, e) we have $L_S f = Tf$. Thus (3.2) holds.

Conversely, suppose that (3.1) or equivalently (3.2) holds. As T is an averaging operator, $Tf \cdot Tg = T(Tf \cdot g)$, which combined with (3.2) gives

$$T((L_S f - Tf) \cdot g) = 0. \tag{3.3}$$

Taking $g = P_\pm e$ where $P_\pm$ are the band projections onto the bands generated by $(L_S f - Tf)^\pm$ in (3.4) gives

$$0 = T((L_S f - Tf) \cdot P_\pm e) = T(P_\pm(L_S f - Tf)) = T((L_S f - Tf)^\pm). \tag{3.4}$$

The strict positivity of T now gives that $(L_S f - Tf)^\pm = 0$, hence $L_S f = Tf$. □

4 Application to Conditional Weak Mixing

We recall from [6] the definition of conditional weak mixing in Riesz spaces. Various other types of mixing in Riesz spaces were studied in [10].

Definition 4.1 (Weak Mixing) The conditional expectation preserving system (E, T, S, e) is said to be weakly mixing if, for all band projections P and Q on E,

$$\frac{1}{n}\sum_{k=0}^{n-1}\left|T\left(\left(S^k Pe\right)\cdot Qe\right) - TPe \cdot TQe\right| \to 0, \tag{4.1}$$

in order, as $n \to \infty$.

From [6] we have (4.1) is equivalent to

$$\frac{1}{n}\sum_{k=0}^{n-1}\left|T\left(\left(S^k f\right)\cdot g\right) - Tf \cdot Tg\right| \to 0 \tag{4.2}$$

for all $f, g \in E_e$, in order as $n \to \infty$.

Combining Theorem 3.5 with Lemma 2.7 and (4.2), we have the following.

Corollary 4.2 *If the conditional expectation preserving system (E, T, S, e), with T strictly positive, is conditionally weak mixing then it is ergodic.*

Proof If (E, T, S, e) is conditionally weak mixing, then by (4.2),

$$\frac{1}{n}\sum_{k=0}^{n-1}\left|T\left(\left(S^k f\right)\cdot g\right) - Tf \cdot Tg\right| \to 0,$$

in order, as $n \to \infty$, for all $f, g \in E_e$, which, by Lemma 2.7 gives

$$\frac{1}{n}\sum_{k=0}^{n-1}T\left(\left(S^k f\right)\cdot g\right) - Tf \cdot Tg = \frac{1}{n}\sum_{k=0}^{n-1}\left(T\left(\left(S^k f\right)\cdot g\right) - Tf \cdot Tg\right) \to 0,$$

in order, as $n \to \infty$, for all $f, g \in E_e$. That is

$$\frac{1}{n}\sum_{k=0}^{n-1}T\left(\left(S^k f\right)\cdot g\right) \to Tf \cdot Tg,$$

in order, as $n \to \infty$, for all $f, g \in E_e$, giving that (E, T, S, e) is ergodic, by Theorem 3.5. □

5 Application to Processes with Conditionally Independent Shifts

We refer the reader to Vardy and Watson [16], Kuo, Vardy and Watson [9], and Kuo, Labuschagne and Watson [8], for background on T-conditional independence. A summary of T-conditional independence, based on [9], follows.

Let E be a Dedekind complete Riesz space with weak order unit, say, e, and T be a conditional expectation operator on E with $Te = e$. Band projections P and Q on E are said to be T-conditionally independent (independent with respect to T), if

$$TPTQe = TPQe = TQTPe. \tag{5.1}$$

We note, from [5, Lemma 4.2], that band projections P and Q on E are T-conditionally independent if and only if $TPTQe = TPQe$ (or equivalently $TPQe = TQTPe$), i.e. assuming both equalities in (5.1) is unnecessary. In the f-algebra setting of E_e, using the averaging property of the conditional expectation operator T, the independence condition (5.1) can equivalently be written as $TPe \cdot TQe = TPQe$ from which it is apparent that weak mixing condition (4.1) is an asymptotic (in k) form of T-conditional independence of S^kPe and Qe.

We say that Riesz subspaces E_1 and E_2 of E are T-conditionally independent if all band projections P and Q are T-conditionally independent where $Pe \in E_1$ and $Qe \in E_2$. For $(E_\lambda)_{\lambda\in\Lambda}$ a family of order closed, Dedekind complete Riesz subspaces of E with $R(T) \subset E_\lambda$, for each $\lambda \in \Lambda$, we say that the family is T-conditionally independent if, for each pair of disjoint non-empty index sets $\Lambda_1, \Lambda_2 \subset \Lambda$, we have that $\left\langle\bigcup_{\lambda\in\Lambda_1} E_\lambda\right\rangle$ and $\left\langle\bigcup_{\lambda\in\Lambda_2} E_\lambda\right\rangle$ are T-conditionally independent. A sequence $(f_n)_{n\in\mathbb{N}} \subset E$ is said to be T-conditionally independent if the family of closed, Dedekind complete Riesz subspaces $\langle\{f_n\} \cup R(T)\rangle$, for each $n \in \mathbb{N}$, is T-conditionally independent. Here, for F an non-empty subset of E, $\langle F\rangle$ denotes the closed (under order limits in E) Riesz subspace of E generated by F.

We recall the Strong Law of Large Numbers from [8, Theorem 4.8].

Theorem 5.1 (The Strong Law of Large Numbers) *If (E, T, S, e) is a conditional expectation preserving system with $\mathcal{E}_S = E$ and the sequence $\left(S^j f\right)_{j\in\mathbb{N}}$ is independent with respect to T, for each $f \in E$, then $T = L_S$.*

Restricting our attention to E_e, with $(S_n f)_{n\in\mathbb{N}}$ independent with respect to T, for each $f \in E_e$, we have, from Theorem 5.1, that $T|_{E_e} = L_S|_{E_e}$. Hence, from Theorems 3.2 and 5.1, we have the following corollary.

Corollary 5.2 *Let (E, T, S, e) be a conditional expectation preserving system with $\mathcal{E}_S = E$ and the sequence $\left(S^j f\right)_{j\in\mathbb{N}}$ T-conditionally independent, for each $f \in E$, then (E, T, S, e) is ergodic.*

Acknowledgments Many thanks to the referees for their suggestions and corrections.

B.A. Watson was supported in part by the Centre for Applicable Analysis and Number Theory and by National Research Foundation of South Africa grant IFR170214222646 with grant no. 109289.

References

1. C.D. Aliprantis, K.C. Border, *Infinite Dimensional Analysis: A Hitchhiker's Guide*, 2nd edn. (Springer, Berlin, 1999)
2. Y. Azouzi, M. Trabelsi, L^p Spaces with respect to conditional expectation on riesz spaces. J. Math. Anal. Appl. **447**, 798–816 (2017)
3. D.H. Fremlin, *Measure Theory*, vol. 3 (Torres Fremlin, 2000)
4. J.J. Grobler, Jensen's and martingale inequalities in Riesz spaces. Indagationes Math. **25**, 275–295 (2014)
5. J.J. Grobler, The Kolmogorov-Čentsov theorem and Brownian motion in vector lattices. J. Math. Anal. Appl. **410**, 891–901 (2014)
6. W.-C. Kuo, J.M. Homann, B.A. Watson, A Koopman-von Neumann type Theorem on the Convergence of Cesàro Means in Riesz Spaces. Proc. Amer. Math. Soc. Series B **8**, 75–85 (2021)
7. W.-C. Kuo, C.C.A. Labuschagne, B.A. Watson, Conditional expectations on Riesz spaces. J. Math. Anal. Appl. **303**, 509–521 (2005)
8. W.-C. Kuo, C.C.A. Labuschagne, B.A. Watson, Ergodic theory and the strong law of large numbers on Riesz spaces. J. Math. Anal. Appl. **325**, 422–437 (2007)
9. W.-C. Kuo, J.J. Vardy, B.A. Watson, Bernoulli processes in Riesz spaces. In *Ordered Structures and Applications*, pp. 263–274 (2016)
10. W.-C. Kuo, M.J. Rogans, B.A. Watson, Mixing inequalities in Riesz spaces. J. Math. Anal. Appl. **456**, 992–1004 (2017)
11. W.-C. Kuo, D.F. Rodda, B.A. Watson, Hájek-Rényi-chow maximal inequality and a strong law of large numbers in Riesz spaces. J. Math. Anal. Appl. **481**, 123462 (2020)
12. P. Meyer-Nieberg, *Banach Lattices*. (Springer, Berlin, 1991)
13. K. Petersen, *Ergodic Theory*. (Cambridge University, Cambridge, 1983)
14. P.C. Shields, *Ergodic Theory of Discrete Sample Paths*. (American Mathematical Society, Providence, 1996)
15. G. Stoica, Limit laws for martingales in vector lattices. J. Math. Anal. Appl. **476**, 715–719 (2019)
16. J.J. Vardy, B.A. Watson, Markov processes on Riesz spaces. Positivity **16**, 373–391 (2012)
17. L.M. Venter, P. van Eldik, Universally Complete Riesz Spaces and f-algebras. South African J. Sci. **84**, 343–346 (1987)
18. B.A. Watson, An Andô-Douglas type theorem in Riesz spaces with a conditional expectation. Positivity **13**, 543–558 (2009)
19. A.C. Zaanen, *Introduction to Operator Theory in Riesz Spaces*. (Springer, Berlin, 1997)
20. A.C. Zaanen, *Riesz Spaces II* (North-Holland, Amsterdam, 1983)

Multiplicative Representation of Real-Valued bi-Riesz Homomorphisms on Partially Ordered Vector Spaces

Anke Kalauch and Onno van Gaans

Dedicated to Coenraad C.A. Labuschagne, with fond memories of his mathematics and his good company.

Abstract We show that a bi-Riesz homomorphism from a product of two partially ordered vector spaces into the reals is the product of two Riesz homomorphisms. Further, we investigate kernels of differences of Riesz* homomorphisms.

Keywords Bi-Riesz homomorphism · Partially ordered vector space · pre-Riesz space · Riesz completion · Riesz homomorphism · Riesz* homomorphism

Mathematics Subject Classification (2010) 06F20, 47B60, 47B65

1 Introduction

For Riesz homomorphisms in Riesz spaces, the following interesting results are stated in the book of Kusraev and Kutateladze [9, 3.12.A.2. and 3.5.4.].

Theorem 1 *If E_1, E_2 are Riesz spaces and $b\colon E_1 \times E_2 \to \mathbb{R}$ is a bi-Riesz homomorphism, then there are Riesz homomorphisms $\sigma : E_1 \to \mathbb{R}$ and $\tau : E_2 \to \mathbb{R}$ such that for every (u, v) in $E_1 \times E_2$ we have $b(u, v) = \sigma(u)\tau(v)$.*

Here, b is called a *bi-Riesz homomorphism* if b is bilinear and for every $x_1 \in E_1^+$ and $x_2 \in E_2^+$ the maps $E_2 \to \mathbb{R}$, $v \mapsto b(x_1, v)$ and $E_1 \to \mathbb{R}$, $u \mapsto b(u, x_2)$ are Riesz homomorphisms.

A. Kalauch (✉) · O. van Gaans
Fakultät Mathematik, Institut für Analysis, Dresden, Germany
e-mail: Anke.Kalauch@tu-dresden.de

O. van Gaans
Mathematical Institute, Leiden University, Leiden, The Netherlands
e-mail: vangaans@math.leidenuniv.nl

E. Kikianty et al. (eds.), *Positivity and its Applications*, Trends in Mathematics,
https://doi.org/10.1007/978-3-030-70974-7_10

Theorem 2 *Let E be a Riesz space and $\varphi\colon E \to \mathbb{R}$ an order bounded linear functional. Then the following are equivalent.*

(i) *There are Riesz homomorphisms $\varphi_1, \varphi_2\colon E \to \mathbb{R}$ such that $\varphi = \varphi_1 - \varphi_2$.*
(ii) *The null space* $\ker\varphi$ *of φ is a Riesz subspace.*

From (ii) it follows, in particular, that the canonical embedding

$$h\colon \ker\varphi \to E, \quad x \mapsto x$$

is a Riesz homomorphism.

As Riesz homomorphisms are also introduced on partially ordered vector spaces by Buskes and van Rooij [4], it is natural to ask whether statements as in the above two theorems are also valid for Riesz homomorphisms on partially ordered vector spaces. We will give an affirmative answer for the result in Theorem 1 and discuss statements similar to the one in Theorem 2 by means of examples and counterexamples.

To illustrate Theorem 1, consider the following simple example.

Example 3 A bilinear map $b\colon \mathbb{R}^n \times \mathbb{R}^n \to \mathbb{R}$ can be represented by means of the standard basis $\left\{e^{(1)}, \ldots, e^{(n)}\right\}$ in $\mathbb{R}^n$ as

$$b(u, v) = u^{\mathrm{T}} A v,$$

where $A := (a_{ij})$ with $a_{ij} := b\left(e^{(i)}, e^{(j)}\right)$. A functional $\sigma\colon \mathbb{R}^n \to \mathbb{R}$ is a Riesz homomorphism if and only if there is $r \in [0, \infty)$ and $i \in \{1, \ldots, n\}$ such that $\sigma\colon u \mapsto r u_i$ (cf. also Proposition 8 below). From Theorem 1 it follows that b is a bi-Riesz homomorphism if and only if there is $r \in [0, \infty)$ and a permutation matrix P such that $A = rP$. This is a special case of Corollary 14 below.

Bi-Riesz homomorphisms and multi-Riesz homomorphisms are considered in the literature in particular on f-algebras, see, e.g., [1, 2]. They also appear in [3, Theorem 33].

2 Preliminaries

Let X be a real vector space containing a *cone* K, i.e., $K + K \subseteq K$, $\lambda K \subseteq K$ for every $\lambda \geq 0$, and $K \cap (-K) = \{0\}$. The cone K induces a partial order $\leq$ in X by $x \leq y$ if $y - x \in K$. We call (X, K) a *partially ordered vector space*. We mostly assume that (X, K) is *directed*, meaning that $X = K - K$. The space (X, K) is called *Archimedean* if for every $x, y \in X$ with $nx \leq y$ for all $n \in \mathbb{N}$ we have $x \leq 0$. A linear subspace D of X is *order dense* in X if for every $x \in X$ we have

$$x = \inf\{d \in D; d \geq x\}.$$

This concept appears, e.g., in [4, p. 360]. A linear map $i\colon X \to Y$, where X and Y are partially ordered vector spaces, is called *bipositive* if, for every $x \in X$, $x \geq 0$ is equivalent to $i(x) \geq 0$. Note that every bipositive linear map is injective.

A partially ordered vector space X is called a *pre-Riesz space* if there is a Riesz space Y and a bipositive linear map $i\colon X \to Y$ such that $i[X]$ is order dense in Y. We call (Y, i) a *vector lattice cover* of X. An intrinsic definition of pre-Riesz spaces is given by van Haandel in [10], see also [6, Section 2.2]. Note that every directed Archimedean partially ordered vector space is pre-Riesz. If (Y, i) is a vector lattice cover of X such that no proper Riesz subspace of Y contains $i[X]$, then we call (Y, i) a *Riesz completion* of X. Such a space is unique up to isomorphism (for details see, e.g., [6, Section 2.4]).

For $A \subseteq X$ denote $A^{\mathrm{u}} = \{x \in X;\ \forall a \in A\colon x \geq a\}$ and $A^{\mathrm{l}} = \{x \in X;\ \forall a \in A\colon x \leq a\}$. Riesz* homomorphisms are defined in [10, Definition 5.1 and Corollary 5.4(iv)], Riesz homomorphisms and complete Riesz homomorphisms in [4].

Definition 4 Let X and Y be partially ordered vector spaces. A linear map $h\colon X \to Y$ is called

- a *Riesz* homomorphism* if for every nonempty finite subset F of X one has

$$h\left[F^{\mathrm{ul}}\right] \subseteq h[F]^{\mathrm{ul}},$$

- a *Riesz homomorphism* if for every $x, y \in X$ one has

$$h\left[\{x, y\}^{\mathrm{u}}\right]^{\mathrm{l}} = h[\{x, y\}]^{\mathrm{ul}},$$

- a *complete Riesz homomorphism* if for every nonempty set $A \subseteq X$ we have

$$\inf A = 0 \implies \inf h[A] = 0.$$

If X and Y are pre-Riesz spaces, then every complete Riesz homomorphism is a Riesz homomorphism, every Riesz homomorphism is a Riesz* homomorphism, and every Riesz* homomorphism is positive, see [6, Theorem 2.3.19]. If X is, in addition, a vector lattice, then h is a Riesz homomorphism if and only if for every $u, v \in X$ there exists $h(u) \vee h(v)$ in Y, and $h(u) \vee h(v) = h(u \vee v)$. If X and Y are vector lattices, then the notions of a Riesz homomorphism and a Riesz* homomorphism both coincide with the notion of a Riesz homomorphism from vector lattice theory, see, e.g., [6, Lemma 2.3.2]. Moreover, in this case, h is a complete Riesz homomorphism if and only if h is an order continuous Riesz homomorphism, see [6, Proposition 1.4.5].

Van Haandel observed that Riesz* homomorphisms are exactly those operators between pre-Riesz spaces that can be extended to Riesz homomorphisms between the corresponding Riesz completions, see [10, Theorem 5.6] or [6, Theorem 2.4.11].

Theorem 5 *Let X_1 and X_2 be pre-Riesz spaces with Riesz completions (Y_1, i_1) and (Y_2, i_2), respectively. Let $h\colon X_1 \to X_2$ be a linear map. The following statements are equivalent.*

(i) *h is a Riesz* homomorphism.*
(ii) *There exists a unique Riesz homomorphism $\hat{h}\colon Y_1 \to Y_2$ satisfying $\hat{h} \circ i_1 = i_2 \circ h$.*

The next result is a more general version of the implication from (ii) to (i) in Theorem 5 and will be useful in the proof of Proposition 19 below; it can, e.g., be found in [6, Proposition 2.3.6].

Proposition 6 *Let X_1 and X_2 be partially ordered vector spaces and let $h\colon X_1 \to X_2$ be a linear map. Assume that there exist Riesz spaces Y_1 and Y_2 and bipositive linear maps $i_1\colon X_1 \to Y_1$ and $i_2\colon X_2 \to Y_2$ such that $i_1[X_1]$ is order dense in Y_1. If there exists a Riesz homomorphism $\hat{h}\colon Y_1 \to Y_2$ such that $i_2 \circ h = \hat{h} \circ i_1$, then h is a Riesz* homomorphism.*

The following result can be found in [5, Corollary 1], see also [6, Proposition 2.3.28].

Theorem 7 *Let (X, K) be a directed partially ordered vector space and let $\varphi\colon X \to \mathbb{R}$ be a positive linear map. Then the following statements are equivalent.*

(i) *φ is a Riesz homomorphism.*
(ii) *For every linear map $\psi\colon X \to \mathbb{R}$ with $0 \le \psi \le \varphi$ there is $\lambda \in \mathbb{R}_+$ such that $\psi = \lambda\varphi$.*

Next we consider order unit spaces and characterize functionals that are Riesz* homomorphisms. Let (X, K) be an Archimedean partially ordered vector space with order unit u, equipped with the order unit norm $\|x\|_u := \inf\{\lambda \in (0, \infty);\ -\lambda u \le x \le \lambda u\}$ for $x \in X$, see, e.g., [6, Section 1.5.3]. Every order unit space is pre-Riesz. We construct a specific vector lattice cover. The functional representation of X is given by means of the weakly-$*$ compact convex set $\Sigma := \{\varphi \in X';\ \varphi[K] \subseteq [0, \infty),\ \varphi(u) = 1\}$ and the set Λ of the extreme points of Σ. The weak-$*$ closure $\overline{\Lambda}$ of Λ in Σ is (with the weak-$*$ topology) a compact Hausdorff space, and the map

$$\Phi\colon X \to \mathrm{C}(\overline{\Lambda}), \quad x \mapsto (\varphi \mapsto \varphi(x)) \tag{1}$$

is a bipositive linear map, and hence injective (for details see, e.g., [6, Section 2.5]). In [7] it is shown that $(\mathrm{C}(\overline{\Lambda}), \Phi)$ is a vector lattice cover of X, see also [6, Theorem 2.5.9]. We recall the statement in [6, Proposition 2.5.5].

Proposition 8 *Let X be an order unit space and let $\varphi \in \Sigma$.*

(a) *One has $\varphi \in \Lambda$ if and only if φ is a Riesz homomorphism.*
(b) *One has $\varphi \in \overline{\Lambda}$ if and only if φ is a Riesz* homomorphism.*

Recall that two elements x and y in a pre-Riesz space (X, K) are *disjoint*, denoted $x \perp y$, if $\{x + y, x - y\}^{\mathrm{u}} = \{x - y, -x + y\}^{\mathrm{u}}$. If (Y, i) is a vector lattice cover of X, then $x \perp y$ if and only if $i(x) \perp i(y)$, see, e.g., [6, Proposition 4.1.4]. Let X and Y be partially ordered vector spaces. A linear map $h\colon X \to Y$ is called *disjointness preserving* if for every $x, y \in X$ with $x \perp y$ one has $h(x) \perp h(y)$. Note that every Riesz* homomorphism is a positive disjointness preserving operator.

3 Representation of bi-Riesz Homomorphisms

In this section, let (X_1, K_1), (X_2, K_2) be partially ordered vector spaces and $b\colon X_1 \times X_2 \to \mathbb{R}$ a bilinear map. We will deal with bi-Riesz homomorphisms b, i.e. for every $x_1 \in K_1$ and $x_2 \in K_2$ the maps $X_2 \to \mathbb{R}$, $v \mapsto b(x_1, v)$ and $X_1 \to \mathbb{R}$, $u \mapsto b(u, x_2)$ are Riesz homomorphisms. We generalize Theorem 1 in Theorem 11 below. First we show a technical statement which is more general. For $X_1^* := \{f\colon X_1 \to \mathbb{R};\ f \text{ linear}\}$, we consider sets $\Phi \subseteq X_1^*$ with the property

$$\text{for every } \varphi \in \Phi \text{ and } \lambda \geq 0 \text{ one has } \lambda\varphi \in \Phi. \tag{2}$$

Theorem 9 *Let (X_1, K_1), (X_2, K_2) be directed partially ordered vector spaces and let $\Phi \subseteq X_1^*$ satisfy (2). Let a bilinear map $b\colon X_1 \times X_2 \to \mathbb{R}$ be such that for every $x_1 \in K_1$ the map $X_2 \to \mathbb{R}$, $v \mapsto b(x_1, v)$ is a Riesz homomorphism and for every $x_2 \in K_2$ the map $X_1 \to \mathbb{R}$, $u \mapsto b(u, x_2)$ is in Φ. Then there are a Riesz homomorphism $\tau\colon X_2 \to \mathbb{R}$ and a functional $\sigma\colon X_1 \to \mathbb{R}$ in Φ such that for every (u, v) in $X_1 \times X_2$ we have $b(u, v) = \sigma(u)\tau(v)$.*

Proof In this proof, for $x \in K_1$ we denote $\tau_x := b(x, \cdot)$. Then τ_x is a Riesz homomorphism.

If $b = 0$, then there is nothing to prove. Let $x \in K_1$ be such that $\tau_x \neq 0$. Let $u \in K_1$ and consider the corresponding τ_u. We show that there is a number $\gamma(u) \geq 0$ such that $\tau_u = \gamma(u)\tau_x$. Indeed, let $w := x + u$, then $\tau_w = \tau_x + \tau_u$. So, $\tau_w \geq \tau_x \geq 0$ and $\tau_w \geq \tau_u \geq 0$. As X_2 is directed, by Theorem 7 there are $r, s \in \mathbb{R}_+$ such that $\tau_x = r\tau_w$ and $\tau_u = s\tau_w$. As $\tau_x \neq 0$ and, hence, $r \neq 0$, we can set $\gamma(u) := \frac{s}{r}$ and obtain $\tau_u = \gamma(u)\tau_x$. We put $\tau := \tau_x$.

So far, we have a map $\gamma\colon K_1 \to \mathbb{R}_+$, $u \mapsto \gamma(u)$, and for every $(u, v) \in K_1 \times X_2$ we get $b(u, v) = \tau_u(v) = \gamma(u)\tau(v)$. Moreover, $\gamma(x) = 1$. We show that γ is additive and positively homogeneous. Indeed, let $v \in X_2$ be such that $\tau(v) \neq 0$. Let $u, w \in K_1$ and $\lambda \in \mathbb{R}_+$. Then

$$\gamma(\lambda u + w)\tau(v) = b(\lambda u + w, v) = \lambda b(u, v) + b(w, v) = \lambda\gamma(u)\tau(v) + \gamma(w)\tau(v),$$

and dividing by $\tau(v)$ yields the statement. We apply the Kantorovich theorem given in [8], see also [6, Theorem 1.2.5]. As X_1 is directed, there is a linear map $\sigma\colon X_1 \to \mathbb{R}$ that extends γ, where for every $w, z \in K_1$ we have $\sigma(w - z) = \gamma(w) - \gamma(z)$.

We show that for every $(u, v) \in X_1 \times X_2$ we have $b(u, v) = \sigma(u)\tau(v)$. Indeed, let $w, z \in K_1$ be such that $u = w - z$. Then

$$\begin{aligned} b(u, v) &= b(w - z, v) = b(w, v) - b(z, v) = \gamma(w)\tau(v) - \gamma(z)\tau(v) \\ &= \sigma(w - z)\tau(v) = \sigma(u)\tau(v). \end{aligned}$$

Finally, for fixed $v \in K_2$ with $\tau(v) \neq 0$ we can write

$$\sigma(u) = \tfrac{1}{\tau(v)} b(u, v).$$

As $b(\cdot, v) \in \Phi$ and Φ satisfies (2), we obtain $\sigma \in \Phi$. □

The choice $\Phi := X_1^*$ yields the following.

Corollary 10 *Let (X_1, K_1), (X_2, K_2) be directed partially ordered vector spaces. Let a bilinear map $b\colon X_1 \times X_2 \to \mathbb{R}$ be such that for every $x_1 \in K_1$ the map $X_2 \to \mathbb{R}$, $v \mapsto b(x_1, v)$ is a Riesz homomorphism. Then there are a Riesz homomorphism $\tau\colon X_2 \to \mathbb{R}$ and a functional $\sigma\colon X_1 \to \mathbb{R}$ such that for every (u, v) in $X_1 \times X_2$ we have $b(u, v) = \sigma(u)\tau(v)$.*

Similar corollaries can be obtained if for Φ one takes, e.g., the sets

$$\begin{aligned} &K_1^*, \; K_1^* - K_1^*, \; \{\varphi \in X_1^*;\ \varphi \text{ is order bounded}\}, \\ &\{\varphi \in X_1^*;\ \varphi \text{ is disjointness preserving}\}, \end{aligned}$$

or an intersection of the last one with one of the others. Moreover, Φ can also be one of the sets

$$\begin{aligned} &\{\varphi \in X_1^*;\ \varphi \text{ is a Riesz* homomorphism}\}, \\ &\{\varphi \in X_1^*;\ \varphi \text{ is a Riesz homomorphism}\}, \text{ or} \\ &\{\varphi \in X_1^*;\ \varphi \text{ is a complete Riesz homomorphism}\}. \end{aligned}$$

If we take $\Phi := \{\varphi \in X_1^*;\ \varphi \text{ is a Riesz homomorphism}\}$ and apply Theorem 9, we obtain the following representation of bi-Riesz homomorphisms, generalizing Theorem 1 to directed partially ordered vector spaces.

Theorem 11 *Let (X_1, K_1), (X_2, K_2) be directed partially ordered vector spaces. If $b\colon X_1 \times X_2 \to \mathbb{R}$ is a bi-Riesz homomorphism, then there are Riesz homomorphisms $\sigma\colon X_1 \to \mathbb{R}$ and $\tau\colon X_2 \to \mathbb{R}$ such that for every (u, v) in $X_1 \times X_2$ we have $b(u, v) = \sigma(u)\tau(v)$.*

We apply the above results to order unit spaces.

Corollary 12 *Let $(X_1, K_1, \|\cdot\|_{u_1})$ and $(X_2, K_2, \|\cdot\|_{u_2})$ be order unit spaces with the functional representations $(\mathrm{C}(\overline{\Lambda_1}), \Phi_1)$ and $(\mathrm{C}(\overline{\Lambda_2}), \Phi_2)$, respectively, corresponding to (1). Let $b\colon X_1 \times X_2 \to \mathbb{R}$ be a bilinear map.*

(a) *If b is such that for every $x_1 \in K_1$ the map $b(x_1, \cdot)$ is a Riesz homomorphism and for every $x_2 \in K_2$ the map $b(\cdot, x_2)$ is a Riesz* homomorphism, then there are $\sigma \in \overline{\Lambda_1}$, $\tau \in \Lambda_2$ and $r \in [0, \infty)$ such that for every (u, v) in $X_1 \times X_2$ we have $b(u, v) = r\sigma(u)\tau(v)$.*

(b) *If b is a bi-Riesz homomorphism, then there are $\sigma \in \Lambda_1$, $\tau \in \Lambda_2$ and $r \in [0, \infty)$ such that for every (u, v) in $X_1 \times X_2$ we have $b(u, v) = r\sigma(u)\tau(v)$.*

Proof For a proof of (a), we combine Theorem 9 with Proposition 8, where we use $\Phi := \{\varphi \in X_1^*;\ \varphi \text{ is a Riesz* homomorphism}\}$.

The statement in (b) is a direct consequence of Theorem 11 and Proposition 8(a). □

Note that in Corollary 12 the representation is in fact given by means of point evaluations in the functional representations. We slightly change the point of view and study order dense subspaces of spaces of continuous functions. In [11], van Imhoff characterized Riesz* homomorphisms between such spaces as weighted composition operators, see also [6, Theorem 5.1.14]. We need the following special case.

Proposition 13 *Let Ω be a nonempty compact Hausdorff space, X an order dense subspace of $\mathrm{C}(\Omega)$ and $\tau: X \to \mathbb{R}$ a linear functional. The following statements are equivalent.*

(i) *τ is a Riesz* homomorphism.*

(ii) *There are $\omega \in \Omega$ and $r \in [0, \infty)$ such that for every $x \in X$ one has $\tau(x) = rx(\omega)$.*

Corollary 14 *Let Ω_1 and Ω_2 be nonempty compact Hausdorff spaces and let X_1 and X_2 be order dense subspaces of $\mathrm{C}(\Omega_1)$ and $\mathrm{C}(\Omega_2)$, respectively. Let a bilinear map $b\colon X_1 \times X_2 \to \mathbb{R}$ be such that for every $x_1 \in K_1$ the map $b(x_1, \cdot)$ is a Riesz homomorphism and for every $x_2 \in K_2$ the map $b(\cdot, x_2)$ is a Riesz* homomorphism. Then there are $\omega_1 \in \Omega_1$, $\omega_2 \in \Omega_2$ and $r \in [0, \infty)$ such that for every $(u, v) \in X_1 \times X_2$ one has*

$$b(u, v) = ru(\omega_1)v(\omega_2).$$

In particular, this is true for a bi-Riesz homomorphism b.

Proof We apply Theorem 9 with

$$\Phi := \{\varphi \in X_1^*;\ \varphi \text{ is a Riesz* homomorphism}\}.$$

Then there are a Riesz homomorphism $\tau: X_2 \to \mathbb{R}$ and a Riesz* homomorphism $\sigma: X_1 \to \mathbb{R}$ such that for every (u, v) in $X_1 \times X_2$ we have $b(u, v) = \sigma(u)\tau(v)$. We represent τ and σ by means of Proposition 13, which yields the existence of ω_1, ω_2 and r, as requested. □

Next we observe that the decomposition in Theorem 9 is essentially unique.

Proposition 15 *Assume that the conditions of Theorem 9 are satisfied and that b is nonzero. If* $\tau, \tau' \colon X_2 \to \mathbb{R}$ *are Riesz homomorphisms and* $\sigma, \sigma' \colon X_1 \to \mathbb{R}$ *are in* Φ *such that for every* $(u, v) \in X_1 \times X_2$ *we have* $b(u, v) = \sigma(u)\tau(v) = \sigma'(u)\tau'(v)$*, then there exists* $\alpha \in \mathbb{R}$ *with* $\alpha > 0$ *such that* $\sigma' = \alpha\sigma$ *and* $\tau' = \frac{1}{\alpha}\tau$.

Proof For $v, v' \in K_2$ with $\tau(v) \neq 0$ and $\tau'(v') \neq 0$, we have $\tau(v+v') \geq \tau(v) > 0$ and $\tau'(v+v') \geq \tau'(v') > 0$, as τ and τ' are positive. Let

$$\alpha = \frac{\tau(v+v')}{\tau'(v+v')},$$

which satisfies $\alpha > 0$. For every $u \in X_1$ we have

$$\sigma'(u) = \frac{1}{\tau'(v+v')} b(u, v+v') = \alpha \frac{1}{\tau(v+v')} b(u, v+v') = \alpha\sigma(u),$$

hence $\sigma' = \alpha\sigma$.

Finally, choose $u \in X_1$ such that $\sigma(u) \neq 0$. Then for every $v \in X_2$ we get

$$\tau'(v) = \frac{1}{\sigma'(u)} b(u, v) = \frac{1}{\alpha}\frac{1}{\sigma(u)} b(u, v) = \frac{1}{\alpha}\tau(v).$$

□

The essential uniqueness of the decomposition has the following consequence.

Corollary 16 *Let* (X, K) *be a directed partially ordered vector space and let* $b \colon X \times X \to \mathbb{R}$ *be a bi-Riesz homomorphism such that* $b(u, v) = b(v, u)$ *for every* $u, v \in X$*. Then there exists a Riesz homomorphism* $\varphi \colon X \to \mathbb{R}$ *such that for every* $u, v \in X$ *we have* $b(u, v) = \varphi(u)\varphi(v)$.

Proof According to Theorem 11 there exist Riesz homomorphisms $\sigma, \tau \colon X \to \mathbb{R}$ such that for all $u, v \in X$ we have $b(u, v) = \sigma(u)\tau(v)$. Then also $b(u, v) = \tau(u)\sigma(v)$ for every $u, v \in X$, due to the symmetry assumption. If b is zero, we choose $\varphi := 0$. Otherwise, Proposition 15 yields an $\alpha \in \mathbb{R}$ with $\alpha > 0$ such that $\sigma = \alpha\tau$. Then for every $u, v \in X$ we have $b(u, v) = \sqrt{\alpha}\tau(u)\sqrt{\alpha}\tau(v)$, hence we choose $\varphi := \sqrt{\alpha}\tau$. □

For further research, we list some open questions. Problem (I) is motivated by Theorem 5, problem (II) by Theorem 11, and problem (III) by Kusraev and Kutateladze [9, Theorem 3.12.A.3.].

(I) Let X_1, X_2 be pre-Riesz spaces, (Y_1, i_1), (Y_2, i_2) their Riesz completions, respectively, Y a vector lattice, and $B \colon X_1 \times X_2 \to Y$ a bilinear map. Under which conditions on B does there exist a bi-Riesz homomorphism $\hat{B} \colon Y_1 \times Y_2 \to Y$ such that for every $u \in X_1$ and $v \in X_2$ one has $B(u, v) = \hat{B}(i_1(u), i_2(v))$?

(II) Consider the setting of Theorem 11, but assume that $b(x_1, \cdot)$ and $b(\cdot, x_2)$ are both Riesz* homomorphisms instead of Riesz homomorphisms. Do there exist corresponding Riesz* homomorphisms σ and τ representing b?

(III) Let X_1, X_2 be directed partially ordered vector spaces, Y an Archimedean vector lattice, $Y^{\mathbf{u}}$ its universal completion and $B\colon X_1 \times X_2 \to Y$ a bi-Riesz homomorphism. Do there exist Riesz homomorphisms $S\colon X_1 \to Y^{\mathbf{u}}$ and $T\colon X_2 \to Y^{\mathbf{u}}$ such that for every $u \in X_1$ and $v \in X_2$ one has $B(u, v) = S(u)T(v)$?

Concerning Problem (I), we have the following partial result.

Proposition 17 *Let (X_1, K_1) and (X_2, K_2) be pre-Riesz spaces and (Y_1, i_1), (Y_2, i_2) their Riesz completions, respectively. Let $b\colon X_1 \times X_2 \to \mathbb{R}$ be a bilinear map such that for every $x_1 \in K_1$ the map $b(x_1, \cdot)$ is a Riesz homomorphism and for every $x_2 \in K_2$ the map $b(\cdot, x_2)$ is a Riesz* homomorphism. Then there exists a bi-Riesz homomorphism $\hat{b}\colon Y_1 \times Y_2 \to \mathbb{R}$ such that for every $x_1 \in X_1$, $x_2 \in X_2$ one has $b(x_1, x_2) = \hat{b}(i(x_1), i(x_2))$.*

Proof For the set $\Phi := \{\varphi \in X_1^*;\ \varphi \text{ is a Riesz* homomorphism}\}$ we apply Theorem 9. Then there exist a Riesz homomorphism $\tau\colon X_2 \to \mathbb{R}$ and a Riesz* homomorphism $\sigma\colon X_1 \to \mathbb{R}$ such that for every (u, v) in $X_1 \times X_2$ we have $b(u, v) = \sigma(u)\tau(v)$. By Theorem 5 there are Riesz homomorphisms $\hat{\tau}\colon Y_2 \to \mathbb{R}$ and $\hat{\sigma}\colon Y_1 \to \mathbb{R}$ extending τ and σ, respectively. Set $\hat{b}\colon Y_1 \times Y_2 \to \mathbb{R}$, $(w, z) \mapsto \hat{\sigma}(w)\hat{\tau}(z)$. Then $\hat{b}$ has the requested properties. □

4 Kernels of Differences of Riesz Homomorphisms

We discuss statements as in Theorem 2 in the setting of a pre-Riesz space X. It is natural to ask for properties of $\ker\varphi$ if $\varphi\colon X \to \mathbb{R}$ is the difference of two Riesz* homomorphisms. More precisely, we deal with the question under which conditions the inclusion map $h\colon \ker\varphi \to X$, $x \mapsto x$, is a Riesz* homomorphism. First we give an example which shows that this is not true, in general. We use [6, Example 4.4.18].

Example 18 We consider in $X := \mathbb{R}^3$ the cone K that is the positive-linear hull of

$$\left\{\begin{pmatrix}1\\0\\1\end{pmatrix}, \begin{pmatrix}0\\1\\1\end{pmatrix}, \begin{pmatrix}-1\\0\\1\end{pmatrix}, \begin{pmatrix}0\\-1\\1\end{pmatrix}\right\}.$$

(X, K) is an Archimedean directed partially ordered vector space. By Proposition 8, a functional $\varphi\colon X \to \mathbb{R}$ is a Riesz homomorphism if and only if it is a positive multiple of one of the functionals

$$\varphi^{(1)} = (-1, -1, 1)\,, \ \varphi^{(2)} = (1, -1, 1)\,, \ \varphi^{(3)} = (1, 1, 1)\,, \ \varphi^{(4)} = (-1, 1, 1)\,.$$

Here, Riesz homomorphisms and Riesz* homomorphisms coincide. By [6, Theorem 2.6.2], the Riesz completion of (X, K) is given by $(\mathbb{R}^4, \Phi)$, where

$$\Phi\colon \mathbb{R}^3 \to \mathbb{R}^4, \quad x \mapsto \left(\varphi^{(1)}(x), \varphi^{(2)}(x), \varphi^{(3)}(x), \varphi^{(4)}(x)\right)^{\mathrm{T}}.$$

By [6, Proposition 2.3.27], Φ is a complete Riesz homomorphism.

In both of the following cases (a) and (b), we choose φ such that $\ker\varphi$ is a vector lattice with the cone $K_0 := K \cap \ker\varphi$. If the inclusion map $h\colon \ker\varphi \to X$, $x \mapsto x$, would be a Riesz* homomorphism, then $\Phi \circ h\colon \ker\varphi \to \mathbb{R}^4$ as well, as the composition of Riesz* homomorphisms is a Riesz* homomorphism, see, e.g., [6, Proposition 2.3.21]. As $\ker\varphi$ and $\mathbb{R}^4$ are vector lattices, $\Phi \circ h$ would be a Riesz homomorphism. By this technique, we show in both cases that h is not a Riesz* homomorphism and, therefore, not a Riesz homomorphism. Note that in (a), $\ker\varphi$ is an order ideal, whereas in (b) it is not, see [6, Theorem 4.3.22].

(a) Let $\varphi := \varphi^{(1)}$. Then

$$\ker\varphi = \left\{s\begin{pmatrix}1\\0\\1\end{pmatrix} + t\begin{pmatrix}0\\1\\1\end{pmatrix};\, s, t \in \mathbb{R}\right\},$$

and K_0 is a face[1] of K. To see that $\Phi\circ h$ is not a Riesz homomorphism, consider $x^{(1)} := (1, 0, 1)^{\mathrm{T}}$ and $x^{(2)} := (0, 1, 1)^{\mathrm{T}}$ in $(\ker\varphi, K_0)$, where we get $x^{(1)} \vee x^{(2)} = (1, 1, 2)^{\mathrm{T}}$. Hence $(\Phi \circ h)\left(x^{(1)} \vee x^{(2)}\right) = (0, 2, 4, 2)^{\mathrm{T}}$. On the other hand, $(\Phi \circ h)\left(x^{(1)}\right) = (0, 2, 2, 0)^{\mathrm{T}}$ and $(\Phi \circ h)\left(x^{(2)}\right) = (0, 0, 2, 2)^{\mathrm{T}}$, so

$$(\Phi \circ h)\big(x^{(1)}\big) \vee (\Phi \circ h)\big(x^{(2)}\big) = (0, 2, 2, 2)^{\mathrm{T}} \neq (\Phi \circ h)\big(x^{(1)} \vee x^{(2)}\big).$$

(b) Let $\varphi := \varphi^{(3)} - \varphi^{(1)} = (2, 2, 0)$. Then

$$\ker\varphi = \left\{s\begin{pmatrix}1\\-1\\0\end{pmatrix} + t\begin{pmatrix}0\\0\\1\end{pmatrix};\, s, t \in \mathbb{R}\right\}$$

is not an order ideal, since it contains the interior point $(0, 0, 1)^{\mathrm{T}}$ of K. To show that $\Phi\circ h$ is not a Riesz homomorphism, take the elements $x^{(1)} := \left(-\frac{1}{2}, \frac{1}{2}, 1\right)^{\mathrm{T}}$ and $x^{(2)} := \left(\frac{1}{2}, -\frac{1}{2}, 1\right)^{\mathrm{T}}$ in $(\ker\varphi, K_0)$, where we have $x^{(1)} \vee x^{(2)} = (0, 0, 2)^{\mathrm{T}}$ and $(\Phi \circ h)\left(x^{(1)} \vee x^{(2)}\right) = (2, 2, 2, 2)^{\mathrm{T}}$. But $(\Phi \circ h)\left(x^{(1)}\right) = (1, 0, 1, 2)^{\mathrm{T}}$ and $(\Phi \circ h)\left(x^{(2)}\right) = (1, 2, 1, 0)^{\mathrm{T}}$, so

$$(\Phi \circ h)\big(x^{(1)}\big) \vee (\Phi \circ h)\big(x^{(2)}\big) = (1, 2, 1, 2)^{\mathrm{T}} \neq (\Phi \circ h)\big(x^{(1)} \vee x^{(2)}\big).$$

[1] A cone $K_0 \subseteq K$ is called a *face* of K if for every $y \in K_0$ and $x \in K$ with $x \leq y$ one has $x \in K_0$.

The next statement provides a class of pre-Riesz spaces where a partial result in the spirit of Theorem 2 is true.

Proposition 19 *Let Ω be a nonempty compact Hausdorff space and let X be a uniformly dense subspace of $\mathrm{C}(\Omega)$. Let $\varphi\colon X \to \mathbb{R}$ be linear. If there exist Riesz* homomorphisms $\varphi_1, \varphi_2\colon X \to \mathbb{R}$ such that $\varphi = \varphi_1 - \varphi_2$, then the inclusion map $h\colon \ker\varphi \to X$, $x \mapsto x$, is a Riesz* homomorphism.*

Proof Denote $X_0 := \ker\varphi$. As X is uniformly dense in $\mathrm{C}(\Omega)$, it is order dense in $\mathrm{C}(\Omega)$, see [6, Corollary 1.6.5]. By Proposition 13, there are $\omega_1, \omega_2 \in \Omega$ and $r_1, r_2 \in [0, \infty)$ such that for every $x \in X$ we have $\varphi_1(x) = r_1 x(\omega_1)$ and $\varphi_2(x) = r_2 x(\omega_2)$. We may assume that $r_2 \geq r_1$, as otherwise we could consider $-\varphi$ instead of φ. Moreover, if $r_1 = r_2 = 0$, then $\ker\varphi = X$ and $h\colon X \to X$ is then the identity map, which is a Riesz* homomorphism. Thus, in the remainder of the proof we may also assume that $r_2 > 0$.

Define $\psi_1, \psi_2\colon \mathrm{C}(\Omega) \to \mathbb{R}$ by $\psi_1\colon y \mapsto r_1 y(\omega_1)$ and $\psi_2\colon y \mapsto r_2 y(\omega_2)$. Furthermore, let

$$Y_0 := \{y \in \mathrm{C}(\Omega);\ \psi_1(y) - \psi_2(y) = 0\} = \{y \in \mathrm{C}(\Omega);\ r_1 y(\omega_1) = r_2 y(\omega_2)\}.$$

Then $X_0 \subseteq Y_0$. Let $i\colon X \to \mathrm{C}(\Omega)$, $j\colon X_0 \to Y_0$ and $k\colon Y_0 \to \mathrm{C}(\Omega)$ denote the inclusion maps. We have $i \circ h = k \circ j\colon X_0 \to \mathrm{C}(\Omega)$. Since ψ_1 and ψ_2 are Riesz homomorphisms, Theorem 2 yields that Y_0 is a Riesz subspace of $\mathrm{C}(\Omega)$, and therefore k is a Riesz homomorphism. We intend to apply Proposition 6 (with $X_1 := X_0$, $X_2 := X$, $Y_1 := Y_0$, $Y_2 := \mathrm{C}(\Omega)$, $i_1 := j$, $i_2 := i$ and $\hat{h} := k$). For this, we want to show that X_0 is order dense in Y_0. We do so by showing that X_0 is uniformly dense in Y_0. Indeed, let $y \in Y_0$ and let $\varepsilon > 0$. As X is uniformly dense in $\mathrm{C}(\Omega)$, there is $x \in X$ such that $\|x - y\| < \frac{\varepsilon r_2}{5(1+r_1+r_2)}$. Then

$$\begin{aligned} |r_1 x(\omega_1) - r_2 x(\omega_2)| &= |r_1 (x(\omega_1) - y(\omega_1)) - r_2 (x(\omega_2) - y(\omega_2))| \\ &\leq (r_1 + r_2)\frac{\varepsilon r_2}{5(1+r_1+r_2)} < \frac{\varepsilon r_2}{5}. \end{aligned}$$

If $\omega_1 \neq \omega_2$, then by Urysohn's Lemma there is $v \in \mathrm{C}(\Omega)$ with

$$0 \leq v \leq 1, \quad v(\omega_1) = 0 \text{ and } v(\omega_2) = 1.$$

(If $\omega_1 = \omega_2$, then we can take $r_1 = 0$ and $v\colon \Omega \to \mathbb{R}$, $\omega \mapsto 1$.) Take $w \in X$ such that $\|v - w\| < \frac{1}{4}$. Then $\|w\| \leq \|w - v\| + \|v\| < \frac{5}{4}$. As $w(\omega_1) - v(\omega_1) < \frac{1}{4}$, $v(\omega_2) - w(\omega_2) < \frac{1}{4}$, and $r_1 \leq r_2$, we get

$$\begin{aligned} r_2 w(\omega_2) - r_1 w(\omega_1) &> r_2 v(\omega_2) - \tfrac{r_2}{4} - r_1 v(\omega_1) - \tfrac{r_1}{4} \\ &> r_2 - \tfrac{r_2}{4} - \tfrac{r_1}{4} \geq \tfrac{r_2}{2}. \end{aligned}$$

Put

$$z := x - \frac{r_2 x(\omega_2) - r_1 x(\omega_1)}{r_2 w(\omega_2) - r_1 w(\omega_1)} w.$$

Then $z \in X$ and $r_2 z(\omega_2) - r_1 z(\omega_1) = 0$, so $z \in X_0$. Moreover,

$$\|y - z\| \leq \|y - x\| + \frac{|r_2 x(\omega_2) - r_1 x(\omega_1)|}{r_2 w(\omega_2) - r_1 w(\omega_1)} \|w\|$$

$$< \frac{\varepsilon}{5} + \frac{\varepsilon r_2}{5} \frac{2}{r_2} \frac{5}{4} < \varepsilon.$$

Thus, X_0 is uniformly dense in Y_0. It follows that X_0 is order dense in Y_0. Now we apply Proposition 6 and obtain that h is a Riesz* homomorphism. □

It remains to discuss the converse implication in Proposition 19. Assume that E is a Riesz space and $\varphi: E \to \mathbb{R}$ an order bounded linear functional such that $h: \ker\varphi \to E$, $x \mapsto x$, is a Riesz homomorphism. This can be ensured, e.g., by choosing φ such that $\ker\varphi$ is order dense in E; then by [6, Proposition 2.3.27] the map h is even a complete Riesz homomorphism. We give an example of such a φ which is not the difference of two Riesz* homomorphisms.

Example 20 Let $E = \mathrm{C}[-1, 1]$ and $\varphi: E \to \mathbb{R}$, $x \mapsto \int_{-1}^{0} x(t)\mathrm{d}t - \int_0^1 x(t)\mathrm{d}t$. Then φ is order bounded and $\ker\varphi$ is order dense in E. So, the map $h: \ker\varphi \to E$, $x \mapsto x$, is a complete Riesz homorphisms. But φ is not a difference of two Riesz* homomorphism (cf. Proposition 13).

Acknowledgments The authors thank A. Kusraev for interesting discussions. At a workshop in Dresden in 2019, he pointed the results in Theorems 1 and 2 out to us and asked the question whether similar statements are valid for Riesz homomorphisms on partially ordered vector spaces. This inspired the present work.
Anke Kalauch is grateful for support by the Deutsche Forschungsgemeinschaft (grant CH 1285/5-1, Order preserving operators in problems of optimal control and in the theory of partial differential equations).

References

1. K. Boulabiar, Some aspects of Riesz multimorphisms, Indag. Mathem. N.S. **13**(4), 419–432 (2002)
2. K. Boulabiar, M.A. Toumi, Lattice bimorphisms on f-algebras. Algebra Univ. **48**, 103–116 (2002)
3. Q. Bu, G. Buskes, A.G. Kusraev, Bilinear maps on products of vector lattices: a survey, in *Positivity*, ed. by K. Boulabiar, G. Buskes, A. Triki (Birkhäuser, Basel etc., 2007), pp. 97–126
4. G. Buskes, A.C.M. van Rooij, The vector lattice cover of certain partially ordered groups. J. Austral. Math. Soc. (Series A) **54**, 352–367 (1993)
5. A. Hayes, Indecomposable positive additive functionals. J. London Math. Soc. **41**, 318–322 (1966)
6. A. Kalauch, O. van Gaans, *Pre-Riesz Spaces* (Walter de Gruyter GmbH, Berlin/Boston, 2019)

7. A. Kalauch, B. Lemmens, O. van Gaans, Riesz completions, functional representations, and anti-lattices. Positivity **18**(1), 201–218 (2014)
8. L.V. Kantorovich, Linear operators in semi-ordered spaces. Math. Sbornik **49**, 209–284 (1940)
9. A.G. Kusraev, S.S. Kutateladze, *Boolean valued analysis: selected topics*. Trends in science 6 (Southern Mathematical Institute, Vladikavkaz, 2014)
10. M. van Haandel, *Completions in Riesz Space Theory*, Ph.D. thesis (University of Nijmegen, Nijmegen, 1993)
11. H. van Imhoff, Riesz* homomorphisms on pre-Riesz spaces consisting of continuous functions. Positivity **22**(2), 425–447 (2018)

Orthogonality: An Antidote to Kadison's Anti-Lattice Theorem

Anil Kumar Karn

Abstract In this article, we introduce the notions ortho-infimum and ortho-supremum of a pair of self-adjoint elements in a general C*-algebra with the help of algebraic orthogonality as non-commutative analogues of infimum and supremum respectively. In a commutative C*-algebra, ortho-infimum coincides with infimum and ortho-supremum coincides with supremum. We explore an order theoretic aspect of the algebraic orthogonality in a C*-algebra and prove that its order theoretic equivalent notion, namely, absolute ∞-orthogonality coincides with lattice orthogonality in AM-spaces.

Keywords Algebraic orthogonality · Ortho-infimum · Ortho-supremum · Absolute ∞-orthogonality · Absolute order unit space

Mathematics Subject Classification (2010) Primary 46L05; Secondary 46B40, 46B42

1 Introduction

Order structure is an essential component of C*-algebra theory. Using order theoretic techniques, Gelfand and Naimark proved in [2] that every C*-algebra can be characterized as a $*$-subalgebra of $B(H)$, the bounded operators on some complex Hilbert space H. For the commutative case, they proved that a unital commutative C*-algebra is isometrically *-isomorphic to $C(X, \mathbb{C})$ for a suitable compact, Hausdorff space X.

A. K. Karn (✉)
School of Mathematical Sciences, National Institute of Science Education and Research, HBNI, Bhubaneswar, India
e-mail: anilkarn@niser.ac.in

E. Kikianty et al. (eds.), *Positivity and its Applications*, Trends in Mathematics,
https://doi.org/10.1007/978-3-030-70974-7_11

In 1941, Kakutani characterized a unital AM-space, up to isometric lattice isomorphism, as $C(X, \mathbb{R})$ for a suitable compact, Hausdorff space X [4]. Now, in the light of Gelfand Naimark theorem, we note that the self-adjoint part of a (unital) commutative C*-algebra is a vector lattice.

In 1951, Kadison proved that if H is a complex Hilbert space and if S and T are bounded self-adjoint operators on H, then $\inf\{S, T\}$ exists in $B(H)_{sa}$ if and only if S and T are comparable [3]. This is known as Kadison's "anti-lattice theorem". Further, in the same year, Sherman proved that the self-adjoint part of a C*-algebra A is a vector lattice if and only if it is commutative [12]. In particular, a vector lattice structure can not be expected in a general C*-algebra.

In this short note, we make an attempt to establish that the order structure of a C*-algebra is not completely anti-lattice, as the term coined by Kadison suggests. We introduce the notions ortho-infimum and ortho-supremum of a pair of self-adjoint elements in a general C*-algebra. These notions extend the notions of infimum and supremum respectively. In this context, algebraic orthogonality plays a central role. We show that in a commutative C*-algebra, the ortho-infimum of a pair of self-adjoint elements coincides with the infimum of these elements. Similarly, the ortho-supremum of a pair of self-adjoint elements coincides with the supremum of these elements. We further explore an order theoretic aspect of algebraic orthogonality in a C*-algebra and prove that its order theoretic equivalent notion, namely, absolute ∞-orthogonality coincides with lattice orthogonality on AM-spaces. This observation justifies the notion of absolute order unit spaces which include both AM-spaces as well as the self-adjoint parts of unital C*-algebras as examples.

This article may be seen as a prequel of [6, 7].

2 Substitutes for Infimum and Supremum

Let A be a C*-algebra. For $a, b \in A^+$, we say that a is *algebraically orthogonal* to b if $ab = 0$. In this case, we write $a \perp^a b$. Algebraically orthogonal pairs of positive elements play an important role in the theory of C*-algebras. For example, it follows from the functional calculus that every self-adjoint element $a \in A_{sa}$ has a unique decomposition: $a = a^+ - a^-$ in A^+, with $a^+ a^- = 0$. By the functional calculus again, we also get $|a| = a^+ + a^-$.

In the literature, the algebraic orthogonality has been defined for a pair of general elements of a C*-algebra. Here, we shall revisit the notion in the light of its order theoretic characterization. We begin with the following observation.

Lemma 2.1 *Let A be a C*-algebra and let $ab = 0$ for some $a, b \in A^+$. Then $cd = 0$ whenever $0 \le c \le a$ and $0 \le d \le b$.*

Proof Let $0 \le c \le a$. Since $x \mapsto z^* x z$ maps positive elements to positive elements [10, Proposition 1.3.5], we get that $0 \le bcb \le bab = 0$ so that $bcb = 0$. It follows that $\|c^{\frac{1}{2}} b\|^2 = \|bcb\| = 0$. Thus $c^{\frac{1}{2}} b = 0$ so that $cb = 0$. Now, by the same arguments, we may further conclude that $cd = 0$ whenever $0 \leq c \leq a$ and $0 \leq d \leq b$. □

Proposition 2.2 *Let A be a C^*-algebra. For $a, b \in A_{sa}$, the following statements are equivalent:*

1. $|a||b| = 0$;
2. a^+, a^-, b^+, b^- *are mutually algebraically orthogonal;*
3. $|a \pm b| = |a| + |b|$.

This result is proved in a more general set-up using order theoretic techniques [9, Proposition 2.4]. Here we provide a C*-algebraic proof.

Proof

(1) implies (2): Let $|a||b| = 0$. As $0 \leq a^+, a^- \leq |a|$ and $0 \leq b^+, b^- \leq |b|$, a repeated use of Lemma 2.1 yields that a^+, a^-, b^+, b^- are mutually algebraically orthogonal.

(2) implies (1): Let a^+, a^-, b^+, b^- be mutually algebraically orthogonal. Then $(a^+ + a^-)(b^+ + b^-) = 0$. That is, $|a||b| = 0$.

(2) implies (3): Again, let a^+, a^-, b^+, b^- be mutually algebraically orthogonal. Then

$$(a^+ + b^+)(a^- + b^-) = 0$$

and

$$(a^+ + b^-)(a^- + b^+) = 0.$$

Thus

$$\begin{aligned} |a + b| &= |a^+ - a^- + b^+ - b^-| \\ &= |(a^+ + b^+) - (a^- + b^-)| \\ &= a^+ + b^+ + a^- + b^- = |a| + |b| \end{aligned}$$

and

$$\begin{aligned} |a - b| &= |a^+ - a^- - b^+ + b^-| \\ &= |(a^+ + b^-) - (a^- + b^+)| \\ &= a^+ + b^- + a^- + b^+ = |a| + |b|. \end{aligned}$$

(3) implies (2): Finally, assume that $|a \pm b| = |a| + |b|$. Then

$$|(a^+ - a^-) \pm (b^+ - b^-)| = a^+ + a^- + b^+ + b^-$$

so that $(a^+ + b^+) \perp^a (a^- + b^-)$ and $(a^+ + b^-) \perp^a (a^- + b^+)$. Thus a^+, a^-, b^+, b^- are mutually algebraically orthogonal.

□

We say that $a, b \in A_{sa}$ are *algebraically orthogonal*, if $|a||b| = 0$. We further extend this notion to a general pair of elements in A in the following sense. Let $a, b \in A$. We say that a is *algebraically orthogonal* to b, if $\left|\begin{bmatrix} 0 & a \\ a^* & 0 \end{bmatrix}\right| \left|\begin{bmatrix} 0 & b \\ b^* & 0 \end{bmatrix}\right| = 0$ in $M_2(A)^+$.

Note that $\left|\begin{bmatrix} 0 & x \\ x^* & 0 \end{bmatrix}\right| = \begin{bmatrix} |x^*| & 0 \\ 0 & |x| \end{bmatrix}$ for any $x \in A$. Thus a is algebraic orthogonal to b if and only if $|a||b| = 0$ and $|a^*||b^*| = 0$. The following result relates this definition with the standard one of algebraic orthogonality available in the literature.

Proposition 2.3 *Let A be a C^*-algebra and let $a, b \in A$. Then $a^*b = 0$ if and only if $|a^*||b^*| = 0$. In other words, a is algebraically orthogonal to b if and only if $ab^* = 0 = a^*b$. In particular, for $a, b \in A_{sa}$, we have a is algebraically orthogonal to b if and only if $ab = 0$.*

The following proof was suggested to the author by Antonio M. Peralta.

Proof Without any loss of generality, we may assume that $\|a\| = 1 = \|b\|$. First, we assume that $a^*b = 0$. Then $aa^*bb^* = 0$ so that $|a^*||b^*| = 0$.

Conversely, assume that $|a^*||b^*| = 0$. Then for any $m, n \in \mathbb{N}$ we have $|a^*|^{\frac{1}{m}} |b^*|^{\frac{1}{n}} = 0$. Since $x^{\frac{1}{m}} \to r(x)$ in the SOT in A^{**} for any $x \in A^{**+}$, we may conclude that $r(|a^*|)r(|b^*|) = 0$. Thus

$$a^*b = a^*r(|a^*|)r(|b^*|)b = 0.$$

Now, the other statements follow easily from here. □

This confirms with the traditional definition. Let $a, b \in A$. We say that a is *algebraically orthogonal* to b, if $ab^* = 0 = a^*b$. In this case, we write $a \perp^a b$.

Remark 2.4 An alternative proof of the converse, with simpler arguments, was suggested by the referee which I produce below with my thanks to the anonymous referee. Let $|a^*||b^*| = 0$. Then $aa^*bb^* = |a^*|^2|b^*|^2 = 0$. Thus

$$|a^*b|^4 = ((a^*b)^*(a^*b))^2 = b^*aa^*bb^*aa^*b = 0$$

so that $a^*b = 0$.

Now, we present a result which generalizes the notions of infimum and supremum.

Theorem 2.5 *Let A be a C^*-algebra and let $a, b \in A_{sa}$.*

1. *There exists a unique $c \in A_{sa}$ such that*
   (a) $c \le a, c \le b$; *and*
   (b) $(a - c) \perp^a (b - c)$.
2. *There exists a unique $d \in A_{sa}$ such that*
   (a) $a \le d, b \le d$; *and*
   (b) $(d - a) \perp^a (d - b)$.

Proof

(1) Put $a - b = x$. Then $x \in A_{sa}$. By the functional calculus, there exist unique $x^+, x^- \in A^+$ with $x^+x^- = 0$ such that $x = x^+ - x^-$ and $|x| := (x^2)^{\frac{1}{2}} = x^+ + x^-$. Set $c = \frac{1}{2}(a + b - |a - b|)$. Then $c \in A_{sa}$,

$$a - c = \frac{1}{2}(a - b + |a - b|) = x^+ \in A^+,$$

and

$$b - c = \frac{1}{2}(b - a + |a - b|) = x^- \in A^+$$

so that $(a - c)(b - c) = x^+x^- = 0$. Thus $(a - c) \perp^a (b - c)$.

Next, let $c_1 \in A_{sa}$ such that $c_1 \le a, c_1 \le b$; and $(a - c_1)(b - c_1) = 0$. Put $a - c_1 = a_1$ and $b - c_1 = b_1$. Then $a_1, b_1 \in A^+$ with $a_1 b_1 = 0$. Also $a_1 - b_1 = a - b = x$. Thus by the functional calculus, we get $a_1 = x^+$ and $b_1 = x^-$. Now, it follows that

$$c = a - x^+ = a - a_1 = c_1.$$

Now, (2) can be proved by replacing a, b and c in (1) with $-a, -b$ and $-c$ respectively.

□

Definition 2.6 Let A be a C*-algebra and let $a, b \in A_{sa}$. We define

$$a \dot{\wedge} b := \frac{1}{2}(a + b - |a - b|)$$

as the *ortho-infimum* of a and b. Similarly, we define

$$a \dot{\vee} b := \frac{1}{2}(a + b + |a - b|)$$

as the *ortho-supremum* of a and b.

We prove that these notions coincide with the usual notions of infimum and supremum respectively, in the case of a vector lattice. Let $(L, L^+, \wedge, \vee)$ be a vector lattice. Recall that a pair of elements $x, y \in L$ is said to be orthogonal, (we write, $x \perp^\ell y$), if $|x| \wedge |y| = 0$. Here $|u| := u \vee (-u)$ for all $u \in L$.

Corollary 2.7 *Let V be a vector lattice and let $x, y \in V$.*

1. *There exists a unique element $u = x \wedge y \in V$ such that*
   (a) $u \leq x, u \leq y$; *and*
   (b) $(x - u) \perp^\ell (y - u)$.
2. *There exists a unique element $v = x \vee y \in V$ such that*
   (a) $x \leq v, y \leq v$; *and*
   (b) $(v - x) \perp^\ell (v - y)$.

Proof Note that

$$\frac{1}{2}(x + y - |x - y|) = x \wedge y$$

and

$$\frac{1}{2}(x + y + |x - y|) = x \vee y.$$

Therefore, the proof of Theorem 2.5 can be replicated. □

Thus, in a vector lattice, the notions ortho-infimum and ortho-supremum coincide with the notions of infimum and supremum, respectively. In other words, the notions of ortho-infimum and ortho-supremum extend the notions of infimum and supremum respectively to the self-adjoint part of a general C*-algebra.

The ortho-infimum and the ortho-supremum enjoy some interesting properties. We list some of the properties discussed in [7].

Remark 2.8 Let A be a C*-algebra.

1. The mapping $\dot\wedge : A_{sa} \times A_{sa} \to A_{sa}$ satisfies the following conditions:
   (a) $a \dot\wedge a = a$ for all $a \in A_{sa}$.
   (b) $a \dot\wedge b = b \dot\wedge a$ for all $a, \in A_{sa}$.
   (c) $(a \dot\wedge b) + c = (a + c) \dot\wedge (b + c)$ for all $a, b, c \in A_{sa}$.
   (d) $k(a \dot\wedge b) = (ka) \dot\wedge (kb)$ for all $a, b \in A_{sa}$ and $k \geq 0$.
   (e) If $a \dot\wedge b = a$, then $(c \dot\wedge a) \dot\wedge b = c \dot\wedge (a \dot\wedge b)$ for all $c \in A_{sa}$.
2. The mapping $\dot\vee : A_{sa} \times A_{sa} \to A_{sa}$ also satisfies a similar set of conditions.
3. The following conditions are equivalent:
   (a) $a \dot\vee b = \sup\{a, b\} := a \vee b$ for all $a, b \in A_{sa}$.
   (b) $\dot\vee$ is associative in A_{sa}.
   (c) $a \dot\wedge b = \inf\{a, b\} := a \wedge b$ for all $a, b \in A_{sa}$.

(d) $\dot{\wedge}$ is associative in A_{sa}.
(e) A_{sa} is a vector lattice.
(f) A is commutative.

Example Now we provide some counter-examples to show that, in general, the ortho-infimum differ in nature from infimum. Similar examples can be found for the ortho-supremum also.

1. In general, the ortho-infimum of a pair of positive elements in a C*-algebra need not be positive. Consider the C*-algebra $A = M_2(\mathbb{C})$, $a = \begin{bmatrix} 1 & 1 \\ 1 & 1 \end{bmatrix}$ and $b = \begin{bmatrix} 2 & 0 \\ 0 & 0 \end{bmatrix}$.
   Then $2(a\dot{\wedge}b) = \begin{bmatrix} 3-\sqrt{2} & 1 \\ 1 & 1-\sqrt{2} \end{bmatrix} \notin M_2(\mathbb{C})^+$.
2. In general, $\dot{\wedge}$ is not associative. Consider the C*-algebra $A = M_2(\mathbb{C})$, $a = \begin{bmatrix} 1 & 1 \\ 1 & 1 \end{bmatrix}$, $b = \begin{bmatrix} 2 & 0 \\ 0 & 0 \end{bmatrix}$ and $c = \begin{bmatrix} 1 & 0 \\ 0 & 0 \end{bmatrix}$. Then
   $$4((a\dot{\wedge}b)\dot{\wedge}c) = \begin{bmatrix} 4-\sqrt{2} & \sqrt{2} \\ \sqrt{2} & -\sqrt{2} \end{bmatrix}$$
   and
   $$(2\sqrt{5})(a\dot{\wedge}(b\dot{\wedge}c)) = \begin{bmatrix} 2\sqrt{5}-2 & \sqrt{5}-1 \\ \sqrt{5}-1 & \sqrt{5}-3 \end{bmatrix}.$$
   Thus $(a\dot{\wedge}b)\dot{\wedge}c \neq a\dot{\wedge}(b\dot{\wedge}c)$.

In the case of projections, the positivity of ortho-infimum is a strong statement.

Theorem 2.9 *Let p and q be (orthogonal) projections in a unital C^*-algebra A. Then $p\dot{\wedge}q \in A^+$ if and only if p and q commute. In this case, $p\dot{\wedge}q = pq$ (and hence a projection in A).*

Proof First, assume that $p\dot{\wedge}q \in A^+$. Then $|p-q| \leq p+q$. As p and q are projections, it follows from [1, Corollary 2.2] that $|p-q|$ commutes with p and q. Thus $|p-q|^2 \leq (p+q)^2$ so that $a := pq+qp \in A^+$. Now $(1-p)a(1-p) = 0$ whence $a(1-p) = 0$. Therefore, $a = ap = pa$. As $ap = pqp + qp$ and $pa = pq + pqp$, we have $pq = qp$.

Conversely, let $pq = qp$. Then $pq \in A^+$ so that $|p-q|^2 \leq (p+q)^2$. Thus $|p-q| \leq p+q$ and we have $p\dot{\wedge}q \in A^+$.

The last assertion follows from [7, Theorem 4.8]. □

Remark 2.10 The above result fails in the case of non-projection, positive elements. To see this, let $A = M_2(\mathbb{C})$ and consider $a = \begin{bmatrix} 9 & 6 \\ 6 & 27 \end{bmatrix}$ and $b = \begin{bmatrix} 4 & -6 \\ -6 & 32 \end{bmatrix}$. Then

$(a-b)^2 = 169I_2$ so that $|a-b| = 13I_2$. Also $a+b = \begin{bmatrix} 13 & 0 \\ 0 & 59 \end{bmatrix}$ so that $a\dot{\wedge}b = \begin{bmatrix} 0 & 0 \\ 0 & 23 \end{bmatrix} \in M_2(\mathbb{C})^+$. However, it is routine to check that $ab \neq ba$.

Replacing p by $(1-p)$ and q by $(1-q)$, we get a dual result.

Corollary 2.11 *Let p and q be (orthogonal) projections in a unital C*-algebra A. Then $p\dot{\vee}q \leq 1$ if and only if p and q commute.*

3 Orthogonality and Norm

In the previous section, we showed that we can extrapolate the notions of infimum and supremum, with the help of algebraic orthogonality in general C*-algebras. Thus it is interesting as well as beneficial to explore algebraic orthogonality in terms of order and norm. More so as it is defined as a zero product, that is, it is multiplicative in nature. Whereas vector lattices do not have pre-assigned multiplication. In this section, we shall discuss an order theoretic equivalent form of algebraic orthogonality.

Recall that a vector lattice V with a norm $\|\cdot\|$ is called an AM-space, if $(V, \|\cdot\|)$ is a Banach space and the following conditions hold:

1. $\|u\| \leq \|v\|$ whenever $u, v \in V$ with $|u| \leq |v|$; and
2. $\|u \vee v\| = \max\{\|u\|, \|v\|\}$ for all $u, v \in V^+$.

A positive element $e \in V^+$ is called an *order unit* for V, if for each $v \in V$, there exists $k > 0$ such that $ke \pm v \in V^+$. If, in addition,

$$\|v\| = \inf\{k > 0 : ke \pm v \in V^+\}$$

for all $v \in V$, then V is called a unital AM-space. (For details on AM-spaces, please refer to any book on Banach lattices, for example [11].)

Definition 3.1 Let V be a real normed linear space. For $u, v \in V$ we say that u is *∞-orthogonal* to v, (we write, $u \perp_\infty v$), if $\|u + kv\| = \max\{\|u\|, \|kv\|\}$, for all $k \in \mathbb{R}$ [5, Section 2].

Definition 3.2 Let A be a C*-algebra. For $a, b \in A^+$, we say that a is *absolutely ∞-orthogonal* to b, (we write $a \perp_\infty^a b$), if $c \perp_\infty d$ whenever $0 \leq c \leq a$ and $0 \leq d \leq b$ [6, Definition 4.5].

If V is an AM-space, then V is isometrically order isomorphic to A_{sa} for some commutative C*-algebra A. Thus the notion of absolute ∞-orthogonality makes sense in V^+ as well. In fact, absolute ∞-orthogonality can be defined in a more general set up [6, Definition 4.5]. It was proved in [7, Theorem 2.1], (see also [6,

Theorem 4.3 and Conjecture 4.4]), that in a C*-algebra A, we have $\perp^a = \perp^a_\infty$ on A^+. For an AM-space, we have a stronger result.

Proposition 3.3 *In an AM-space V, we have $\perp^\ell = \perp^a_\infty$ on V^+.*

Though this result can be deduced from [6, Theorem 4.2] using Kakutani's theorem for AM-spaces, we give a direct order theoretic proof.

Proof Let $u, v \in V^+$. First, we assume that $u \perp^\ell v$, that is, $u \wedge v = 0$. Let $0 \leq u_1 \leq u$, $0 \leq v_1 \leq v$ and $k \in \mathbb{R}$. First, we note that

$$|u_1| \wedge |kv_1| = u_1 \wedge (|k|v_1) = 0.$$

Since $x \wedge y = \frac{1}{2}(x + y - |x - y|)$ for all $x, y \in V$, we have

$$|u_1 - |k|v_1| = u_1 + |k|v_1 - 2u_1 \wedge (|k|v_1) = u_1 + |k|v_1 = |u_1 + |k|v_1|.$$

Further, as $x \vee y - x \wedge y = |x - y|$ for all $x, y \in V$, we find that

$$|u_1 + kv_1| = u_1 + |k|v_1 = u_1 \vee (|k|v_1).$$

Since V is an AM-space, we get that

$$\|u_1 + kv_1\| = \||u_1 + kv_1|\| = \|u_1 \vee (|k|v_1)\| = \max\{\|u_1\|, \|kv_1\|\}.$$

Thus $u \perp^a_\infty v$.

Conversely, assume that $u \perp^a_\infty v$. Put $w = u \wedge v$. Then $0 \leq w \leq u$ and $w \leq v$. Thus, by assumption, $w \perp_\infty w$. But then $w = 0$. Hence $u \perp^\ell v$. □

If we write $\perp$ as a generalization of $\perp^a$ in a C*-algebra and $\perp^\ell$ in a vector lattice, we are motivated for the following definition. (See [7] with special attention to Remark 3.3.)

Definition 3.4 Let (V, V^+) be a real ordered vector space. Assume that $\perp$ is a binary relation in V such that for $u, v, w \in V$, we have

1. $u \perp 0$;
2. $u \perp v$ implies $v \perp u$;
3. $u \perp v$ and $u \perp w$ imply $u \perp (kv + w)$ for all $k \in \mathbb{R}$;
4. For each $u \in L$, there exist unique $u^+, u^- \in L^+$ with $u^+ \perp u^-$ such that $u = u^+ - u^-$.
   Let us put $u^+ + u^- := |u|$.
5. If $u \perp v$ and if $|w| \leq |v|$, then $u \perp w$.

Then V is called an *absolutely ordered vector space*.

A normed version of this notion is included in the following result. For this purpose, we extend the notion of absolute orthogonality to order unit spaces. Let (V, e) be an order unit space. For $u, v \in V^+$, we say that u is *absolutely* ∞*-orthogonal* to v, (we write $u \perp_\infty^a v$), if $u_1 \perp_\infty v_1$ whenever $0 \le u_1 \le u$ and $0 \le v_1 \le v$.

Theorem 3.5 *Let* (V, e) *be an order unit space. Then the following two sets of conditions are equivalent:*

1. (a) *For each* $u \in V$, *the exists a unique pair* $u^+, u^- \in V^+$ *with* $u^+ \perp_\infty^a u^-$ *such that* $u = u^+ - u^-$;
   Set $|u| := u^+ + u^-$.
   (b) *If* $u, v, w \in V^+$ *with* $u \perp_\infty^a v$ *and* $u \perp_\infty^a w$, *then we have* $u \perp_\infty^a |v \pm w|$.
2. V *is an absolutely ordered vector space in which* $\perp = \perp_\infty^a$ *on* V^+.

In this case, $(V, |\cdot|, e)$ is called an *absolute order unit space*. In [7], a more general normed version of absolutely ordered vector spaces was introduced [7, Definition 3.8].

Proof Clearly, condition (2) implies conditions (1)(a) and (1)(b). Let us now assume that (1)(a) and (1)(b) hold. For $u, v \in V$, we define $u \perp v$, if $|u| \perp_\infty^a |v|$. Then $\perp = \perp_\infty^a$ on V^+. Also, the following facts follow immediately from the definition of $\perp_\infty^a$:

(i) For each $u \in V$, there exist unique $u^+, u^- \in V^+$ with $u^+ \perp u^-$ such that $u = u^+ - u^-$;
(ii) $u \perp 0$ for all $u \in V$;
(iii) $u \perp v$ implies $v \perp u$;
(iv) $u \perp v$ implies $u \perp kv$ for any $k \in \mathbb{R}$; and
(v) if $u \perp v$ and $|w| \le |v|$, then $u \perp w$.

Thus it only remains to prove that $u \perp (v + w)$ whenever $u \perp v$ and $u \perp w$. Assume that $u \perp v$ and $u \perp w$. Then $|u| \perp_\infty^a |v|$ and $|u| \perp_\infty^a |w|$. Then by the definition of $\perp_\infty^a$, we may deduce that $|u| \perp_\infty^a \{v^+, v^-, w^+, w^-\}$. Now, by a repeated use of (1)(b), we obtain that $|u| \perp_\infty^a |v + w|$ for

$$v + w = v^+ - v^- + w^+ - w^- = (v^+ + w^+) - (v^- + w^-).$$

Thus $u \perp (v + w)$. Hence V is an absolutely ordered vector space. □

In some recent works, we have been able to extend some of the properties of operator algebras to absolute (matrix) order unit spaces [7–9]. We hope to present absolute (matrix) order unit spaces as a non-commutative analogue of unital AM-spaces.

Acknowledgments The author is thankful to Antonio M. Peralta for his input on orthogonality in C*-algebras. The author is also thankful to the referee for some valuable suggestions and comments.

References

1. J. Avron, R. Seiler, B. Simon, The index of a pair of projections. J. Funct. Anal. **120**, 220–237 (1994)
2. I.M. Gelfand, M.A. Naimark, On the embedding of normed rings into the ring of operators in Hilbert space. Mat. Sb. **12**, 87–105 (1943)
3. R.V. Kadison, Order properties of bounded self-adjoint operators. Proc. A. M. S. **2**(3), 505–510 (1951)
4. S. Kakutani, Concrete representation of abstract (M)- spaces. Ann. Math. **42**(4), 994–1024 (1941)
5. A.K. Karn, Orthogonality in ℓ_p spaces and its bearing on ordered Banach spaces. Positivity **18**, 223–234 (2014)
6. A.K. Karn, Orthogonality in C*-algebras. Positivity **20**(03), 607–620 (2016)
7. A.K. Karn, Algebraic orthogonality and commuting projections in operator algebras. Acta Sci. Math. (Szeged) **84**, 323–353 (2018)
8. A.K. Karn, A. Kumar, *Isometries of Absolute Order Unit Spaces* (2019). (Published online in Positivity). https://doi.org/10.1007/s11117-019-00731-y
9. A.K. Karn, A. Kumar, *Partial Isometries in an Absolute Order Unit Space*. Preprint (2019). https://arxiv.org/abs/1912.05792
10. G.K. Pedersen, *C^*-algebras and their Automorphism Groups* (Academic Press, New York, 1979)
11. H.H. Schaefer, *Banach Lattices and Positive Operators* (Springer, Berlin, 1974)
12. S. Sherman, Order in operator algebras. Am. J. Math. **73**(1), 227–232 (1951)

Binary Relations in Mathematical Economics: On Continuity, Additivity and Monotonicity Postulates in Eilenberg, Villegas and DeGroot

M. Ali Khan and Metin Uyanik

Abstract This chapter examines how *positivity* and *order* play out in two important questions in mathematical economics, and in so doing, subjects the postulates of *continuity, additivity* and *monotonicity* to closer scrutiny. Two sets of results are offered: the first departs from Eilenberg's necessary and sufficient conditions on the topology under which an anti-symmetric, complete, transitive and continuous binary relation exists on a topologically connected space; and the second, from DeGroot's result concerning an additivity postulate that ensures a complete binary relation on a σ-algebra to be transitive. These results are framed in the registers of order, topology, algebra and measure-theory; and also beyond mathematics in economics: the exploitation of Villegas' notion of *monotonic continuity* by Arrow-Chichilnisky in the context of Savage's theorem in decision theory, and the extension of Diamond's impossibility result in social choice theory by Basu-Mitra. As such, this chapter has also a synthetic and expository motivation, and can be read as a plea for inter-disciplinary conversations, connections and collaboration.

Keywords Continuity · Additivity · Monotonicity · Ordered space · Weakly ordered space

Mathematics Subject Classification (2010) 91B55, 37E05

M. A. Khan (✉)
Department of Economics, Johns Hopkins University, Baltimore, MD, USA
e-mail: akhan@jhu.edu

M. Uyanik
School of Economics, University of Queensland, Brisbane, QLD, Australia
e-mail: m.uyanik@uq.edu.au

E. Kikianty et al. (eds.), *Positivity and its Applications*, Trends in Mathematics,
https://doi.org/10.1007/978-3-030-70974-7_12

It has often happened that a theory designed originally as a tool for the study of a physical problem came subsequently to have purely mathematical interest. When that happens the theory is generalized way beyond the point needed for applications, the generalizations make contact with other theories (frequently in completely unexpected directions), and the subject becomes established as a new part of pure mathematics. Physics is not the only external source of mathematical theories; other disciplines (such as economics and biology) can play a similar role.[1]

Halmos (1956)

It is also possible that algebra, as a separate discipline within mathematics may not survive. The 20th century was a period of unification, with algebra invading other areas of math, and they counter-invading it. If I am engaged in studying a family of functions on multi-dimensional manifolds, those families having a group structure, am I working in analysis (the functions), topology (the manifolds) or algebra (the groups)?[2]

Derbyshire (2006)

1 Introduction

In this chapter revolving around the ideas of *positivity* and *order* in mathematical economics, one can do worse than begin with Garett Birkhoff's review of Eilenberg [33]: it is well-worth quoting in full.

> An "ordered topological space" is, in effect, a simply ordered set whose topology is obtainable by a weakening of its intrinsic topology. The author proves that a topological connected space X can be ordered if and only if the subset of its square X^2 obtained by deleting the diagonal of points (x, x) is not connected; the same condition also characterizes those connected locally connected separable topological spaces which are homeomorphic with subsets of the linear continuum.

In this, his paper on "ordered topological spaces," Eilenberg [33] is justly celebrated for posing two questions of seminal importance for economic theory. First, can a continuous binary relation on a set be represented by a continuous function on the same set? Second, what are the conditions on the set under which a complete and continuous relation is necessarily transitive? Both questions, the second perhaps more than the first, investigate how technical topological conditions, assumed for tractability, necessarily translate into behavioral consequences. However, Eilenberg limited himself to the study of anti-symmetric relations, and thereby to studying agency in a context wherein distinct elements in the choice set are necessarily preferred one to another, a kind of extreme decisiveness. It remained for [20, 22] to place the first question,[3] and for [81, 82] the second, in a setting where the symmetric part of the given binary relation is not an equality, which is to say, the set

[1]Halmos [44, p.419]. The part of pure mathematics so created does not (and need not) pretend to solve the physical problem from which it arises; it must stand and fall on its own merits.

[2]Derbyshire [28, p.319].

[3]In his reproduction of Debreu's theorem [Proposition 1] on the sufficiency of connectedness of a choice set in a finite-dimensional Euclidean space, [65] for example, observes that "Debreu credits a paper of Eilenberg's [33] as containing the mathematical essence of [his] Proposition 1."

of indifferent elements of the relation are not singletons. They and their followers have by now given rise to a rich and mature body of work.

Eilenberg also asked, and answered, two other questions that seem to have had less traction in economic theory, at least in the way that they were initially posed. He asked for conditions on the topology under which there exist "nice" relations (in the sense of being anti-symmetric, transitive, complete and continuous) on a given set, and furthermore, turning the matter on its head, how such relations disallow sets that are "rich" in the meaning endowed to the term through the topological and/or algebraic structures on the set over which they are defined. We shall think of these as Eilenberg's third and fourth questions. Both questions are again "natural" ones. The third is in some sense analogous[4] to the question concerning conditions on a topology under which non-constant continuous functions exist. If the topology is too "sparse" then every continuous function is necessarily constant, and every reflexive, transitive and continuous relation is necessarily trivial in sense that no element is preferred to another. In the context of his fourth question, Eilenberg showed that the existence of a "nice" relation defined on a connected, locally connected and separable space necessarily renders the space to be a linear continuum. These results then are a testimony to the mutual imbrication of assumptions on a relation and the space on which the relation is defined, a two-way relationship that in recent work, Khan-Uyanik [59] see and study as the Eilenberg-Sonnenschein (ES) research program.

In terms of the third and fourth questions concerning "nice" relations, to be sure topologists have understood this mutual imbrication very well. Thus, for there to be a rich supply of continuous linear functions, the topology on the common domain of the functions must, of necessity, satisfy some properties, and cannot be too sparse. Alternatively, the only continuous functions on a set endowed with an indiscrete topology are the constant functions; and digging a little deeper, there is a plethora of (say) non-locally convex spaces with no continuous function at all other than the zero function. The question of the existence of a supporting hyperplane is explicitly studied by [63] in the context of an algebraic structure, and in the context of topological vector spaces, Kalton-Peck-Roberts [56, p. vii] write:

> The role of the Hahn-Banach theorem may be said to be that of a universal simplifier whereby infinite-dimensional arguments can be reduced to the scalar case by the use of the ubiquitous linear functional. Thus the problem with non-locally convex spaces is that of "getting off the ground."

The point is that there is some *hiddenness* in the mutual interaction of a function and set that needs to be flushed out. In terms of the origins, [86] studies the problem of determining the most general class of topological spaces in which non-constant real-valued continuous functions exist. Hewitt's [52] provides an example of a countable, connected Urysohn space[5] in which every continuous function is constant. Following Hewitt's work, there are results on the class of topological

---

[4]Eilenberg's third question is entirely analogous to the existence of a *one-to-one* continuous function since he requires the anti-symmetry property. In Sect. 3 we introduce a result for binary relations that is analogous to the existence of a non-constant continuous function.

[5]A topological space in which any two distinct points can be separated by closed neighborhoods.

spaces on which every continuous function is constant; see [18] for the original paper, and the following, for example, for more modern work: [50, 68, 92] and [53].

The question is of substantive consequence for functional analysis but also beyond it for economic theory and mathematical economics. In terms of this register, the problem gets translated into the question of the sustaining of technologically efficient program as value maximization programs. Majumdar [72] furnishes a complete characterization and refers to his result as follows:

> One should recall that a major motivation behind research in this area comes from the need to determine whether efficient allocations can be attained by the use of a price mechanism in a decentralized system achieving economy of information and utilizing individual incentives. The implications of any result on complete characterization should be seriously considered in this context, and as far as [the result] goes, they seem to be somewhat negative in character. The equivalence established indicates that, in general, one would need a family of price systems to specify an efficient program. Indeed, the applicability of the criterion is rather restricted since one has to know too many prices.[6]

It is then to this literature that we connect Eilenberg's third and fourth questions. We see him asking this question: rather than the existence of a function from a "nice" class of functions, does there exist a binary relation from a "nice" class of binary relations? And the first contribution of this chapter is that it generalizes Eilenberg's answer to this question by relaxing connectedness and anti-symmetry assumptions: in a nutshell, we do to Eilenberg in this context what Sonnenschein did in another and Debreu did in yet another. This is to say that we generalize Eilenberg's result by dropping the anti-symmetry assumption, and then extend the generalization to k-connected spaces, and then to a setting that substitutes k-connectedness with local-connectedness. Finally, we note that our first two results can be analogously generalized to general preferences, and connect our results to the literature on the non-existence of non-constant continuous functions.

But we also make another connection that has been missed in the economic literature. This is the application of our results on the existence of a "nice" preference relation to Diamond's [29] impossibility theorem: what this economic literature sees an impossibility result, we see simply as a question of the existence of a nice binary relation where the adjective nice has been given a meaning and an elaboration in terms of intergenerational equity. In introducing his own paper, Zame [98, p. 188] documents the trajectory of this substantial economic literature.

> Diamond [29] shows that a complete transitive preference relation that displays intergenerational equity and respects the Pareto ordering cannot be continuous in the topology induced by the supremum norm. Basu-Mitra [7] show that such a preference relation—whether continuous or not—cannot be represented by a (real-valued) utility function. On the other

---

[6]Majumdar continues, "But being a complete characterization,[the result] provides a new angle from which the difficulties faced by the earlier approaches can be viewed and tends to suggest that simpler criteria involving fewer price systems, in particular, the use of just one price system as is typically the case, may be incapable of isolating the set of efficient programs unless restrictive assumptions on technology are introduced." For further work on the problem, see [79], and [83], and the references to his chapters in [34].

hand, Svensson [84] proves that such preference relations do exist. Fleurbaey-Michel [37], Hara-Suzumura-Xu [46], Basu-Mitra [8], and Bossert-Suprumont-Suzumura [15] provide further results, both positive and negative.

Moreover, as already illustrated by Toranzo-Hervés-Beleso [85], there are continuous, complete and transitive relations on non-separable spaces which are not representable. This connection that we make is important in that it sights Eilenberg [33] as one of the originating papers of this substantial economic literature. This concludes our discussion of the first substantive section, Sect. 3, of the paper.

Section 4 of the paper returns to Eilenberg's second question: to find a suitable topological condition which ensures the transitivity of a complete, reflexive and continuous binary relations. Khan-Uyanik [59, 60] frame this question in settings that remain squarely remain within the purely topological register, but go considerably beyond Eilenberg. In his consideration of the relationship[7] however, Sonnenschein [81] move to a setting that also embrace linear structures. In a complementary result, [39] show the existence of a mixture-continuous, anti-symmetric, transitive and complete relation defined on a mixture space renders the setting to be isomorphic to either a greater-than-or-equal-to relation, or its inverse, defined on the interval! These results are of substantive consequence for social science since they pertain directly to the formalization of human agency. It mandates that in a sufficiently rich choice set, an agent in an economy, or a player in a game, cannot be simultaneously consistent (transitive) and extremely decisive (anti-symmetric and complete); or to put the matter in a contra-positive way, the choice-set of a sufficiently rational agent in the sense of satisfying the above two *desiderata* must of necessity be sparse and impoverished: a linear continuum in the case of Eilenberg and an interval in the case of [39]. Note that these results, while bearing obvious implication for results on the representation of binary relations, belong to an entirely different register. They concern the dove-tailing and mutual imbrication of a set of assumptions on one object for those on a different but not unrelated object.

The second contribution of this chapter is to make a further move from the register of mixture-spaces to a more abstract algebraic one. Our point of departure now is Villegas [89, 90]: this work studied countably additive qualitative probability representations and showed that given a finite additive qualitative probability, monotone continuity is necessary and sufficient for a countably additive representation[8] It remained for DeGroot [27] to flush out the abstract algebraic register grounding this result. The contribution of Sect. 4 below is (i) to introduce an sharper additivity postulate, one supplemented by monotone continuity postulates, on abstract algebraic structures that are analogous to Villegas' additivity postulate,

[7]It is worth noting that both Eilenberg [33] and Sonnenschein [81] limit their attention to one way of the two-way relationship, in that they examine the implication of assumptions on the choice set on the properties of the relation defined on that set; the *backward* direction exploring the implication of the properties of a class of preferences on the choice set over which they are defined is the signature of the Khan-Uyanik's work.

[8]See Krantz-Luce-Suppes-Tversky [66, Section 5.4.2] for further discussion; also see [61].

(ii) to obtain an equivalence result between additivity and transitivity without referring to completeness or continuity of the binary relation, (iii) to show that under additivity, different variants of the monotonicity concepts are equivalent, (iv) to relate our result to Villegas, DeGroot, de Finetti, Arrow and Chichinisky. In particular we highlight the *hiddenness* and *redundancy* of the transitivity assumption as these *desiderata* are emphasized in [59]. As such, it contributes to the depth and maturity of the ES program. This concludes our discussion of Sect. 4 of the paper.

We began this this introduction by reading Eilenberg [33] as a text revolving around four questions concerning binary relations: leaving Sect. 2 for notational and conceptual preliminaries, we shall focus on the third and the fourth in Sect. 3, and on the second in Sect. 4. The reader may well wonder at our silence about the (first) question that mathematical economists and economic theorists know him by. As mentioned, this pertains to the representation of a binary relation by a function, of a continuous relation by a continuous function, of a monotonic relation by a monotonic function, and of a concave relation by a quasi-concave function. We state the question in this elaborated baroque way simply to allude with the river of work that has accumulated in mathematical economics and mathematical psychology on this question. But to keep to Eilenberg's parameters except that of his singleton indifference sets, we quote from Beardon [9, p. 3].

> In this expository essay we consider how much of the theory can be developed from a purely topological perspective. We focus on those ideas which provide a link between utility theory and topology, and we leave the economic interpretations to others. Briefly, we give priority to results that seem to be topologically important, so we pay more attention to the quotient space of indifference classes than is usual, and more attention to the order topology than other topologies.

We refer the reader to the above article and to [14], the book of which it is a chapter, register our attunement with it, and move on.[9]

2 Mathematical and Conceptual Preliminaries

Let X be a set. A subset $\succcurlyeq$ of $X \times X$ denote a *binary relation* on X. We denote an element $(x, y) \in\ \succcurlyeq$ as $x \succcurlyeq y$. The *asymmetric part* $\succ$ of $\succcurlyeq$ is defined as $x \succ y$ if $x \succcurlyeq y$ and $y \not\succcurlyeq x$, and its *symmetric part* $\sim$ is defined as $x \sim y$ if $x \succcurlyeq y$ and $y \succcurlyeq x$. The inverse of $\succcurlyeq$ is defined as $x \preccurlyeq y$ if $y \succcurlyeq x$. Its asymmetric part $\prec$ is defined analogously and its symmetric part is $\sim$. We provide the descriptive adjectives pertaining to a relation in a tabular form for the reader's convenience in the Table 1.

Let $\succcurlyeq$ be a binary relation on a set X. For any $x \in X$, let $A_{\succcurlyeq}(x) = \{y \in X | y \succcurlyeq x\}$ denote the *upper section* of $\succcurlyeq$ at x and $A_{\preccurlyeq}(x) = \{y \in X | y \preccurlyeq x\}$ its *lower section* at

[9]The reader interested in this (first) Eilenberg question can also see [16, 48, 77, 96]; we shall return to Wold [96] in Section 5.

Table 1 Properties of binary relations

Reflexive	$x \succsim x \ \forall x \in X$
Complete	$x \succsim y$ or $y \succsim x \ \forall x, y \in X$
Non-trivial	$\exists x, y \in X$ such that $x \succ y$
Transitive	$x \succsim y \succsim z \Rightarrow x \succsim z \ \forall x, y, z \in X$
Semi-transitive	$x \succ y \sim z \Rightarrow x \succ z$ and $x \sim y \succ z \Rightarrow x \succ z \ \forall x, y, z \in X$
Anti-symmetric	$x \succsim y$ and $y \succsim x \Rightarrow x = y \forall x, y \in X$

x. Now assume X is endowed with a topology. We say $\succsim$ is *continuous* if its upper and lower sections are closed at all $x \in X$ and the upper and lower sections of its asymmetric part $\succ$ are open at all $x \in X$.

A topological space X is said to be *connected* if it is not the union of two non-empty, disjoint open sets. The space X is *disconnected* if it is not connected. A subset of X is connected if it is connected as a subspace. We say X is *locally connected* if for all $x \in X$, every open neighborhood of x contains a connected and open set containing x. A *component* of a topological space is a maximal connected set in the space; that is, a connected subset which is not properly contained in any connected subset. For any natural number k, a topological space is k-*connected* if it has at most k components.[10] The concept of k-connectedness provides a quantitative measure of the degree of disconnectedness of a topological space. It is easy to see that 1-connectedness is equivalent to connectedness and that any k-connected space is l-connected for all $l \geq k$.

3 On the Existence of a Continuous Binary Relation

Eilenberg [33, Theorem I] provides a necessary and sufficient condition for the existence of an anti-symmetric, complete, transitive and continuous binary relation on a connected topological space X. In this section we start with introducing a generalization of Eilenberg's result to k-connected spaces, and then show that when the space is locally connected, then cardinality of the components of the space does not matter. We continue by presenting a result which eliminates the anti-symmetry requirement in Eilenberg's theorem. We end the section with a brief discussion of our results.

Before presenting our result, we need the following notation: for any set X, define

$$P(X) = \{(x, y) \in X \times X : x \neq y\}.$$

[10]See [59] for a detailed discussion on k-connectedness.

3.1 On Ordered Topological Spaces

Eilenberg [33] calls a topological space *ordered* if there exists an anti-symmetric, complete, transitive and continuous binary relation on it. He then presents

Theorem (Eilenberg) *A connected topological space X which contains at least two elements can be ordered if and only if $P(X)$ is disconnected.*

The following theorem generalizes Eilenberg's theorem to k-connected spaces.

Theorem 3.1 *For any natural number k, a k-connected topological space X can be ordered if and only if $P(C)$ is disconnected for each non-singleton component C of X.*

Proof of Theorem 3.1 Let $\{C_i\}_{i=1}^{\ell}$ be the collection of the components of X where $\ell \leq k$. First note that Theorem (Eilenberg) implies that for each component C_i of X which contains at least two elements, there exists an anti-symmetric, complete, transitive and continuous binary relation $\succcurlyeq_i$ on C_i if and only if $P(C_i)$ is disconnected.

In order to prove the forward direction, assume $\succcurlyeq$ is an anti-symmetric, complete, transitive and continuous binary relation on X. Then, for each C_i, the restriction of $\succcurlyeq$ on C_i, defined as $\succcurlyeq_i = \succcurlyeq \cap (C_i \times C_i)$, is an anti-symmetric, complete, transitive and continuous binary relation on C_i. Then, Theorem (Eilenberg) implies that $P(C_i)$ is disconnected for all non-singleton C_i.

In order to prove the backward direction, assume $P(C_i)$ is disconnected for each non-singleton C_i. It follows from Theorem (Eilenberg) that there exists an anti-symmetric, complete, transitive and continuous binary relation $\succcurlyeq_i$ on every non-singleton C_i. If C_i is a singleton, then define $\succcurlyeq_i = C_i \times C_i$. Then define a binary relation $\succcurlyeq$ on X as follows: $\bigcup_{i=1}^{\ell} \succcurlyeq_i \subseteq \succcurlyeq$, and for all $i > j$, $C_i \times C_j \subseteq \succcurlyeq$. Then, $\succcurlyeq$ is anti-symmetric, complete, and transitive. Since each C_i is closed in X, therefore $\succcurlyeq_i$ has closed sections in both C_i and X, and hence $\succcurlyeq$ has closed sections. Therefore, $\succcurlyeq$ is continuous. □

Theorem 3.2 *A locally connected topological space X can be ordered if and only if $P(C)$ is disconnected for each non-singleton component C of X.*

Proof of Theorem 3.2 Let $\{C_i\}_{i\in I}$ be the collection of the components of X. The proof of the forward direction is identical to the proof of the forward direction of Theorem 3.1. In order to prove the backward direction, assume $P(C_i)$ is disconnected for each non-singleton C_i. It follows from Theorem (Eilenberg) that there exists an anti-symmetric, complete, transitive and continuous binary relation $\succcurlyeq_i$ on each non-singleton C_i. If C_i is a singleton, then define $\succcurlyeq_i = C_i \times C_i$. The well-ordering theorem (Munkres [76, Theorem, p.65]) implies that there exists an anti-symmetric, complete and transitive binary relation $\hat{\succcurlyeq}$ on I. Then define a binary relation $\succcurlyeq$ on X as follows: $\bigcup_{i\in I} \succcurlyeq_i \subseteq \succcurlyeq$, and for all $i \hat{\succ} j$, $C_i \times C_j \subseteq \succcurlyeq$. Then, $\succcurlyeq$ is anti-symmetric, complete, and transitive. Since X is locally connected, each C_i

is both open and closed. Then, each $\succcurlyeq_i$ has closed sections in both C_i and X. Note that for all C_i and all $x \in C_i$,

$$A_{\succcurlyeq}(x) = A_{\succcurlyeq_i}(x) \cup \left(\bigcup_{j \hat{\succ} i} C_j \right) = A_{\succcurlyeq_i}(x) \cup \left(\bigcap_{i \hat{\succcurlyeq} j} C_j^c \right).$$

Then, it follows from C_i is open for all $i \in I$ that $\succcurlyeq$ has closed upper sections. An analogous argument implies that $\succcurlyeq$ has closed lower sections. Therefore, $\succcurlyeq$ is continuous. □

3.2 On Weakly Ordered Topological Spaces

This subsection provides a necessary and sufficient condition for the existence of a non-trivial, complete, transitive and continuous binary relation on a connected topological space. This result is analogue to Theorem (Eilenberg), except that the binary relation is not necessarily anti-symmetric.

Definition 3.3 A topological space is *weakly ordered* if there exists a non-trivial, complete, transitive and continuous binary relation on it.

Theorem 3.4 *A connected topological space X which contains at least two elements can be weakly ordered if and only if $P(X| \sim)$ is disconnected for some equivalence relation $\sim$ on X.*

Proof of Theorem 3.4 Let X be a topological space with at least two elements. Assume there exists a non-trivial, complete, transitive and continuous binary relation $\succcurlyeq$ on X. Let $\sim$ denote the symmetric part of $\succcurlyeq$. Since X is connected, the quotient space $X| \sim$ is connected. It is easy to show that the induced binary relation $\hat{\succcurlyeq}$ on $X| \sim$, defined as $([x], [y]) \in \hat{\succcurlyeq}$ if and only if $(x', y') \in \succcurlyeq$ for all $x' \in [x]$ and all $y' \in [y]$, is non-trivial, anti-symmetric, complete, transitive and continuous. Then, it follows from Theorem (Eilenberg) that $P(X| \sim)$ is disconnected.

In order to prove the backward direction, assume there exists an equivalence relation $\hat{\sim}$ on X such that $P(X|\hat{\sim})$ is disconnected. Then, $X|\hat{\sim}$ contains at least two elements. Since $X|\hat{\sim}$ is connected, it follows from Theorem (Eilenberg) that there exists an anti-symmetric, complete, transitive and continuous binary relation $\hat{\succcurlyeq}$ on $X|\hat{\sim}$. Define a binary relation $\succcurlyeq$ on X as $(x, y) \in \succcurlyeq$ if and only if $([x], [y]) \in \hat{\succcurlyeq}$. Then the symmetric part $\sim$ of $\succcurlyeq$ is identical to $\hat{\sim}$. It follows from $A_{\hat{\succcurlyeq}}([x])$ and $A_{\hat{\preccurlyeq}}([x])$ are closed in $X|\hat{\sim}$ and the definition of the quotient topology that the sections

$$A_{\succcurlyeq}(x) = \bigcup_{[y] \hat{\succcurlyeq} ([x])} [y] \text{ and } A_{\preccurlyeq}(x) = \bigcup_{[y] \hat{\preccurlyeq} [x]} [y]$$

of $\succcurlyeq$ are closed in X, hence $\succcurlyeq$ is continuous. The non-triviality, completeness and transitivity of $\succcurlyeq$ directly follow from its construction. □

Note that in an ordered space, the indifference relation $\sim$ in Theorem 3.4 is assumed to be the *equality* relation. Hence, as expected, the requirement for the existence of an order is stronger than the requirement for that of a weak order. The following example illustrates a weakly ordered topological space which cannot be ordered.

Example Let $X = [0, 2]$ and the following define a basis for the topology on X: $[0, x)$ for all $x \in (1, 2]$, $(x, 2]$ for all $x \in [1, 2)$, and (x, y) for all $x, y \in [1, 2]$. Note that the smallest closed set containing any point in $[0, 1]$ is $[0, 1]$. It is clear that X is connected. Since the topology is not Hausdorff, Eilenberg [33, 1.4] implies that there does not exist an anti-symmetric, complete and continuous binary relation on X. However, the following is a non-trivial, complete, transitive and continuous binary relation on X: $(x, y) \in \preccurlyeq$ for all $x, y \in [0, 1]$, and $(x, y) \in \preccurlyeq$ for all $x, y \in X$ with $x < y$.

Finally, the methods of proofs of Theorems 3.1 and 3.2 can be used to provide generalizations of this result to disconnected spaces.

3.3 Discussion of the Results

We can apply our results to the literature on the non-existence of a non-constant function on topological spaces as follows. First, note that every non-constant continuous function induces a non-trivial, complete, transitive and continuous binary relation. Therefore, by Hewitt's [52] result we know that there does not exist a non-trivial, complete, transitive and continuous relation. Moreover, note that the space in Hewitt's paper is countable, hence separable, and connected. Therefore, every non-trivial, complete, transitive and continuous relation has a non-constant, continuous real-valued representation. Therefore, Theorem 3 provides an equivalence condition for the existence of a non-constant function in Hewitt's setting. Hence, Theorem 3 may provide a new perspective on Hewitt's theorem and on the subsequent work in this line of work.

Moreover, [43, 54, 62, 74] provide countable spaces that are connected and satisfy the Hausdorff separation axiom. Since continuous functions take connected sets to connected sets, therefore there cannot exist a non-constant continuous function on these spaces. We next show that there does not exist a continuous, non-trivial, semi-transitive relation with a transitive symmetric part on these spaces. First, by appealing to the current authors' earlier work,[11] any such relation is complete and transitive. Since the space is countable, it is separable. Therefore, it follows

[11] See Khan-Uyanik [59, Theorem 2]. We refer the reader to [60, 88] for generalizations to bi-preference structures and general parametrized topological spaces.

from Debreu [20, Theorem I] that there exists a continuous real-valued function representing the binary relation. Since the relation is non-trivial, therefore the function is non-constant. This furnishes us a contradiction.

The literature has focused on the existence, or non-existence, of a non-constant continuous function. For binary relations, different continuity postulates has been introduced and used in mathematical economics. The existence of a non-trivial binary relation satisfying different continuity assumptions may be of interest; see [87] for an extended discussion on the continuity postulate.[12]

4 On the Additivity Postulate

In this section we provide two results on the implications of the additivity postulate. We first show that a strong form of additivity postulate is equivalent to the transitivity postulate. Then we define three monotone continuity postulates on partially ordered sets, inspired by the pioneering work of Villegas on qualitative probability, and then show that under the additivity postulate, the three continuity postulates are equivalent. We end this section by relating our results to the antecedent literature.

4.1 *Additivity and Transitivity: A Two-Way Relationship*

We first present a result on the relationship between additivity and strong additivity.

Definition 4.1 A binary relation $\succcurlyeq$ on an Abelian group $(X, +)$ is called *additive* if for all $x, y, z \in X$, $x \succcurlyeq y$ implies $x + z \succcurlyeq y + z$. Moreover, we say $\succcurlyeq$ is *strongly additive* if for all $x_1, x_2, y_1, y_2 \in X$, $x_i \succcurlyeq y_i$ for $i = 1, 2$ implies $x_1 + x_2 \succcurlyeq y_1 + y_2$.

Proposition 4.2 *Every reflexive and strongly additive relation on an Abelian group is additive.*

Proof of Proposition 4.2 Assume $\succcurlyeq$ is strongly additive relation on an Abelian group $(X, +)$. Pick $x, y, z \in X$ such that $x \succcurlyeq y$. Then $z \succcurlyeq z$, by reflexivity, and strong additivity of $\succcurlyeq$ imply $x + z \succcurlyeq y + z$. Hence $\succcurlyeq$ is additive. □

Along with this observation, the next result shows that when a reflexive binary relation is transitive, the two additivity postulates are equivalent. Moreover, it shows that the transitivity of the relation is implied by strong additivity.

[12]There is a literature on different continuity postulates for functions; see Ciesielski-Miller's [19] recent survey on this.

Theorem 4.3 *An additive binary relation $\succcurlyeq$ on an Abelian group $(X, +)$ is transitive if and only if it is strongly additive.*

Proof of Theorem 4.3 Let $\succcurlyeq$ be an additive binary relation on an Abelian group $(X, +)$. Assume $\succcurlyeq$ is transitive. Pick $x_1, x_2, y_1, y_2 \in X$ such that $x_i \succcurlyeq y_i$ for $i = 1, 2$. Then it follows from additivity that $x_1+x_2 \succcurlyeq y_1+x_2$ and $x_2+y_1 \succcurlyeq y_2+y_1$. Then commutativity of $+$ and transitivity of $\succcurlyeq$ implies that $x_1 + x_2 \succcurlyeq y_1 + y_2$.

Now assume $\succcurlyeq$ is strongly additive. Pick $x, y, z \in X$ such that $x \succcurlyeq y \succcurlyeq z$. Then strong additivity implies $x + y \succcurlyeq y + z$. Then additivity of $\succcurlyeq$ imply $x + y + (-y) \succcurlyeq y + z + (-y)$. Therefore $x \succcurlyeq z$. □

The following is a direct corollary of Proposition 4.2 and Theorem 4.3.

Corollary 4.4 *Every reflexive and strongly additive relation on an Abelian group is transitive.*

4.2 Implications of Additivity for Monotone Continuity

Let $(X, \geq)$ be a partially ordered set. We say X is *order-complete* if every non-empty subset of X with an upper bound has a least upper bound. Note that a poset X is order-complete if and only if every non-empty subset of X with a lower bound has a greatest lower bound; see Fremlin (3.14B, vol3I).

Definition 4.5 Let $(X, \geq)$ be an order-complete poset and $\succcurlyeq$ a binary relation on X. Consider the following monotone continuity axioms for $\succcurlyeq$.

[C1′] For all $y \in X$ and all bounded below sequence $\{x_i\}_{i\in\mathbb{N}}$ in X, $x_i \geq x_{i+1}$ and $x_i \succcurlyeq y$ for all i imply $\inf\{x_i\}_{i\in\mathbb{N}} \succcurlyeq y$.

[C2′] For all $y \in X$ and all bounded above sequence $\{x_i\}_{i\in\mathbb{N}}$ in X, $x_{i+1} \geq x_i$ and $y \succcurlyeq x_i$ for all i imply $y \succcurlyeq \sup\{x_i\}_{i\in\mathbb{N}}$.

[C3′] For all $y \in X$ and all bounded above sequence $\{x_i\}_{i\in\mathbb{N}}$, $x_{i+1} \geq x_i$ and $y \prec \sup\{x_i\}_{i\in\mathbb{N}}$ imply there exists an integer $N > 0$ such that, for $i \geq N$, we have $y \prec x_i$.

Theorem 4.6 *For any complete and strongly additive binary relation on an Abelian group which is also an order-complete poset, the continuity axioms* C1′, C2′, *and* C3′ *are equivalent.*

Proof of Theorem 4.6 Let $(X, +, \geq)$ be an order-complete poset on an Abelian group and $\succcurlyeq$ a complete and strongly additive binary relation on X. It follows from Corollary 4.4 that $\succcurlyeq$ is transitive.

First, we show that C1′ is equivalent to C2′. Note that additivity implies $x \succcurlyeq y$ if and only if $-y \succcurlyeq -x$. Assume C1′. Pick a bounded above sequence $\{x_i\}_{i\in\mathbb{N}}$ and y in X such that $x_{i+1} \geq x_i$ and $y \succcurlyeq x_i$ for all i. Then, $-x_i \geq -x_{i+1}$ and $-x_i \succcurlyeq -y$ for all i. It follows form C1′ that $\inf\{-x_i\}_{i\in\mathbb{N}} \succcurlyeq -y$. We now show that additivity implies $\inf\{-x_i\}_{i\in\mathbb{N}} = -\sup\{x_i\}_{i\in\mathbb{N}}$. Define $\underline{x} = \inf\{-x_i\}_{i\in\mathbb{N}}$ and $\bar{x} = -\sup\{x_i\}_{i\in\mathbb{N}}$

Assume towards a contradiction that $\underline{x} > \bar{x}$. By definition, $\underline{x} \leq -x_i$ for all i. Then additivity implies $-\underline{x} \geq x_i$ for all i. Then $-\underline{x}$ is an upper bound of $\{x_i\}_{i\in\mathbb{N}}$, hence $-\underline{x} \geq \bar{x}$. This contradicts the assumption that $\underline{x} > \bar{x}$. An analogous argument yields a contradiction for $\underline{x} < \bar{x}$. Therefore, $\underline{x} = \bar{x}$. Then $-\sup\{x_i\}_{i\in\mathbb{N}} \not\succcurlyeq -y$, hence by additivity, $y \not\succcurlyeq \sup\{x_i\}_{i\in\mathbb{N}}$. Therefore, C2′ holds. The proof of the converse relationship is analogous.

We next show that C2′ is equivalent to C3′. Assume C2′. Assume towards a contradiction that there exists a bounded above $\{x_i\}_{i\in\mathbb{N}}$ and y in X such that $x_{i+1} \geq x_i$ and $y \prec \sup\{x_i\}_{i\in\mathbb{N}}$, but for all $N > 0$, there exists $j \geq N$ such that $y \not\succcurlyeq x_i$. Then there exists a subsequence $\{x_{ik}\}_{k\in\mathbb{N}}$ such that for all k, $x_{i_{k+1}} \geq x_{i_k}$ and $y \not\succcurlyeq x_{i_k}$. Then C2′ implies $y \not\succcurlyeq \sup\{x_{i_k}\}_{k\in\mathbb{N}}$. It is easy to see that $\sup\{x_{i_k}\}_{k\in\mathbb{N}} = \sup\{x_i\}_{i\in\mathbb{N}}$. This contradicts the assumption that $y \prec \sup\{x_i\}_{i\in\mathbb{N}}$. Hence C3′ holds. The converse relationship immediately follows from the definitions. □

4.3 *Discussion of the Results*

Villegas [89] introduced the following additivity concept for binary relations on a σ-algebra.

Definition 4.7 A preference relation $\succcurlyeq$ on a σ-algebra $\mathcal{X}$ on a set X is *Villegas-additive* if for all $A_1, A_2, B_1, B_2 \in \mathcal{X}$ with $A_1 \cap A_2 = B_1 \cap B_2 = \emptyset$, $A_i \not\succcurlyeq B_i$ for $i = 1, 2$ implies $A_1 \cup A_2 \not\succcurlyeq B_1 \cup B_2$. If, in addition, $A_1 \succ B_1$ or $A_2 \succ B_2$, then $A_1 \cup A_2 \succ B_1 \cup B_2$.

First note that the union operation is similar to the additivity operation[13] but it does not satisfy all properties the addition in an Abelian group satisfies. Moreover, the usual additivity assumption is neither stronger nor weaker than Villegas-additivity: the latter imposes restriction on a smaller class of elements whereas additivity does not impose a restriction on the strict relation. DeGroot [27, Theorem 1, p. 71] followed Villegas and proved a result analogous to Theorem 4.3 where the space is a σ-algebra with the usual inclusion relation.

Theorem (DeGroot) *Every complete and Villegas-additive binary relation on a σ-algebra is transitive.*

We next apply our results to de Finetti's expected utility representation theorem. Let $X = \mathbb{R}^n$ which is endowed with the usual topological, algebraic and order structures. A real valued function u is called *monotone* if for all $x, y \in \mathbb{R}^n$ such that $x > y$ (i.e. $x_i \geq y_i$ for all i and $x \neq y$), $u(x) > u(y.)$ A preference relation $\succcurlyeq$ on $\mathbb{R}^n$ is *monotone* if for all $x, y \in \mathbb{R}^n$, if $x > y$, then $x \succ y$. The following theorem is due to [25, 26].[14]

[13]Note that Fishburn [36, p. 336] calls Villegas-additivity the *additivity axiom*.

[14]See Wakker [91, Theorem A.2.1, p.161] for the statement and further details.

Theorem (de Finetti) *Let $\succcurlyeq$ be a binary relation on $\mathbb{R}^n$. The following are equivalent.*

(a) *The binary relation $\succcurlyeq$ is complete, transitive, additive and continuous.*
(b) *There exist positive $(p_i)_{i=1}^n$, summing to one, such that $u(x) = \sum_i p_i x_i$ represents $\succcurlyeq$.*

The equivalence theorem of de Finetti can be restated as

Corollary 4.8 *Let $\succcurlyeq$ be a binary relation on $\mathbb{R}^n$. The following are equivalent.*

(a) *The binary relation $\succcurlyeq$ is complete, strongly additive and continuous.*
(b) *There exist positive $(p_i)_{i=1}^n$, summing to one, such that $u(x) = \sum_i p_i x_i$ represents $\succcurlyeq$.*

Therefore, we can drop the transitivity assumption in de Finetti's theorem by replacing additivity with strong additivity, which are equivalent in the presence of the transitivity postulate. We can also drop the completeness assumption; see [87] for a detailed exposition on the hiddenness and redundancy in mathematical economics.[15]

We next move to monotone continuity. The second subsection above is an attempt to understand this postulate introduced in Villegas, DeGroot, Arrow and Chichilnisky in order to study qualitative/subjective probability. As we illustrate above, monotone continuity neither requires any topological property on the choice set, nor uses the structure of the unit interval, unlike the continuity assumption of [51]. Hence, an investigation of the relationship between monotone continuity with the other continuity postulates, and its applications may be of interest, and our definitions and results can be considered as the first step for such investigation.[16] Villegas [89] and DeGroot [27] provide the following monotone continuity postulates for binary relations defined on σ-algebras.

Definition 4.9 Let $\mathcal{X}$ be a σ-algebra on a set and $\succcurlyeq$ a binary relation on $\mathcal{X}$. We define the following monotone continuity axioms for $\succcurlyeq$.

[C1] For all sets $\{A_i\}_{i\in\mathbb{N}}$, B in $\mathcal{X}$, $A_1 \supseteq A_2 \supseteq \cdots$ and $A_i \succcurlyeq B$ for all i imply $\bigcap_i A_i \succcurlyeq B$.
[C2] For all sets $\{A_i\}_{i\in\mathbb{N}}$, B in $\mathcal{X}$, '$A_1 \subseteq A_2 \subseteq \cdots$ and $B \succcurlyeq A_i$ for all i imply $B \succcurlyeq \bigcup_i A_i$.

[15] See also Krantz-Luce-Suppes-Tversky [66, Section 5.4.2] for an interesting discussion on hiddenness and redundancy. Moreover, it may be of interest to generalize this result to groupoids or semigroups; see Fishburn [35, Chapter 11].

[16] See [40, 59] for a discussion on the relationship among different continuity postulates. Moreover, [71] shows that on a σ-algebra monotone continuity implies the Arhimedeanity and solvability postulates, which are introduced by Luce [69]. Finally, DeGroot's assumption SP_5 "There exists a random variable which has a uniform distribution on the interval [0, 1]" may have some relevance to the existence of a nice relation, or a continuous function. We leave this question for future work.

[C3] For all sets $\{A_i\}_{i \in \mathbb{N}}$, B in $\mathcal{X}$, $A_1 \subseteq A_2 \subseteq \cdots$ and $B \prec \bigcup_i A_i$ imply there exists an integer $N > 0$ such that, for $i \geq N$, we have $B \prec A_i$.

The following is a result analogous to Theorem 4.6 above for the special case of σ-algebras.

Theorem 4.10 *For any complete and Villegas-additive binary relation on a σ-algebra, the monotone continuity postulates* C1, C2 *and* C3 *are equivalent.*

The equivalence between C2 and C3 is due to Villegas [89, Theorem] and between C1 and C2 is due to DeGroot [27, Theorem 5].

Villegas [89, 90] studied countably additive qualitative probability representation and showed that given a finitely additive qualitative probability, monotone continuity is necessary and sufficient for countably additive representation; see Krantz-Luce-Suppes-Tversky [66, Section 5.4.2] for further discussion. In particular, the following result is quoted.[17]

Theorem (Villegas) *A finitely additive probability representation of a structure of qualitative probability, on a σ-algebra, is countably additive if and only if the structure is monotonically continuous.*

Finally, the following monotone continuity postulate is due to [3]; also see [1, 2, 4].

Definition 4.11 Given a and b, where $a \succ b$, a consquence c and a vanishing sequence $\{E_i\}$, suppose sequence of actions satisfy the conditions that (a^i, s) yield the same consequences as (a, s) for all $s \in E_i^c$, and the consequence c for all $s \in E_i$, while (b^i, s) yield the same consequences as (b, s) for all $s \in E_i^c$, and the consequence c for all $s \in E_i$. Then, for all i sufficiently large, $a^i \succ b$ and $a \succ b^i$.

Chichilnisky [17] interpreted Arrow's definition as follows and showed that it is equivalent to the continuity postulate C1.

Definition 4.12 Let $\mathcal{X}$ be a σ-algebra on a set and $\succcurlyeq$ a binary relation on X. We call $\succcurlyeq$ satisfies *Monotone Continuity Axiom 4* (C4) if for all $\{A_i\}_{i \in \mathbb{N}}$, F, G in $\mathcal{X}$, $A_1 \supseteq A_2 \supseteq \cdots$, $\bigcap_{i=1}^{\infty} A_i = \emptyset$ and $F \succ G$ imply there exists $N > 0$ such that altering arbitrarily the events F and G on the set A_i, where $i > N$, does not alter the ranking of the events, namely $F' \succ G'$, where F' and G' are the altered events.

5 Order and Positivity in Mathematical Economics

We began this essay with Halmos' take on how applied mathematics transits to pure mathematics; and Derbyshire's take on how an important sub-field, with increasing importance, gets incorporated into the larger field of which it is a part, and thereby

[17] For definitions, we refer the reader to [66].

changes the identity of the larger field and loses its own. In this concluding section[18] to this chapter on binary relations in mathematical economics, we read, against the grain, these two texts and their claims on the incorporation of *positivity* and *order-theoretic methods* in Walrasian general equilibrium theory.

In classical Walrasian general equilibrium theory, as brought to fruition in [5, 21, 64, 73, 78], the agents in the economy are categorized as *consumers* and *producers,* with the former parametrized by preferences (a binary relation) defined on a (consumption) set and endowments being elements of such a set; and the latter drawing their signature simply by having an access to a *production set.*[19] The vernacular of *order* and *positivity* is relevant in so far as it is relevant to its constituent conceptions of a consumer and a producer. The idea of *monotonicity* enters the theory of production through the assumption of *free disposal,* an assumption delineated by Debreu [20] in the context of a production set, say in a ordered normed space whose positive cone has a non-empty interior.

> The assumption of free disposal for the technology means that if an input-output combination is possible, so is one where one where some outputs are smaller or some inputs larger; it is implied that a surplus can be freely disposed of. With this assumption, if the production set is non-empty, it has an interior point.

It is the existence of an interior that proves crucial for the sustainablity of technologically efficient production plans and Pareto optimal allocations through individual value and profit maximization.[20]

As far as the theory of the consumer is concerned, the ideas of *order* and *positivity* enter through the assumption of *monotonicity* of preferences which gets translated into "more is always preferred to less." To be sure, it factors into the Eilenberg

[18]In his participation in the composition of this section, Khan should like to acknowledge his indebtedness to conversations with Malcolm King, and Niccolò Urbinati, and to JJ Grobler's inspiring talk titled *101 years of vector lattice theory: A general form of integral: PJ Daniell (1918)* at the Conference. He should also like to acknowledge the stimulus received from Schliesser's readings of [38].

[19]Our choice of these four texts, to the exclusion of all others, should perhaps be justified. The Walrasian general equilibrium model is referred to in the economics mainstream as the Arrow-Debreu or the Arrow-Debreu-McKenzie model, and this naming facilitated a homogeneous monolithic view and added to the confusion and to an unfortunate haste in canonization; see Footnote 22 below. In this connection, the interested reader can also see Düppe-Weintraub [31, p. 204] who argue that the 1954 papers of Arrow-Debreu and McKenzie "by being applied, interpreted, shaped, and reshaped ... came to symbolize a new intellectual culture in American economics and help[ed] reconstruct the body of economic knowledge (p. 204)," a claim contested in [57] who urged the inclusion of Uzawa, Nikaido and Gale also as fellow-pioneers of what we are calling here "Walrasian general equilibrium theory." A confounding factor in this is that many of the pioneers of Walrasian general equilibrium theory were also pioneers of linear and non-linear programming; Uzawa breing one of the leaders. For this line of work, see [30], and the recent application of Uzawa's consequential extension of the Kuhn-Tucker-Karush theorem in [58].

[20]We invite the reader to compare Debreu's definition with corresponding definitions of the concept in the five texts to which Footnote 19 refers. The idea of "free disposal" is intimately tied to the non-negativity of prices; see [45], and compare [21] and [73] on this issue.

questions regarding binary relations with which we began the introduction. Thus Arrow-Hahn [5, p. 106] write:

> Wold seems to have been the first to see the need of specifying assumptions under which the representation of the continuous utility functions exists. Wold assumed that the [consumption set] is the entire non-negative orthant [of finite-dimensional Euclidean space] and that preference is strictly monotone in each commodity. A very considerable generalization, based on a mathematical paper by Eilenberg [33], was achieved with the deeper methods of [20]; he assumed only the continuity of preferences and the connectedness of the [consumption set] (a property weaker than convexity).

This is an important passage: its irony lies in the fact that it comes from two of the more distinguished and senior Walrasian theorists at the time who could not refrain from drawing arbitrary and needless distinctions between mathematicians and economists, and between mathematical and economic papers, and thereby in sighting [33], and bracketing it at the same time. The point is that Eilenberg and Wold were independent pioneers of what later assumed the identity of an important subfield of "choice and decision theory."[21]

But returning to trajectories being implicitly charted by Halmos and to Derbyshire, the point is that the monotonicity assumption for consumers in the Walrasian conception comes rather late in its development: it is not there, for example, in [21], or in [73], or the term even indexed in [5].[22] The more important question, however is where the subject is in terms of these, their trajectories. This is a question that merits an investigation of its own, and is outside the scope of this technical essay: it suffices to make two observations. With respect to Halmos, classical Walrasian general equilibrium theory has neglected, by its very definitional conception, interdependencies between the parametrizations of what it sees as the relevant agents in the economy; and classical game theory, again by virtue of its definitional conception, has neglected the market in its formalizations. The applied problems of our time cry out for a formalization of these interdependencies in what perhaps ought to be a synthetic view of both subjects. Thus even after 70

[21] We can recommend [35, 41, 42, 70, 75] and their references for this subject, which branches off also into mathematical psychology.

[22] This also suggests how much a reader of Walrasian general equilibrium theory loses by ascribing to it a monolithic conception. Each of these pioneers had their own ways of looking at their subject. In this connection one may also refer the interested reader to McKenzie's conception of production in his McKenzie [73, Section 2.8, pages 77–82] on an "Economy of Activities." It is also perhaps worth noting that Debreu's resistance to the monotonicty assumption on consumers may be due to his having relaxed the montonicity assumption in [96]. To the authors knowledge, his first recourse to the assumption is in connection with the Debreu-Scarf theorem in 1963, and to be sure the auumption irrevocably enters into the field with Aumann and his Israeli school of Walrasian theory; see [24] for the relevant papers and references. As emphasized in [57], the erasure of the production sector can also be ascribed to this school, and it becomes folded into the iideological divide between the "two Cambridges," those of the UK and the US. Foucault's [38] emphasis on "governmentality" in the formulation of *perfect competition* and its normative properties is clearly relevant here. What we are referring to as the Israeli School has also been referred to as the Belgian-Israeli School and includes besides Aumann, Kannai and Schmeidler Dreze, Mertens, Gabszswicz, Vind, Kurz, Neyman, Shitovitz, Greenburg, Peleg and Einy, to take a random order.

years, mathematical economics (including game theory) has very much retained its dependence on both economics *and* mathematics. This is to say that it has remained pure *and* applied. As to Derbyshire on algebra, in terms of the algebraic approach to these subjects, it has yet to be incorporated into both Walrasian general equilibrium theory and in non-cooperative game theory. In the authors' judgement, this cannot but be a fruitful task.

There is another, perhaps narrower, way to view the substance of these results. The question of the "right" commodity space for general Walrasian general equilibrium, or the "right setting" of the individual action sets in game theory, has not been explicitly posed. There has been little need to do so. Given the substantive questions at issue, the economic or game-theoretic formulations assume a strong-enough structure on the payoff functions and the choice sets by setting them either in a finite-dimensional Euclidean space, or in the context of game theory, a finite number of actions, to allow the question to be investigated and determinatively answered. When this rather arbitrary limitation is removed, the question becomes of consequence, and notions of *order* and *positivity* began to take on colours that one may not have previously imagined.[23]

This introduction has framed the results to follow as stemming from Eilenberg's [33] seminal work. In this connection we observe that it is an interesting curiosum in the history of ideas that a piece of work entirely peripheral to an author's *oeuvre,* written almost as a fragmentary passing thought, proves to be of such decisive and sustainable consequence in what may have been perceived at the time of its writing to be an unrelated discipline. Eilenberg's paper, along with Kakutani's [55] fixed point theorem, coincidentally published in the same year, may well be two canonical examples.[24] In any case, as far as mathematical economics is concerned, the belated recognition of this pioneering paper by Debreu [20, 23] and Sonnenschein [81, 82] has subsequently waned, and it is only recent work that has re-emphasized Eilenberg's work and given it importance under the rubric of what it refers to as the Eilenberg-Sonnenschein program.[25] It is a source of satisfaction to the authors that its importance can also be delineated in a chapter on a book on *positivity* and *order*, and with an explicit inter-disciplinary thrust.[26]

[23]This investigation remains an ongoing project of Khan and Urbinati, and in his talk in Pretoria, Khan made some room to expand at some length to report on Nikaido's contributions to this question in keeping with this project.

[24]1941 was a particular productive years for Eilenberg, especially given the standards of the time: in addition to the paper being discussed here, he published at least six other papers; see Derbyshire [28, p.302] for his 1940 meeting with Saunders MacLane, and his subsequent involvement with algebraic topology. His paper with Wilder on "uniform local connectedness and contractability" was to follow an year later, and the generalization of Kakutani's fixed point theorem with Montgomery, 5 years later.

[25]Barring Debreu and Sonnenschein, neglect of [33] goes back to [96, 97] and their followers.

[26]The reader has surely noted that DeGroot and Villegas mathematical statisticians. For the former's priority regarding the notion of *monotonic continuity,* the reader is referred to the reviews of Chacon (MR167588) and Good (MR215325).

Acknowledgments This work was initiated during Khan's visit to the Department of Economics, University of Queensland, July 27 to August 13, 2018. In addition to the hospitality of the Department, Khan also thanks Rabee Tourky for emphasizing the importance of Eilenberg [33] during a most pleasant visit at the trimester program "Stochastic Dynamics in Economics and Finance" held by Hausdorff Research Institute for Mathematics (HIM) in August 2013. The authors are still trying to track down Lerner (1907). They thank an anonymous referee and the editors of the Conference Proceedings for their comments, and also Youcef Askoura, Ying Chen, Aniruddha Ghosh and Eddie Schlee, Eric Schliesser and Niccollò Urbinati for conversation and collaboration. This paper draws its basic conception and composition from an invited plenary talk titled "The Role of Positivity in Mathematical Economics: Monotonicity and Free-Disposal in Walrasian Equilibrium Theory," and delivered by Khan at *Positivity X* held in Pretoria, July 8–12, 2019. He thanks Jan Harm and his team of Organizers for the invitation, and for their indispensable help regarding the logistics. He also thanks Jacek Banaciak, Bernard Cornet, Jacobus Grobler, Malcolm, King, Sonja Mouton and Asghar Ranjbari for stimulating conversation and encouragement after his talk, and Rajiv Vohra for his close reading of Section 5.

References

1. K.J. Arrow, *Aspects of the Theory of Risk-bearing* (Yrjö Jahnsson Foundation, Helsinki, 1965)
2. K.J. Arrow, Exposition of the theory of choice under uncertainty. Synthese **16**(3/4), 253–269 (1966)
3. K.J. Arrow, *Essays in the Theory of Risk-Bearing* (Amsterdam: North-Holland, 1971)
4. K.J. Arrow, Exposition of the theory of choice under uncertainty, in *Decision and Organization: A Volume in Honor of Jacob Marschak*, ed. by C.B. McGuire, R. Radner (North-Holland, Amsterdam, 1972)
5. K.J. Arrow, F.H. Hahn, *General Competitive Analysis* (North-Holland, Amsterdam, 1971)
6. G.B. Asheim, T. Mitra, B. Tungodden, Sustainable recursive social welfare functions. Econ. Theory **49**(2), 267–291 (2012)
7. K. Basu, T. Mitra, Aggregating infinite utility streams with intergenerational equity: the impossibility of being paretian. Econometrica **71**(5), 1557–1563 (2003)
8. K. Basu, T. Mitra, Possibility theorems for equitably aggregating infinite utility streams, in *Intergenerational Equity and Sustainability* (Springer, Berlin, 2007), pp. 69–84
9. A.F. Beardon, Topology and preference relations, in *Mathematical Topics on Representations of Ordered Structures and Utility Theory: Essays in Honor of Professor Ghanshyam B. Mehta*, ed. by G. Bosi, M.J. Campión, J.C. Candeal, E. Indurain (Springer, Switzerland, 2020), pp. 3–21
10. A.F. Beardon, G.B. Mehta, The utility theorems of wold, debreu, and arrow-hahn. Econometrica **62**(1), 181–186 (1994)
11. G. Birkhoff, *Lattice Theory (Revised Edition)*, vol. XXV (American Mathematical Society, New York, 1948)
12. G. Bosi, G. Herden, The structure of useful topologies. J. Math. Econ. **82**, 69–73 (2019)
13. G. Bosi, M. Zuanon, Continuity and continuous multi-utility representations of nontotal preorders: some considerations concerning restrictiveness, in *Mathematical Topics on Representations of Ordered Structures and Utility Theory: Essays in Honor of Professor Ghanshyam B. Mehta*, ed. by G. Bosi, M.J. Campión, J.C. Candeal, E. Indurain (Springer, Switzerland, 2020), pp. 213–236
14. G. Bosi, M.J. Campión, J.C. Candeal, E. Indurain, *Mathematical Topics on Representations of Ordered Structures and Utility Theory* (Springer, Switzerland, 2020)
15. W. Bossert, Y. Sprumont, K. Suzumura, Ordering infinite utility streams. J. Econ. Theory **135**(1), 579–589 (2007)
16. D.S. Bridges, G.B. Mehta, *Representations of Preference Orderings* (Springer, Berlin, 1995)

17. G. Chichilnisky, The foundations of statistics with black swans. Math. Social Sci. **59**(2), 184–192 (2010)
18. E.W. Chittenden, On general topology and the relation of the properties of the class of all continuous functions to the properties of space. Trans. Am. Math. Soc. **31**(2), 290–321 (1929)
19. K.C. Ciesielski, D. Miller, A continuous tale on continuous and separately continuous functions. Real Anal. Exch. **41**(1), 19–54 (2016)
20. G. Debreu, Valuation equilibrium and pareto optimum. Proc. Natl. Acad. Sci. **40**(8), 588–592 (1954)
21. G. Debreu, *Theory of Value: An Axiomatic Analysis of Economic Equilibrium*, vol. 17 (Yale University Press, Connecticut, 1959)
22. G. Debreu, Topological methods in cardinal utility theory, in *Mathematical Methods in the Social Sciences*, ed. by K.J. Arrow, S. Karlin, P. Suppes (Stanford University, California, 1960) pp. 16–26
23. G. Debreu, Continuity properties of paretian utility. Int. Econ. Rev. **5**(3), 285–293 (1964)
24. G. Debreu, *Mathematical Economics: Twenty Papers of Gerard Debreu*, vol. 4. Econometric Society Monographs (Cambridge University, Cambridge, 1983)
25. B. de Finetti, La prévision: ses lois logiques, ses sources subjectives, in *Annales de l'institut Henri Poincaré*, vol. 7 (1937), pp. 1–68
26. B. de Finetti, *Theory of Probability*, vol. 1 (Wiley, New York, 1974)
27. M.H. DeGroot, *Optimal Statistical Decisions* (McGraw-Hill, New York, 1970)
28. J. Derbyshire, *Unknown Quantity: A Real and Imaginary History of Algebra* (Plume Books, New York, 2006)
29. P.A. Diamond, The evaluation of infinite utility streams. Econometrica **33**(1), 170–177 (1965)
30. R. Dorfman, P.A. Samuelson, R.M. Solow, *Linear Programming and Economic Analysis* (McGraw-Hill, New York, 1958)
31. T. Düppe, E.R. Weintraub, *Finding Equilibrium: Arrow, Debreu, McKenzie and the Problem of Scientific Credit* (Princeton University, New Jersey, 2014)
32. B. Dushnik, E.W. Miller, Partially ordered sets. Am. J. Math. **63**(3), 600–610 (1941)
33. S. Eilenberg, Ordered topological spaces. Am. J. Math. **63**(1), 39–45 (1941)
34. M. Faber, *Studies in Austrian Capital Theory, Investment and Time* (Springer, New York, 1986)
35. P.C. Fishburn, *Mathematics of Decision Theory* (The Hauge, Mouton, 1972)
36. P.C. Fishburn, The axioms of subjective probability. Stat. Sci. **1**(3), 335–345 (1986)
37. M. Fleurbaey, P. Michel, Intertemporal equity and the extension of the ramsey criterion. J. Math. Econ. **39**(7), 777–802 (2003)
38. M. Foucault, *The Birth of Biopolitics: Lectures at the College de France* (Palgrave MacMillan, New York, 2008). French edition (2004)
39. T. Galaabaatar, M. Ali Khan, M. Uyanik, Completeness and transitivity of preferences on mixture sets. Math. Social Sci. **99**, 49–62 (2019)
40. A. Ghosh, M. Ali Khan, M. Uyanik, *Solvability and Continuity: On Bridging Two Communities* (Johns Hopkins University, Baltimore, mimeo, 2020)
41. I. Gilboa, *Uncertainty in Economic Theory* (Cambridge University, Cambridge, 2004)
42. I. Gilboa, *Theory of Decision Under Uncertainty* (Cambridge University, Cambridge, 2009)
43. S.W. Golomb, A connected topology for the integers. Am. Math. Mon. **66**(8), 663–665 (1959)
44. P.R. Halmos, The basic concepts of algebraic logic. Am. Math. Mon. **53**(6), 217–249 (1956)
45. C. Hara, Existence of equilibria in economies with bads. Econometrica **73**(2), 647–658 (2005)
46. C. Hara, K. Suzumura, Y. Xu, On the possibility of continuous, paretian and egalitarian evaluation of infinite utility streams, in *Andrew Young School of Policy Studies Research Paper Series* (2006), pp. 07–12
47. G. Herden, On the existence of utility functions. Math. Social Sci. **17**(3), 297–313 (1989)
48. G. Herden, On the existence of utility functions ii. Math. Social Sci. **18**(2), 107–117 (1989)
49. G. Herden, Topological spaces for which every continuous total preorder can be represented by a continuous utility function. Math. Social Sci. **22**(2), 123–136 (1991)
50. H. Herrlich, Wann sind alle stetigen abbildungen in y konstant? Math. Z. **90**(2), 152–154 (1965)

51. I.N. Herstein, J. Milnor, An axiomatic approach to measurable utility. Econometrica **21**(2), 291–297 (1953)
52. E. Hewitt, On two problems of urysohn. Ann. Math. **47**, 503–509 (1946)
53. S. Iliadis, V. Tzannes, Spaces on which every continuous map into a given space is constant. Can. J. Math. **38**, 1281–1298 (1986)
54. G.J. Jameson, *Topology and Normed Spaces* (Chapman and Hall, London, 1974)
55. S. Kakutani, A generalization of brouwer's fixed point theorem. Duke Math. J. **7**(2), 457–459 (1941)
56. N.J. Kalton, N.T. Peck, J.W. Roberts, *An F-space Sampler*, vol. 89 (Cambridge University, New York, 1984)
57. M.A. Khan, On the finding of an equilibrium: Düppe-weintraub and the problem of scientific credit. J. Econ. Lit. online(1), 1–50 (2020)
58. M.A. Khan, E.E. Schlee, Money-metrics in applied welfare analysis: a saddlepoint rehabilitation. ASU Working Paper **Forthcoming**, 169–171 (2019)
59. M.A. Khan, M. Uyanik, Topological connectedness and behavioral assumptions on preferences: a two-way relationship. Econ. Theory, 1–50 (2019). https://doi.org/10.1007/s00199-019-01206-7
60. M.A. Khan, M. Uyanik, On an extension of a theorem of eilenberg and a characterization of topological connectedness. Topol. Appl. **273**, 107–117 (2020)
61. M.A. Khan, L. Qiao, K.P. Rath, Y. Sun, Modeling large societies: why countable additivity is necessary. J. Econ. Theory **189**, 105102 (2020)
62. A.M. Kirch, A countable, connected, locally connected hausdorff space. Am. Math. Mon. **76**(2), 169–171 (1969)
63. V. Klee, On a question of bishop and phelps. Am. J. Math. **85**(1), 95–98 (1963)
64. T.C. Koopmans, *Three Essays on the State of Economic Science* (McGraw-Hill, New York, 1957)
65. T.C. Koopmans, Representation of preference orderings with independent components of consumption, in *Decision and Organization*, ed. by C.B. McGuire, R. Radner (University of Minnesota Press, Minneapolis, 1972), pp. Chapter 4, 79–100
66. D. Krantz, D. Luce, P. Suppes, A. Tversky, *Foundations of Measurement, Volume I: Additive and Polynomial Representations* (Academic Press, New York, 1971)
67. L. Lauwers, Ordering infinite utility streams comes at the cost of a non-ramsey set. J. Math. Econ., **46**(1), 32–37 (2010)
68. H. Lord, Connectednesses and disconnectednesses. Ann. N. Y. Acad. Sci. **767**(1), 115–139 (1995)
69. D. Luce, Sufficient conditions for the existence of a finitely additive probability measure. Ann. Math. Stat. **38**(3), 780–786 (1967)
70. D. Luce, *Utility of Gains and Losses: Measurement-Theoretical and Experimental Approaches* (Lawrance Erlbaum Associates, New Jersey, 2000)
71. A. Mackenzie, On atom-swarming, archimedeanity, and solvability, in *Working Paper* (2020)
72. M. Majumdar, Efficient programs in infinite dimensictnd spaces: A complete characterization. J. Econ. Theory **7**(4), 355–369 (1974)
73. L.W. McKenzie, *Classical General Equilibrium Theory*, vol. 1 (MIT Press, Cambridge, 2002)
74. G.G. Miller, Countable connected spaces. Proc. Am. Math. Soc. **26**(2), 355–360 (1970)
75. I. Moscati, *Measurinng Utility* (Oxford University, Oxford, 2016)
76. J.R. Munkres, *Topology* (Prentice Hall, New Jersey, 2000)
77. L. Nachbin, *Topology and Order*, vol. 4 (Krieger Pub Co, New Jersey, 1965)
78. H. Nikaido, Convex structures and economic theory, in *Mathematics in Science and Engineering*, vol. 51 (Academic Press, New York, 1968)
79. B. Peleg, M.E. Yaari, Efficiency prices in an infinite-dimensional space. J. Econ. Theory **2**(1), 41–85 (1970)
80. F.D. Sentilles, Bounded continuous functions on a completely regular space. Trans. Am. Math. Soc. **168**, 311–336 (1972)

81. H. Sonnenschein, The relationship between transitive preference and the structure of the choice space. Econometrica **33**(3), 624–634 (1965)
82. H. Sonnenschein, Reply to "a note on orderings". Econometrica **35**(3, 4), 540 (1967)
83. G. Stephan, Competitive finite value prices: a complete characterization, in *Studies in Austrian Capital Theory, Investment and Time*, ed. by M. Faber (Springer, New York, 1986), pp. 173–183
84. L.-G. Svensson, Equity among generations. Econometrica **48**(5), 1251–1256 (1980)
85. M.E. Toranzo, C. Hervés-Beloso, On the existence of continuous preference orderings without utility representations. J. Math. Econ. **24**(4), 305–309 (1995)
86. P. Urysohn, Uber die mächtigkeit der zusammenhängenden mengen. Math. Ann. **94**(1), 262–295 (1925)
87. M. Uyanik, M.A. Khan, The continuity postulate in economic theory: a deconstruction and an integration, in *Working Paper* (2019)
88. M. Uyanik, M.A. Khan, On the consistency and the decisiveness of the double-minded decision-maker. Econ. Lett. **185**, 108657, 1–5 (2019)
89. C. Villegas, On qualitative probability σ-algebras. Ann. Math. Stat. **35**(4), 1787–1796 (1964). Mathematical Reviews (R. V. Chacon): MR167588
90. C. Villegas, On qualitative probability. Am. Math. Mon. **74**(6), 661–669 (1967). Mathematical Reviews (I. J. Good): MR215325
91. P. Wakker, *Additive Representations of Preferences: A New Foundation of Decision Analysis* (Kluwer Academic Publishers, Boston, 1989)
92. A.D. Wallace, Relations on topological spaces, in *General Topology and its Relations to Modern Analysis and Algebra, Proceedings of the symposium held in Prague in September 1961* (Academia Publishing House of the Czechoslovak Academy of Sciences, Prague, 1962), pp. 356–360
93. L.E. Ward, Binary relations in topological spaces. An. Acad. Brasil. Ci **26**, 357–373 (1954)
94. L.E. Ward, Partially ordered topological spaces. Proc. Am. Math. Soc. **5**(1), 144–161 (1954)
95. R.L. Wilder, *Topology of Manifolds* (American Mathematical Society Colloquium Publications XXXII, Berlin, 1949)
96. H. Wold, A synthesis of pure demand analysis, i–iii. Scandinavian Actuarial J. **26**, 85–118, 220–263; 27, 69–120 (1943–1944)
97. H. Wold, L. Jureen, *Demand Analysis* (Wiley, New York, 1953)
98. W.R. Zame, Can intergenerational equity be operationalized? Theor. Econ. **2**(2), 187–202 (2007)

On Fixed Point Theory in Partially Ordered (Quasi-)metric Spaces and an Application to Complexity Analysis of Algorithms

M. D. Mabula, J. J. Miñana, and O. Valero

Dedicated to the memory of Coenraad Labuschagne

Abstract In this paper we prove an existence of fixed point theorem in partially ordered quasi-metric spaces which extends a celebrated result of Nieto and Rodríguez-López. Moreover, we derive from our main result a fixed point theorem for self-mappings in partially ordered metric spaces which improves the Nieto and Rodríguez-López one. Furthermore, we show that our assumptions can not be weakened and we show that our result can not be deduced from the well known Kleene's fixed point theorem. In addition, we focus our attention on the particular case in which the specialization order is under consideration. Finally, an application to complexity analysis of algorithms is presented.

Keywords Partially ordered metric space · Partially ordered quasi-metric space · Fixed point · Specialized partial order · Complexity analysis of algorithms

Mathematics Subject Classification (2010) 47H10, 54E50, 54F05, 68Q25

Mabula was supported by South African National Research Foundation grant N01504-112207.

M. D. Mabula (✉)
Department of Mathematics and Applied Mathematics, University of Pretoria, Pretoria, South Africa
e-mail: mokhwetha.mabula@up.ac.za

J. J. Miñana · O. Valero
Departament de Ciències Matemàtiques i Informàtica, Universitat de les Illes Balears, Palma, Spain
Institut d'Investigació Sanitària Illes Balears (IdISBa), Palma, Spain
e-mail: jj.minana@uib.es; o.valero@uib.es

E. Kikianty et al. (eds.), *Positivity and its Applications*, Trends in Mathematics,
https://doi.org/10.1007/978-3-030-70974-7_13

1 Introduction and Preliminaries

In [10], J.J. Nieto and R. Rodríguez-López proved a fixed point theorem for self-mappings in ordered metric spaces (metric spaces endowed with a partial order relation) which was useful to show the existence and uniqueness of solutions to differential equations. In order to recall such a result let us fix a few pertinent notations. From now on, given a sequence $(x_n)_{n\in\mathbb{N}}$ in a partially ordered metric space $(X, d, \preceq)$, we will set

$$U_{\preceq}(x_n) = \{y \in X : y \text{ is an upper bound of } (x_n)_{n\in\mathbb{N}} \text{ in } (X, \preceq)\};$$
$$L_d(x_n) = \{y \in X : (x_n)_{n\in\mathbb{N}} \text{ is convergent to } y \text{ with respecto to } \tau(d)\};$$
$$UL_{\preceq,d}(x_n) = U_{\preceq}(x_n) \cap L_d(x_n).$$

Following the previous notations, the aforementioned fixed point result can be stated as follows:

Theorem 1.1 *Let $(X, d, \preceq)$ be an ordered metric space such that (X, d) is complete. Let $f : X \to X$ be a monotone mapping. Assume that the following conditions are satisfied:*

(i) *there exists $x_0 \in X$ such that $x_0 \preceq f(x_0)$,*
(ii) *there exists $k \in [0, 1[$ such that $d(f(x), f(y)) \leq kd(x, y)$ for all $y \preceq x$,*
(iii) *if $(x_n)_{n\in\mathbb{N}}$ is an increasing sequence in $(X, \preceq)$ and there exists $x \in L_d(x_n)$, then $x \in U_{\preceq}(x_n)$.*

Then f has a fixed point $x^\star$ such that $x^\star \in UL_{\preceq,d}(f^n(x_0))$.

In the last years fixed point theory in quasi-metric spaces has been shown to be useful in Computer Science (see [6, 7, 13, 15, 17]). Inspired, in part by the utility of fixed point theory in Computer Science, we focus our attention on the possibility of extending Theorem 1.1 to the quasi-metric context. Thus we prove an existence of fixed point theorem in partially ordered quasi-metric spaces (quasi-metric spaces endowed with a partial order). Moreover, we show that our assumptions can not be weakened and we retrieve as a particular case of a new result an improved version of Theorem 1.1 when a partially ordered metric space is under consideration. Furthermore, we focus our attention on the particular case in which the specialization order is under consideration and we show that our result can not be deduced from the well known Kleene's fixed point theorem. Finally, we apply the exposed theory to complexity analysis of algorithms.

We assume that the reader is familiar with the basics on fixed point theory in metric spaces (we refer the reader to [16] for a detailed treatment of the topic). In what follows, we denote by $\mathbb{N}$ and $\mathbb{R}^+$ the set of positive integers and the set of positive real numbers, respectively. We recall some preliminary notions that we will need in our discussion later on.

According to [19], if $(X, \preceq)$ is a partially ordered set and $Y \subseteq X$, then an upper bound for Y in $(X, \preceq)$ is an element $x \in X$ such that $y \preceq x$ for all $y \in Y$. The least upper bound for Y in $(X, \preceq)$, if exists, is an element $z \in X$ which is an upper bound for Y and, in addition, satisfies that $z \preceq x$ provided that $x \in X$ is an upper bound for Y.

Following [6], a quasi-metric on a nonempty set X is a function $d : X \times X \to \mathbb{R}^+$ such that for all $x, y, z \in X$:

(i) $d(x, y) = d(y, x) = 0 \Leftrightarrow x = y$;
(ii) $d(x, z) \leq d(x, y) + d(y, z)$.

The pair (X, d) is called a quasi-metric space. Clearly a metric d on a nonempty set X is a quasi-metric which holds additionally the property (iii) $d(x, y) = d(y, x)$ for all $x, y \in X$.

Each quasi-metric d on a set X induces a T_0 topology $\tau(d)$ on X which has as a base the family of open d-balls $\{B_d(x, r) : x \in X,\ r > 0\}$, where $B_d(x, r) = \{y \in X : d(x, y) < r\}$ for all $x \in X$ and $r > 0$. Thus a sequence $(x_n)_{n\in\mathbb{N}}$ in a quasi-metric space (X, d) is $\tau(d)$-convergent to $x \in X$ provided that $d(x, x_n) \to 0$ as $n \to \infty$ (see [6]). Notice that unlike the metric case, the topology induced by a quasi-metric is only T_0 but not T_2 (Hausdorff) in general. A quasi-metric space (X, d) is called T_1 whenever $d(x, y) = 0 \Leftrightarrow x = y$. Examples 1 and 7 provide instances of quasi-metric spaces whose induced topology is only T_0 and T_2, respectively. A quasi-metric space whose topology is T_1 is given by the pair (X, d), where $X = \{0, 1\}$ and d is the defined by $d(x, x) = 0$ and $d(x, y) = 2^{-(y+1)}$ for all $x, y \in X$ with $x \neq y$.

It must be stressed that every quasi-metric d on a nonempty set X induces in a natural way a metic d^s given by $d^s(x, y) = \max\{d(x, y), d^{-1}(x, y)\}$ for all $x, y \in X$, where d^{-1} is a quasi-metric, called the conjugate quasi-metric of d, defined by $d^{-1}(x, y) = d(y, x)$ for all $x, y \in X$.

2 The Existence of Fixed Point

In this section, we prove an existence result for fixed point theory in partially ordered quasi-metric spaces which extends Theorem 1.1. To this end, let us note that, given a sequence $(x_n)_{n\in\mathbb{N}}$ in a quasi-metric space (X, d), we can consider the subsets $L_d(x_n)$ and $UL_{\preceq,d}(x_n)$ introduced in previous section simply replacing the metric by a quasi-metric in the definition. Next we introduce the following notion which will play a central role in our subsequent discussion. From now on, we will say that a sequence $(x_n)_{n\in\mathbb{N}}$ in a partially ordered quasi-metric space $(X, d, \preceq)$ is $L_{d^{-1}}$-bounded provided that the following is hold: If there exists $x \in X$ such that $x \in L_{d^{-1}}(x_n)$, then there exists $y \in UL_{\preceq,d^s}(x_n)$.

The next example provides an instance of $L_{d^{-1}}$-bounded sequence.

Example 1 Consider the partially ordered quasi-metric space $([0, 1], d_u, \preceq)$ such that d_u is the quasi-metric defined by $d_u(x, y) = \max\{y - x, 0\}$ for all $x, y \in [0, 1]$ and the partial order $\preceq$ is defined as follows: $1 \preceq 1$, $x \preceq 0 \preceq 1$ for all $x \in [0, 1[$

and $x \preceq y \Leftrightarrow x \leq y$ whenever $x, y \in]0, 1[$, $\leq$ stands for the usual order in $\mathbb{R}^+$ restricted to $[0, 1]$. Next let $(x_n)_{n\in\mathbb{N}}$ be the sequence given by $x_n = 1 - \frac{1}{n}$ for all $n \in \mathbb{N}$. Clearly $U_{\preceq}(x_n) = \{0, 1\}$. Moreover, it is obvious that $0, 1 \in L_{d_u^{-1}}(x_n)$ and that $1 \in L_{d_u^s}(x_n)$ and, thus $1 \in UL_{\preceq, d_u^s}(x_n)$.

The next example shows an instance of a sequence which is not $L_{d^{-1}}$-bounded.

Example 2 Consider the partially ordered quasi-metric space $(\mathbb{R}^+, d_l, \preceq_=)$ such that d_l is the quasi-metric defined by $d_l(x, y) = \max\{x - y, 0\}$ and the partial order $\preceq_=$ is defined by $x \preceq_= y \Leftrightarrow x = y$. Let $(x_n)_{n\in\mathbb{N}}$ in $\mathbb{R}^+$ be the sequence given by $x_n = 1 - \frac{1}{n}$ for all $n \in \mathbb{N}$. Then it is not hard to check that $L_{d_l^{-1}}(x_n) = [1, \infty[$ but $U_{\preceq_=}(x_n) = \emptyset$.

In order to introduce our main result we need to fix a notion of continuity for the self-mapping and of completeness for the quasi-metric space. Given a partially ordered quasi-metric space $(X, d, \preceq)$ and $x \in X$, a mapping $f : X \to X$ will say to be $\tau(d)$-$\preceq$-continuos at x provided that $f(y) \in L_d(f^{n+1}(x)_n)$ whenever $y \in UL_{\preceq, d^s}(f^n(x)_n)$. Examples 4, 5, and 7 give instances of $\tau(d)$-$\preceq$-continuous mappings and, in addition, Example 6 gives an instance of a mapping which is not $\tau(d)$-$\preceq$-continuous.

Observe that every continuous mapping at x from $(X, \tau(d^s))$ into $(X, \tau(d))$ is always $\tau(d)$-$\preceq$-continuos at x. However the example below shows that the converse is not true.

Example 3 Consider the partially ordered quasi-metric space $([0, 1], d_u, \leq)$, where (X, d_u) is the quasi-metric space introduced in Example 1. Define the mapping $f : [0, 1] \to [0, 1]$ by $f(x) = \frac{x}{2}$ if $x \in]0, 1]$ and $f(0) = 1$. It is clear that f is $\tau(d_u)$-$\leq$-continuous at 1 but, however, it is not continuous at 1 from $([0, 1], \tau(d^s))$ into $([0, 1], d_u)$.

On account of [11], a quasi-metric space (X, d) is right K-sequentially complete provided that every right K-Cauchy sequence is convergent with respect to $\tau(d)$, where a sequence $(x_n)_{n\in\mathbb{N}}$ is said to be right K-Cauchy if, given a real number $\varepsilon > 0$, there exists $n_0 \in \mathbb{N}$ such that $d(x_m, x_n) < \varepsilon$ for all $m \geq n \geq n_0$.

In the light of the exposed notions we are able to state our main result whose contractive condition is inspired in those assumed in [9, 12, 14].

Theorem 2.1 *Let $(X, d, \preceq)$ be a partially ordered quasi-metric space such that (X, d^{-1}) is right K-sequentially complete and let $f : X \to X$ be a mapping. If there exist $k \in [0, 1[$ and $x_0 \in X$ satisfying that:*

(i) *the sequence $(f^n(x_0))_{n\in\mathbb{N}}$ is $L_{d^{-1}}$-bounded,*
(ii) *f is $\tau(d)$-$\preceq$-continuos at x_0,*
(iii) *for each $n \in \mathbb{N} \cup \{0\}$,*

$$d(f^{n+1}(x_0), f(y)) \leq kd(f^n(x_0), y)$$

for all $y \in U_{\preceq}(f^n(x_0)) \cup \mathcal{O}_n(f, x_0)$, where $\mathcal{O}_n(f, x_0) = \{f^{n+i}(x_0) : i \in \mathbb{N}\}$.

Then f has a fixed point $x^\star$ such that $x^\star \in UL_{\preceq, d^s}(f^n(x_0))$.

Proof By condition (iii) we have that $d(f^n(x_0), f^{n+1}(x_0)) \leq k^n d(x_0, f(x_0))$ for all $n \in \mathbb{N}$. It follows that $(f^n(x_0))_{n\in\mathbb{N}}$ is a right K-Cauchy sequence in (X, d^{-1}). Indeed, let $m, n \in \mathbb{N}$ such that $n \leq m$. Then

$$\begin{aligned} d(f^n(x_0), f^m(x_0)) &\leq d(f^n(x_0), f^{n+1}(x_0)) + \cdots + d(f^{m-1}(x_0), f^m(x_0)) \\ &\leq \left(k^n + \cdots + k^{m-1}\right) d(x_0, f(x_0)) \\ &= \frac{k^{n-1} - k^{m+1}}{1-k} d(x_0, f(x_0)) \\ &\leq \frac{k^{n-1}}{1-k} d(x_0, f(x_0)). \end{aligned}$$

Since (X, d^{-1}) is a right K-sequentially complete quasi-metric space we have that the sequence $(f^n(x_0))_{n\in\mathbb{N}}$ is convergent with respect to $\tau(d^{-1})$. The fact that $(f^n(x_0))_{n\in\mathbb{N}}$ is a $L_{d^{-1}}$-bounded sequence guarantees that there exist $x^\star \in U_{\preceq}(f^n(x_0)_n)$ such that $x^\star \in L_{d^s}(f^n(x_0)_n)$. It remains to prove that $x^\star$ is a fixed point of f. The fact that $x^\star \in U_{\preceq}(f^n(x_0)_n)$ and assumption (iii) guarantee that $d(f^{n+1}(x_0), f(x^\star)) \leq kd(f^n(x_0), x^\star) \leq kd^s(f^n(x_0), x^\star)$ for all $n \in \mathbb{N}$. Hence we deduce that $d(x^\star, f(x^\star)) = 0$, since we have that $\lim_{n\to\infty} d^s(f^n(x_0), x^\star) = 0$ and $d(x^\star, f(x^\star)) \leq d(x^\star, f^n(x_0)) + d(f^n(x_0), f(x^\star)) \leq d^s(f^n(x_0), x^\star) + kd^s(f^{n-1}(x_0), x^\star)$ for all $n \in \mathbb{N}$. The $\tau(d)$-$\preceq$-continuity of f at x_0 yields that $f(x^\star) \in L_d(f^n(x_0))$. It follows that $d(f(x^\star), x^\star) = 0$ because $d(f(x^\star), x^\star) \leq d(f(x^\star), f^n(x_0)) + d(f^n(x_0), x^\star)$ for all $n \in \mathbb{N}$. So we deduce that $d(x^\star, f(x^\star)) = d(f(x^\star), x^\star) = 0$ and, thus, that $f(x^\star) = x^\star$. □

We now show that the conditions assumed in Theorem 2.1 are necessary. Indeed, the next example yields that the right K-sequential completeness is a necessary condition.

Example 4 Consider, the partial order $\preceq_-$ defined on $]0, 1]$ by $x \preceq_- y \Leftrightarrow y \leq x$. Consider, in addition, the partially ordered quasi-metric space $(]0, 1], d_2, \preceq_-)$, where the quasi-metric $d_2 :]0, 1] \times]0, 1] \to \mathbb{R}^+$ is given by

$$d_2(x, y) = \begin{cases} y - x & \text{if } x \leq y \\ 2(x - y) & \text{if } y < x. \end{cases}$$

It is clear that $(]0, 1], d_2^{-1})$ is not right K-sequentially complete, since the sequence $(x_n)_{n\in\mathbb{N}}$, given by $x_n = \frac{1}{n}$ for all $n \in \mathbb{N}$, is right K-Cauchy but is not convergent with respect to $\tau(d_2^{-1})$. Define the mapping $f :]0, 1] \to]0, 1]$ by $f(x) = \frac{x}{2}$ for all $x \in]0, 1]$. Thus the sequence $(f^n(1))_{n\in\mathbb{N}}$ is $L_{d_2^{-1}}$-bounded. Moreover, a straightforward computation shows that f is $\tau(d_2)$-$\preceq_-$-continuos at 1. Furthermore, $d_2(f^{n+1}(1), f(y)) \leq \frac{1}{2} d(f^n(1), y)$ holds trivially whenever $y \in U_{\preceq_-}(f^n(1)) \cup \mathcal{O}_n(f, 1)$. Nevertheless, f has not fixed points.

The next example proves that the $L_{d^{-1}}$-boundness of the sequence $(f^n(x_0))_{n\in\mathbb{N}}$ is also a necessary condition.

Example 5 Consider the partially ordered quasi-metric space $(\mathbb{R}_1^+, d_u, \leq)$, where $\mathbb{R}_1^+ = \mathbb{R}^+ \setminus \{1\}$, $\leq$ denotes the usual partial order on $\mathbb{R}^+$ restricted to $\mathbb{R}_1^+$ and the quasi-metric $d_u : \mathbb{R}_1^+ \times \mathbb{R}_1^+ \to \mathbb{R}^+$ is defined by $d_u(x, y) = \max\{y - x, 0\}$ for all $x, y \in \mathbb{R}_1^+$. It is not hard to check that $(\mathbb{R}_1^+, d_u^{-1})$ is right K-sequentially complete. Consider the mapping $f : \mathbb{R}_1^+ \to \mathbb{R}_1^+$ by $f(x) = \frac{x+1}{2}$ for all $x \in \mathbb{R}_1^+$. An easy verification shows that f is $\tau(d_u)$-$\preceq$-continuous at 0. Moreover, we have that $d_u(f^{n+1}(0), f(y)) \leq \frac{1}{2} d_u(f^n(0), y)$ whenever $y \in U_{\leq}(f^n(0)) \cup \mathbb{O}_n(f, 0)$. However, $0 \in L_{d_u^{-1}}(f^n(0))$ and, in addition, $UL_{\leq, d_u^s}(f^n(0)) = \emptyset$. Hence, the sequence $(f^n(0))_{n\in\mathbb{N}}$ is not $L_{d_u^{-1}}$-bounded. Finally, the mapping f has not fixed points.

The $\tau(d)$-$\preceq$-continuity of the mapping cannot be weakened in Theorem 2.1 in order to guarantee the existence of fixed point.

Example 6 Consider the partially ordered quasi-metric space $([0, 1], d_l, \preceq_-)$, where d_l is the quasi-metric introduced in Example 2 to $[0, 1]$ and $\preceq_-$ is defined on $[0, 1]$ by $x \preceq_- y \Leftrightarrow y \leq x$. Clearly the quasi-metric space $([0, 1], d_l^{-1})$ is right K-sequentially complete. Consider the mapping f introduced in Example 3. Then it is not hard to check that $0 \in L_{d_l^{-1}}(f^n(1))$ and that $0 \in UL_{\preceq_-, d^s}(f^n(1))$. So the sequence $(f^n(1))_{n\in\mathbb{N}}$ is $L_{d_l^{-1}}$-bounded. Besides, $d_l(f^{n+1}(1), f(y)) \leq \frac{1}{2} d_l(f^n(1), y)$. for all $y \in U_{\preceq_-}(f^n(1)) \cup \mathbb{O}_n(f, 1)$. Next we show that f is not $\tau(d_l)$-$\preceq_-$-continuous at 1. Indeed, consider the sequence $(f^n(1))_{n\in\mathbb{N}}$. Clearly $f^n(1) = \frac{1}{2^n}$ for all $n \in \mathbb{N}$ and, thus, $0 \in UL_{\preceq_-, d_l^s}(f^n(1))$. Nonetheless, $f(0) \notin L_{d_l}(f^{n+1}(1))_{n\in\mathbb{N}})$, since $d_l(f(0), f^{n+1}(1)) = 1 - \frac{1}{2^{n+1}}$ for all $n \in \mathbb{N}$. Clearly f has not fixed points.

The contractive condition "$d(f^{n+1}(x_0), f(y)) \leq kd(f^n(x_0), y)$ for all $y \in U_{\preceq}(f^n(x_0)) \cup \mathbb{O}_n(f, x_0)$" is necessary in statement of Theorem 2.1 such as the next shows.

Example 7 Let $([0, 1], d_S, \leq)$ be the partially ordered quasi-metric space where the quasi-metric $d_S : [0, 1] \times [0, 1] \to \mathbb{R}^+$ is given by $d_S(x, y) = \begin{cases} x - y & \text{if } x \geq y \\ 1 & \text{if } x < y. \end{cases}$
Clearly the quasi-metric space $([0, 1], d_S^{-1})$ is right K-sequentially complete. Next consider the mapping $f : [0, 1] \to [0, 1]$ defined by $f(1) = 0$ and $f(x) = \frac{x+1}{2}$ for all $x \in [0, 1[$. The sequence $(f^n(0))_{n\in\mathbb{N}}$ is $L_{d_S^{-1}}$-bounded. Moreover, f is $\tau(d_S)$-$\leq$-continuous at 0.

However, since $d_S(f^n(0), 1) = 1$ and $\lim_{n\to\infty} d_S(f^{n+1}(0), f(1)) = 1$ then does not exist $k \in [0, 1[$ such that $d_S(f^{n+1}(0), f(1)) \leq kd_S(f^n(0), 1)$. Note that $U_{\leq}(f^n(0)) = 1$. Of course, f has not fixed points.

On account of [6], a quasi-metric space (X, d) is Smyth complete provided that every left K-Cauchy sequence is convergent with respect to $\tau(d^s)$, where a sequence

$(x_n)_{n\in\mathbb{N}}$ is said to be left K-Cauchy if, given the real number $\varepsilon > 0$, there exists $n_0 \in \mathbb{N}$ such that $d(x_n, x_m) < \varepsilon$ for all $m \geq n \geq n_0$. It must be pointed out that Theorem 2.1 holds true for Smyth complete partially ordered quasi-metric spaces, since (X, d^{-1}) is right K-sequentially complete when the quasi-metric (X, d) is Smyth complete.

Following [6], every quasi-metric d induces on a non-empty set X a partial order $\preceq_d$, known as specialized order, which is defined by $x \preceq_d y \Leftrightarrow d(x, y) = 0$. In the light of this fact, we obtain a few fixed point theorems from Theorem 2.1, that give a little bit more of information than the aforesaid theorem, when the specialized order is under consideration.

Corollary 2.2 *Let (X, d) be a quasi-metric space such that (X, d^{-1}) is right K-sequentially complete and let $f : X \to X$ be a mapping. If there exist $k \in [0, 1[$ and $x_0 \in X$ satisfying that:*

(i) *the sequence $(f^n(x_0))_{n\in\mathbb{N}}$ is $L_{d^{-1}}$-bounded,*
(ii) *f is $\tau(d)$-$\preceq_d$-continuos at x_0,*
(iii) *for each $n \in \mathbb{N}$, $d(f^{n+1}(x_0), f(y)) \leq kd(f^n(x_0), y)$ for all $y \in U_{\preceq_d}(f^n(x_0)) \cup \mathcal{O}_n(f, x_0)$.*

Then f has a fixed point $x^\star$ such that $x^\star \in UL_{\preceq_d, d^s}(f^n(x_0))$ and $x^\star$ is the least upper bound of $(f^n(x_0))_{n\in\mathbb{N}}$ in $(X, \preceq_d)$.

Proof The existence of a fixed point $x^\star$ such that $x^\star \in UL_{\preceq_d, d^s}(f^n(x_0))$ follows from Theorem 2.1. We only show that $x^\star$ is the least upper bound of $(f^n(x_0))_{n\in\mathbb{N}}$ in $(X, \preceq_d)$. Indeed, assume that there exists $y \in U_{\preceq_d}(f^n(x_0))$. Then $d(f^n(x_0), y) = 0$ for all $n \in \mathbb{N}$. Whence we deduce that $d(x^\star, y) \leq d(x^\star, f^n(x_0)) + d(f^n(x_0), y) = d(x^\star, f^n(x_0))$ for all $n \in \mathbb{N}$. Since $x^\star \in L_{\preceq_d}(f^n(x_0))$ we obtain that there exists $n_0 \in \mathbb{N}$ such that $d(x^\star, f^n(x_0)) < \varepsilon$ for all $n \geq n_0$. Thus we conclude that $d(x^\star, y) = 0$, which implies that $x^\star \preceq_d y$. Whence we have $x^\star$ is the least upper bound of $(f^n(x_0))_{n\in\mathbb{N}}$ in $(X, \preceq_d)$. □

The next result was proved in [9] and it will be useful in the following corollary.

Lemma 2.3 *Let (X, d) be a quasi-metric space. If $x \in X$ and $(x_n)_{n\in\mathbb{N}}$ is an increasing sequence in $(X, \preceq_d)$ such that $x \in L_{d^s}(x_n)$, then x is the least upper bound of $(x_n)_{n\in\mathbb{N}}$.*

Corollary 2.4 *Let (X, d) be a Smyth complete quasi-metric space and let $f : X \to X$ be a mapping. If there exist $k \in [0, 1[$ and $x_0 \in X$ satisfying that:*

(i) *the sequence $(f^n(x_0))_{n\in\mathbb{N}}$ is increasing in $(X, \preceq_d)$,*
(ii) *f is $\tau(d)$-$\preceq_d$-continuos at x_0,*
(iii) *for each $n \in \mathbb{N}$, $d(f^{n+1}(x_0), f(y)) \leq kd(f^n(x_0), y)$ for all $y \in U_{\preceq_d}(f^n(x_0))$.*

Then f has a fixed point $x^\star$ such that $x^\star \in UL_{\preceq_d, d^s}(f^n(x_0))$ and $x^\star$ is the least upper bound of $(f^n(x_0))_{n\in\mathbb{N}}$ in $(X, \preceq_d)$.

Proof Lemma 2.3 guarantees that every increasing sequence in $(X, \preceq_d)$ is always $L_{d^{-1}}$-bounded, whenever (X, d) is Smyth complete. So Corollary 2.2 gives the desired conclusions. □

Corollary 2.5 *Let (X, d) be a Smyth complete quasi-metric space and let $f : X \to X$ be a mapping. If there exists $x_0 \in X$ satisfying that:*

(i) $x_0 \preceq_d f(x_0)$,
(ii) *f is monotone and $\tau(d)$-$\preceq_d$-continuos at x_0,*

Then f has a fixed point $x^\star$ such that $x^\star \in UL_{\preceq_d, d^s}(f^n(x_0))$ and $x^\star$ is the least upper bound of $(f^n(x_0))_{n\in\mathbb{N}}$ in $(X, \preceq_d)$.

Proof Of course the monotony of f and the fact that $x_0 \preceq_d f(x_0)$ provides that the sequence $(f^n(x_0))_{n\in\mathbb{N}}$ is increasing in $(X, \preceq_d)$. Since f is monotone we have that $f(x) \preceq_d f(y)$ whenever $x \preceq_d y$. Whence we have that $f^{n+1}(x_0) \preceq_d f(y)$ for all $n \in \mathbb{N}$ provided that $y \in U_{\preceq_d}(f^n(x_0))$. It follows that $d(f^{n+1}(x_0), f(y)) = 0$ for all $n \in \mathbb{N}$ whenever $y \in U_{\preceq_d}(f^n(x_0))$. Since $d(f^n(x_0), y) = 0$ for all $y \in U_{\preceq_d}(f^n(x_0))$ we deduce that condition (iii) in Corollary 2.4 is also fulfilled. Therefore the aforesaid result provides the desired conclusions. □

Notice that the monotony of the self-mapping and its $\tau(d)$-$\preceq_d$-continuity are not redundant assumptions in Corollary 2.5 such as the next example shows.

Example 8 Consider the quasi-metric space $([0, 1], d_l)$. Let $f : [0, 1] \to [0, 1]$ be the mapping introduced in Example 3. Take $x_0 = 0$. Then all assumptions in Corollary 2.5 are hold except the monotony. Indeed, $0 \preceq_{d_l} 1$ but $f(0) \not\preceq_{d_l} f(1)$. Next consider the quasi-metric space $(\mathbb{R}^+, d_l)$ and define the mapping $f : \mathbb{R}^+ \to \mathbb{R}^+$ by $f(x) = 0$ for all $x \in [0, \frac{1}{2}[$, $f(x) = \frac{x+1}{2}$ for all $x \in [\frac{1}{2}, 1[$ and $f(x) = x + 1$ for all $x \in [1, \infty[$. Fix $x_0 = \frac{1}{2}$. Then all assumptions in Corollary 2.5 are hold except the $\tau(d_l)$-$\preceq_{d_l}$-continuity at $\frac{1}{2}$. Indeed, $1 \in UL_{\preceq_{d_l^s}}(f^n(\frac{1}{2}))$ but $2 = f(1) \notin L_{d_l}(f^n(\frac{1}{2}))$.

Taking into account that every metric space is a quasi-metric space, we derive from our main result, Theorem 2.1, a fixed point theorem for self-mappings in partially ordered metric spaces which improves Theorem 1.1.

Corollary 2.6 *Let $(X, d, \preceq)$ be a partially ordered metric space such that (X, d) is complete and let $f : X \to X$ be a mapping. If there exist $k \in [0, 1[$ and $x_0 \in X$ satisfying that:*

(i) *the sequence $(f^n(x_0))_{n\in\mathbb{N}}$ is L_d-bounded,*
(ii) *f is $\tau(d)$-$\preceq$-continuos at x_0,*
(iii) *for each $n \in \mathbb{N}$, $d(f^{n+1}(x_0), f(y)) \leq kd(f^n(x_0), y)$ for all $y \in U_{\preceq}(f^n(x_0)) \cup \mathcal{O}_n(f, x_0)$*

Then f has a fixed point $x^\star$ such that $x^\star \in UL_{\preceq,d}(f^n(x_0))$.

Notice that, in contrast to Theorem 1.1, the preceding corollary does not need the assumption about the increasing condition of the sequence $(f^n(x_0))_{n\in\mathbb{N}}$, the monotony and continuity of the self-mapping and, in addition, the "contractive condition" is only satisfied for all upper bounds and elements in the sequence $(f^n(x_0))_{n\in\mathbb{N}}$. Observe, in addition, that the property that establish a relationship between order and topology for all increasing convergent sequences can be reduced to the L_d-boundness only for the sequence $(f^n(x_0))_{n\in\mathbb{N}}$. So, in general, all assumptions in the statement of Theorem 1.1 have been weakened in Corollary 2.6.

We end the section showing that Theorem 1.1 can be retrieved as a particular case of Corollary 2.6. To this end, assume that the self-mapping f is monotone and that condition (i) in Theorem 1.1 is hold. Then there exists $x_0 \in X$ such that $x_0 \preceq f(x_0)$ and, hence, the sequence $(f^n(x_0))_{n\in\mathbb{N}}$ is increasing in $(X, \preceq)$. Conditions (ii) and (iii) in Theorem 1.1 warranty that $(f^n(x_0))_{n\in\mathbb{N}}$ is convergent with respect to $\tau(d)$ and that the limit is an upper bound. Whence we have that the sequence $(f^n(x_0))_{n\in\mathbb{N}}$ is L_d-bounded. Again, condition (ii) provides that the self-mapping is $\tau(d)$-$\preceq$-continuos at x_0. Finally, the contractive condition $d(f^{n+1}(x_0), f(y)) \leq kd(f^n(x_0), y)$ is satisfied for all $y \in U_{\preceq}(f^n(x_0)) \cup \mathcal{O}_n(f, x_0)$ because f fulfills the contractive condition given in condition (iii) in Theorem 1.1 which is hold for all order related elements.

3 The Relationship with Kleene Fixed Point Theorem

In Computer Science, the celebrated Kleene fixed point theorem plays a central role (see, for instance, [18]). Let us recall that a partially ordered set $(X, \preceq)$ is called chain-complete if every increasing sequence in X has a least upper bound (see [1]). Of course, a sequence $(x_n)_{n\in\mathbb{N}}$ is said to be increasing whenever $x_n \preceq x_{n+1}$ for all $n \in \mathbb{N}$. In addition, given a partially ordered set $(X, \preceq)$, a mapping $f : X \to X$ is said to be $\preceq$-continuous if the least upper bound of the sequence $(f(x_n))_{n\in\mathbb{N}}$ is $f(x)$ for every increasing sequence $(x_n)_{n\in\mathbb{N}}$ whose least upper bound exists and is x. Taking into account the preceding notions, Kleene's theorem can be states as follows (see [1, 18]).

Theorem 3.1 *Let $(X, \preceq)$ be a chain-complete partially ordered set and let $f : X \to X$ be a $\preceq$-continuous mapping. If there exists $x_0 \in X$ such that $x_0 \preceq f(x_0)$, then f has a fixed point $x^\star$ such that $x^\star \in \{y \in X : x_0 \preceq y\}$ and $x^\star$ is the least upper bound of the sequence $(f^n(x_0))_{n\in\mathbb{N}}$.*

Observe that every Smyth complete quasi-metric space is chain-complete when the specialization order is under consideration (see Proposition 4 in [9]). So it seems natural to try of discerning if Corollary 2.5 can be deduced from Theorem 3.1. Nevertheless this is not the case. Indeed, consider the Smyth complete quasi-metric space $([0, 1], d_u)$ and define the mapping $f : [0, 1] \to [0, 1]$ by $f(0) = 0$ and $f(x) = \frac{x+1}{2}$ for all $x \in]0, 1]$. Then all conditions in Corollary 2.5 are satisfied for $x_0 = 0$ but f is not $\preceq_{d_u}$-continuous.

4 An Application to Complexity Analysis of Algorithms

The complexity of an algorithm is given as the quantity of resources required by the algorithm to get a solution to the problem for which it has been designed. A typical resource is the running time of computing (see, for instance [3])

In most cases there are a few algorithms that are able to get the solution to a posed problem. So, in computer science, one target is to fix which of them solve the problem taken less time. With this aim, a technique for comparing their running time of computing becomes necessary. This comparison is made by means of the asymptotic analysis, in which the running time of an algorithm is denoted by a function $T : \mathbb{N} \rightarrow]0, \infty]$ in such a way that $T(n)$ matches up with the time taken by the algorithm to solve the problem when the input data is of size n. Observe that we exclude 0 from the range of T because an algorithm always takes an amount of time in order to solve the problem for which it has been designed.

Normally the running time does not only depend on the input data size n, but it depends also on the particular input and the distribution of the input data. Therefore, three possible behaviors are usually distinguished when the running time of an algorithm is discussed. Such cases are the so-called best case, the worst case and the average case. The best case and the worst case, for an input of size n, are defined by the minimum and the maximum running time of computing, respectively. The average case for an input of size n is defined by the expected value or average running time of computing over all inputs of size n.

Usually to establish the exact expression of the function which gives the running time of computing of an algorithm is an arduous task. For this reason, normally, the analysis is focused on bounding the running time of computing and, thus, to yield an approximation of it.

One way in which the running time of computing can be approximated is by means of getting an asymptotic lower bound, i.e., giving the Ω asymptotic complexity class. Let us recall such a notion. To this end, from now on, $\leq$ will stand for the usual partial order on $]0, \infty]$ and we set $\mathcal{T}$ as $\mathcal{T} = \{f : \mathbb{N} \rightarrow]0, \infty]\}$.

Consider two functions $f, g \in \mathcal{T}$. Then $f \in \Omega(g)$ if and only if there exist $n_0 \in \mathbb{N}$ and $c \in]0, \infty[$ satisfying $cg(n) \leq f(n)$ for all $n \in \mathbb{N}$ with $n \geq n_0$.

Notice that when $f \in \mathcal{T}$ provides the running time of computing of an algorithm, then the fact that $f \in \Omega(g)$ yields that an asymptotic lower bound of the aforesaid running time is represented by the function g. Hence if the exact expression of the function f is unknown, then the function g gives an approximate information of the running time of computing for each input size n, $f(n)$, in such a way that the algorithm takes a time to process the input data of size n bounded below by the value $g(n)$.

Clearly the set $\mathcal{T}$ becomes a partially ordered set when we endow it with the partial order $\preceq_{\mathcal{T}}$ given by $f \preceq_{\mathcal{T}} g \Leftrightarrow f(n) \leq g(n)$ for all $n \in \mathbb{N}$.

Usually the analysis of the running time of computing of algorithms leads up recurrence equations on $\mathbb{N}$ of the following general type:

$$T(n) = \begin{cases} c_n & \text{if } n \leq n_0, \\ \Phi(n, T(g_1(n)), \ldots, T(g_k(n))) & \text{if } n > n_0 \end{cases} \tag{4.1}$$

where the following facts are assumed:

1. $n_0 \in \mathbb{N}$ is fixed (associated to the input size of the data when the base case occurs).
2. $g_i : \mathbb{N} \to \mathbb{N}$ are unbounded monotone functions with respect to the partial order $\leq$ such that $g_i(n) < n$ for all $n \in \mathbb{N}$ and for all $i = 1, \ldots, k$.
3. $\Phi : \mathbb{N} \times]0, \infty]^k \to]0, \infty]$ is monotone in each of its variables with respect to the partial order $\leq$ and unbounded.

Examples of algorithms whose running time satisfies a recurrence equation of the above introduced type (4.1) are, among others, divide-and-conquer algorithms, as the celebrated Quicksort (worst case) and Mergesort (average case), multiple-size divide-and-conquer algorithms, recursive algorithms as Hanoi Towers Puzzle, Largetwo (average case) and Fibonacci (see, for instance, [2, 3, 5, 8, 17, 20]).

The running time of computing of all aforementioned algorithms is the solution to a recurrence equation of type (4.1) where all functions $\{g_i : i = 1, \ldots, k\}$ fulfill, in addition, either $g_i(n) = n - i$ for all $n \in \mathbb{N}_{n_0}$ or $g_i(n) = \lceil \frac{n_i}{b_i} \rceil$ for all $n \in \mathbb{N}_{n_0}$ with $\mathbb{N}_{n_0} = \{n \in \mathbb{N} : n > n_0\}$. Many methods can be found in the literature in order to obtain asymptotic bounds for those algorithms whose the running time of computing fulfills a recurrence equation of type (4.1). Frequently, such methods are specific for each case under study and are based on tedious and hard arguments coming either from mathematical induction or from calculus involving integrals or limits. A general view of the classical treatment of the topic can be found in [3, 4].

In [13, 15] a method was developed to get asymptotic lower bounds in those cases in which the running time of computing satisfies particular instances of the following recurrence equation which can be retrieved from (4.1):

$$T(n) = \begin{cases} c_n & \text{if } n \leq n_0, \\ \sum_{i=1}^{k} a_i(n) T(g_i(n)) + d(n) & \text{if } n > n_0 \end{cases} \tag{4.2}$$

where the following facts are assumed:

1. $a_i : \mathbb{N}_{n_0} \to]0, \infty[$ are fixed functions for all $i = 1, \ldots, k$.
2. $d \in \mathcal{T}$ with $d(n) < \infty$ for all $n \in \mathbb{N}$.

The aforementioned method is based on the use of the fixed point theory in partially ordered quasi-metric spaces. It must be stressed that such a method does not intend to compete with the standard techniques to analyze the complexity of algorithms based on the classical arguments. The authentic purpose of the method is to introduce a formal treatment of asymptotic complexity by means of really basic and elementary arguments which provide, in some sense, a fixed point theoretical counterpart of the classical techniques. Let us recall the announced method.

4.1 The Fixed Point Method

We first stress that every recurrence equation of type (4.1) has always a unique solution, which represents the running time of computing of the algorithm under consideration, provided the initial conditions $c_1, \ldots, c_n$ and this fact can be proved trivially by means of standard induction arguments (see, for instance [5]). Taking this fact into account, we only need to focus our attention on how we can get an asymptotic lower bound for such a solution without knowing its specific expression. To this end, we will work on the subset $\mathcal{T}_{n_0,c}$ of $\mathcal{T}$ which is given by

$$\mathcal{T}_{n_0,c} = \{f \in \mathcal{T} : f(n) = c_n \text{ for all } n \leq n_0 \text{ and } \sum_{n=1}^{\infty} 2^{-n} \frac{1}{f(n)} < \infty\}.$$

According to [17], the functions belonging to $\mathcal{T}_{n_0,c}$ models the running time of computing of those algorithms whose complexity class is lower than the exponential one, i.e., $f(n) < 2^n$ for all $n \in \mathbb{N}$. Notice that these algorithms are the unique reasonable from a computability point of view. Moreover, given two functions $f, g \in \mathcal{T}_{n_0,c}$, the numerical value

$$d_{\mathcal{T}_{n_0,c}}(f, g) = \sum_{n=1}^{\infty} 2^{-n} \max\{\frac{1}{g(n)} - \frac{1}{f(n)}\}$$

can be interpreted as the relative progress made in lowering the complexity by replacing any algorithm P with running time of computing f by any algorithm Q with running time of computing g. Therefore the condition $d_{\mathcal{T}_{n_0,c}}(f, g) = 0$ provides that the algorithm P is at least as efficient as the algorithm Q, i.e., $f \preceq_{\mathcal{T}} g$. Observe that $d_{\mathcal{T}_{n_0,c}}(f, g) = 0 \Leftrightarrow f(n) \leq g(n)$ for all $n \in \mathbb{N}$ and thus, $\preceq_{d_{\mathcal{T}_{n_0,c}}} = \preceq_{\mathcal{T}}$. Moreover, observe that $d_{\mathcal{T}_{n_0,c}}(f, g) = 0$ implies that $g \in \Omega(f)$. Following [17], the pair $(\mathcal{T}_{n_0,c}, d_{\mathcal{T}_{n_0,c}})$ is a Smyth complete quasi-metric space.

It must be pointed out that the asymmetry of $d_{\mathcal{T}_{n_0,c}}$ plays a central role in order to provide information about the increase of complexity whenever an algorithm is replaced by another one. A metric could be able to yield information on the increase but it, however, would not yield information about which algorithm is more efficient.

Next define the functional $\Psi : \mathcal{T}_{n_0,c} \to \mathcal{T}_{n_0,c}$ by

$$\Psi(f)(n) = \begin{cases} c_n & \text{if } n \leq n_0 \\ \sum_{i=1}^{k} a_i(n) T(g_i(n)) + d(n) & \text{if } n > n_0 \end{cases} \tag{4.3}$$

for all $f \in \mathcal{T}_{n_0,c}$. Of course a function belonging to $\mathcal{T}_{n_0,c}$ is a solution to the recurrence Eq. (4.2) if and only if it is a fixed point of the functional Ψ. A straightforward computation shows that Ψ is monotone with respect to $\preceq_{\mathcal{T}}$.

On account of [15], a functional $\Psi : \mathcal{T}_{n_0,c} \to \mathcal{T}_{n_0,c}$ is called a worsener with respect to $f_0 \in \mathcal{T}_{n_0,c}$ provided that $\Psi^n(f_0) \preceq_{d_{\mathcal{T}_{n_0,c}}} \Psi^{n+1}(f_0)$ for all $n \in \mathbb{N}$. In the light of the preceding notion, the technique to get asymptotic lower bounds can be stated as follows:

Theorem 4.1 *Let $f_T \in \mathcal{T}_{n_0,c}$ be the (unique) solution to a recurrence equation of type (4.2). If there exists $f_0 \in \mathcal{T}_{n_0,c}$ such that $f_0 \preceq_{\mathcal{T}} \Psi(f_0)$, then the functional Ψ associated to (4.2), and given by (4.3), is a worsener with respect to f_0 and $f_T \in \Omega(f_0)$.*

The proof of the preceding result comes down basically to verify that if Ψ is a worsener with respect to f_0, then $UL_{\preceq_{d_{\mathcal{T}_{n_0,c}}}, d^s_{\mathcal{T}_{n_0,c}}}(\Psi^n(f_0)) = \{f_T\}$, where f_T is the unique fixed point of the functional Ψ associated to (4.2). Notice that Ψ is monotone and, thus, $f_0 \preceq_{\mathcal{T}} \Psi^n(f_0)$. So $f_T \in \Omega(f_0)$.

It must be stressed that in order to guarantee the preceding fact, in [13, 15] it was proved that, in those particular cases discussed, the functional Ψ was contractive, i.e., that there exists $k \in [0, 1[$ such that $d^s_{\mathcal{T}_{n_0,c}}(\Psi(g), \Psi(f)) \leq k d^s_{\mathcal{T}_{n_0,c}}(g, f)$ for all $f, g \in \mathcal{T}_{n_0,c}$.

Inspired by the above, and taking into account that only few particular cases of the recurrence of type (4.2) were explored in [13, 15], it seems natural to wonder whether the conditions assumed in the statement of Theorem 4.1 remain valid to assure that the condition "$UL_{\preceq_{d_{\mathcal{T}_{n_0,c}}}, d^s_{\mathcal{T}_{n_0,c}}}(\Psi^n(f_0)) = \{f_T\}$" is still verified when we consider those algorithms whose running time of computing satisfies an instance of the general recurrence of type (4.1).

The following result answers to the preceding question clarifying under what conditions a technique in the spirit of Theorem 4.1 for asymptotic lower bounds can be developed.

In order to introduce the promised answer we consider the functional associated to the recurrence Eq. (4.1), $\Psi_\Phi : \mathcal{T}_{n_0,c} \to \mathcal{T}_{n_0,c}$, given by

$$\Psi_\Phi(f)(n) = \begin{cases} c_n & \text{if } n \leq n_0 \\ \Phi(n, T(g_1(n)), \ldots, T(g_k(n))) & \text{if } n > n_0 \end{cases} \tag{4.4}$$

for all $f \in \mathcal{T}_{n_0,c}$. Observe that Ψ_Φ is monotone with respect to $\preceq_{d_{\mathcal{T}_{n_0,c}}}$ because Φ is assumed to be monotone is each of its variables.

In the light of the preceding we have the following.

Proposition 4.2 *Let $f_T \in \mathcal{T}_{n_0,c}$ be the (unique) solution to a recurrence equation of type (4.1). Let $\Psi_\Phi : \mathcal{T}_{n_0,c} \to \mathcal{T}_{n_0,c}$ be the functional associated to the recurrence equation (4.1) and given by (4.4). If Ψ_Φ is a worsener with respect to any $f_0 \in \mathcal{T}_{n_0,c}$, then the following assertions are equivalent.*

(i) *Ψ_Φ is $\tau(d_{\mathcal{T}_{n_0,c}})$-$\preceq_{d_{\mathcal{T}_{n_0,c}}}$-continuos at f_0.*
(ii) *$UL_{\preceq_{d_{\mathcal{T}_{n_0,c}}}, d^s_{\mathcal{T}_{n_0,c}}}(\Psi^n(f_0)) = \{f_T\}$.*

Proof Let $f_0 \in \mathcal{T}_{n_0,c}$. Assume that Ψ_Φ is a worsener with respect to f_0. It is clear that (ii) implies (i). Next we show that (i) implies (ii). To this end, suppose that there exists $f \in \mathcal{T}_{n_0,c}$ such that $f \in UL_{\preceq_{d_{\mathcal{T}_{n_0,c}}}, d^s_{\mathcal{T}_{n_0,c}}}(\Psi^n_\Phi(f_0))$. Then, by Lemma 2.3, we have that f is the least upper bound of $(\Psi^n_\Phi(f_0))_{n\in\mathbb{N}}$. Moreover, the monotony of Ψ_Φ gives that $\Psi_\Phi(f) \in U_{\preceq_{d_{\mathcal{T}_{n_0,c}}}}(\Psi^n_\Phi(f_0))$. It follows that $d_{\mathcal{T}_{n_0,c}}(\Psi^n_\Phi(f_0), \Psi_\Phi(f)) = 0$ for all $n \in \mathbb{N}$. The $\tau(d_{\mathcal{T}_{n_0,c}})$-$\preceq_{d_{\mathcal{T}_{n_0,c}}}$-continuity at f_0 of Ψ_Φ provides that $\Psi_\Phi(f) \in L_{d_{\mathcal{T}_{n_0,c}}}(\Psi^n_\Phi(f_0))$ and, hence, that $\Psi_\Phi(f) \in L_{d^s_{\mathcal{T}_{n_0,c}}}(\Psi^n_\Phi(f_0))$. Lemma 2.3, again, gives that $\Psi_\Phi(f)$ is the least upper bound of $(\Psi^n_\Phi(f_0))_{n\in\mathbb{N}}$. Therefore $\Psi_\Phi(f) = f$. The uniqueness of fixed point of Ψ_Φ (note that recurrence Eq. (4.1) has a unique solution f_T) gives that $UL_{\preceq_{d_{\mathcal{T}_{n_0,c}}}, d^s_{\mathcal{T}_{n_0,c}}}(\Psi^n(f_0)) = \{f_T\}$. □

In view of the preceding proposition we are able to provide a mathematical method for getting asymptotic lower bounds of the complexity of those algorithms whose running time of computing fulfills the recurrence equation of type (4.1):

Theorem 4.3 *Let $f_T \in \mathcal{T}_{n_0,c}$ be the (unique) solution to a recurrence equation of type (4.1). Let Ψ_Φ be the functional associated to (4.1) and given by (4.4). Assume that the following assertions hold:*

(i) *There exists $f_0 \in \mathcal{T}_{n_0,c}$ such that $f_0 \preceq_{\mathcal{T}} \Psi_\Phi(f_0)$.*
(ii) *Ψ_Φ is $\tau(d_{\mathcal{T}_{n_0,c}})$-$\preceq_{d_{\mathcal{T}_{n_0,c}}}$-continuos at f_0.*

Then $f_T \in \Omega(f_0)$.

Proof Clearly Ψ_Φ is monotone with respect to $\preceq_{\mathcal{T}}$ and, thus, the condition $f_0 \preceq_{\mathcal{T}} \Psi_\Phi(f_0)$ provides that Ψ_Φ is a worsener with respect to $f_0 \in \mathcal{T}_{n_0,c}$. So we have that the sequence $(\Psi^n_\Phi(f_0))_{n\in\mathbb{N}}$ is increasing. So all assumptions in the statement of Corollary 2.5 are satisfied. It follows that $f_T \in UL_{\preceq_{d_{\mathcal{T}_{n_0,c}}}, d^s_{\mathcal{T}_{n_0,c}}}(\Psi^n_\Phi(f_0))$, where f_T is the unique fixed point of Ψ_Φ. By Proposition 4.2, $UL_{\preceq_{d_{\mathcal{T}_{n_0,c}}}, d^s_{\mathcal{T}_{n_0,c}}}(\Psi^n_\Phi(f_0)) = \{f_T\}$. The fact that $f_0 \preceq_{\mathcal{T}} \Psi_\Phi(f_0)$ provides that $f_0 \preceq_{\mathcal{T}} \Psi_\Phi(f_0) \preceq_{\mathcal{T}} \Psi^n_\Phi(f_0) \preceq_{\mathcal{T}} f_T$ and, hence, that $f_T \in \Omega(f_0)$.

□

We end the section stressing that Theorem 4.3 presents an advantage with respect to the approach exposed in [13, 15]. On the one hand, we do not need a contractive condition imposed for all elements of $\mathcal{T}_{n_0,c}$. On the other hand, we have been able to clarify which operators, the $\tau(d_{\mathcal{T}_{n_0,c}})$-$\preceq_{d_{\mathcal{T}_{n_0,c}}}$-continuos, are valid to develop a technique which preserves the essence of those provided in the aforesaid references, that is, the condition

$$UL_{\preceq_{d_{\mathcal{T}_{n_0,c}}}, d^s_{\mathcal{T}_{n_0,c}}}(\Psi^n(f_0)) = \{f_T\}.$$

Finally, our technique has been discussed for a very general recurrence equation instead for a few particular instances.

Acknowledgments J.J. Miñana and O. Valero acknowledge financial support from FEDER/Ministerio de Ciencia, Innovación y Universidades-Agencia Estatal de Investigación/_Proyecto PGC2018-095709-B-C21. This work is also partially supported by Programa Operatiu FEDER 2014-2020 de les Illes Balears, by project PROCOE/4/2017 (Direcció General d'Innovació i Recerca, Govern de les Illes Balears) and by projects ROBINS and BUGWRIGHT2. These two latest projects have received funding from the European Union's Horizon 2020 research and innovation programme under grant agreements No 779776 and No 871260, respectively. This publication reflects only the authors views and the European Union is not liable for any use that may be made of the information contained therein. M.D. Mabula acknowledge financial support from South African National Research Foundation grant N01504-112207. He also acknowledges good hospitality received while visiting O. Valero at Balearic Islands University, in Palma Mallorca, Spain.

References

1. A. Baranga, The contraction principle as a particular case of Kleene's fixed point theorem. Discrete Math. **98**, 75–79 (1991)
2. M. Blum, R. Floyd, V. Pratt, R. Rivest, R. Tarjan, Time bounds for selection. J. Comput. Syst. Sci. **7**, 448–461 (1973)
3. G. Brassard, P. Bratley, *Algorithms: Theory and Practice* (Prentice Hall, New Jersey, 1988)
4. T.H. Cormen, C.E. Leiserson, R.L. Rivest, *Introduction to Algorithms* (MIT Press, New York, 1990)
5. P. Cull, R. Flahive, R. Robson, *Difference Equations: From Rabbits to Chaos* (Springer, New York, 1985)
6. J. Goubault-Larrecq, *Non-Hausdorff Topology and Domain Theory* (Cambridge University, New York, 2013)
7. P. Hitzler, A.K. Seda, *Mathematical Aspects of Logic Programming Semantics* (CRC Press, Boca Raton, 2010)
8. M. Kao, Multiple-size divide-and-conquer recurrences. ACM SIGACT News **28**, 67–69 (1997)
9. M. López-Ramírez, O. Valero, Qualitative versus quantitative fixed point techniques in computer science. Quaest. Math. **41**, 115–127 (2018)
10. J.J. Nieto, R. Rodríguez-López, Contractive mapping theorems in partially ordered sets and applications to ordinary differential equations. Order **22**, 223–239 (2005)
11. I.L. Reilly, P.V. Subrahmanyam, M.K. Vamanamrthy, Cauchy sequences in quasi-pseudo-metric spaces. Mh. Math. **93**, 127–140 (1982)
12. J. Rodríguez-López, S. Romaguera, O. Valero, Denotational semantics for programming languages, balanced quasi-metrics and fixed points. Int. J. Comput. Math. **85**, 623–630 (2008)
13. S. Romaguera, O. Valero, Asymptotic complexity analysis and denotational semantics for recursive programs based on complexity spaces, in *Semantics-Advances in Theories and Mathematical Models*, ed. by M. Afzal, vol. 1 (InTech Open Science, Rijeka, 2012), pp. 99–120
14. S. Romaguera, M.P. Schellekens, O. Valero, The complexity space of partial functions: a connection between complexity analysis and denotational semantics. Int. J. Comput. Math. **88**, 1819–1829 (2011)
15. S. Romaguera, P. Tirado, O. Valero, New results on mathematical foundations of asymptotic complexity analysis of algorithms via complexity space. Int. J. Comput. Math. **89**, 1728–1741 (2012)
16. I.A. Rus, A. Pretusel, G. Petrusel, *Fixed Point Theory* (Cluj University Press, Cluj-Napoca, 2008)
17. M.P. Schellekens, The Smyth completion: a common foundation for denotational semantics and complexity analysis. Electron. Notes Theor. Comput. Sci. **1**, 211–232 (1995)

18. J.E. Stoy, *Denotational Semantics: The Scott-Strachey Approach to Programming Language Theory* (MIT Press, Cambridge, 1977)
19. B.Z. Vulikh, *Introduction to the Theory of Partially Ordered Spaces* (Wolters-Noordhoff Scientific Publications, Groningen, 1967)
20. X. Wang, Q. Fu, A frame for general divide-and-conquer recurrences. Inform. Process. Lett. **59**, 45–51 (1996)

Inheritance Properties of Positive Cones Induced by Subalgebras and Quotients of Ordered Banach Algebras

Kelvin Muzundu

Abstract Let A be a Banach algebra ordered by a positive cone C. We shall describe conditions under which positive cones induced by subalgebras and quotients of A inherit some fundamental properties of the positive cone C.

Keywords Positive cone · Subalgebra · Quotient algebra · Ordered Banach algebra

Mathematics Subject Classification (2010) Primary 06F25; Secondary 46B40

1 Introduction

Throughout A will be a complex Banach algebra with unity $\mathbf{1}$ and F will denote a two-sided ideal of A. If F is a closed ideal of A, the norm on the quotient Banach algebra A/F will be denoted by $|||\cdot|||$. The *spectrum* and *spectral radius* of an element $a \in A$ will be denoted by $\sigma(a)$ and $r(a)$ respectively. The (Jacobson) *radical* of A will be denoted by Rad(A) and A is said to be *semisimple* if Rad(A) $= \{0\}$. An ideal I in A is said to be *inessential* if the spectrum of every element in I is either finite or a sequence converging to zero.

An *ordered Banach algebra* (OBA) is a Banach algebra A containing a subset C, called an *algebra cone*, such that C contains $\mathbf{1}$ and is closed under addition, multiplication, and non-negative scalar multiplication. The ordering $\geq$ induced on A by C is defined by $b \geq a$ if and only if $b - a \in C$. The elements of C are called *positive* and if $a \in C$ then $a \geq 0$. Whenever convenient, we shall write (A, C) to denote a Banach algebra A ordered by an algebra cone C. An algebra cone C is called *normal* if there is a scalar $\alpha > 0$ such that $||a|| \leq \alpha ||b||$ whenever $0 \leq a \leq b$ with respect to C. The spectral radius function is said to be *monotone* with respect to C if $r(a) \leq r(b)$ whenever $0 \leq a \leq b$ with respect to C.

K. Muzundu (✉)
Department of Mathematics and Statistics, University of Zambia, Lusaka, Zambia

E. Kikianty et al. (eds.), *Positivity and its Applications*, Trends in Mathematics,
https://doi.org/10.1007/978-3-030-70974-7_14

There are several fundamental properties that an algebra cone C in a Banach algebra A may be endowed with. The most important ones being normality and monotonicity of the spectral radius, because most of the significant results in the theory of OBAs rely on the algebra cone possessing at least one of these properties. There are two other properties that will play a role in this work, namely closedness and properness of the algebra cone C, which are defined as follows: C is a *closed algebra cone* in A if it is topologically closed in A; C is a *proper algebra cone* if $C \cap -C = \{0\}$. It is well known that if C is normal in A, then it is proper and the spectral radius in (A, C) is monotone.

Let (A, C) be an OBA. If B is a Banach algebra such that $\mathbf{1} \in B$ and $B \subset A$, then $B \cap C$ is an algebra cone in B and so $(B, B \cap C)$ is an OBA. If F is a closed ideal of A and $\pi : A \to A/F$ defined by $\pi(a) = a + F$ is the canonical homomorphism, then $\pi(C)$ is an algebra cone in A/F, so that $(A/F, \pi C)$ is an OBA. We will refer to $B \cap C$ and $\pi(C)$ as algebra cones induced by subalgebras and quotients of A respectively. Algebra cones induced by subalgebras and quotients play a fundamental role in the development of the theory of OBAs.

As a continuation of the development of spectral theory in OBAs, in this paper we describe conditions under which the properties of normality of the algebra cone and monotonicity of the spectral radius function relative to the algebra cone carry over to algebra cones induced by subalgebras and quotients of A. This work was initiated in [10], where Banach algebras ordered by algebra cones were first studied. For more results on spectral theory in OBAs see [1, 4–9].

2 Monotonicity of the Spectral Radius

Let (A, C) be an OBA and B a Banach algebra such that $\mathbf{1} \in B \subset A$. Note that this means that the algebraic operations of B are those of A, while the norms of A and B may be different. If $b \in B$ we obtain from [2, Theorem 3.2.13] that $\sigma(b, A) \subset \sigma(b, B)$ and $\partial(\sigma(b, B)) \subset \partial(\sigma(b, A))$, where ∂S denotes the topological boundary of S. It follows from this that $r(b, A) = r(b, B)$, and so if the spectral radius is monotone in (A, C), then it is monotone in the OBA $(B, B \cap C)$.

We now turn to quotient algebras. As the next example demonstrates, monotonicity of the spectral radius in (A, C) does not generally imply monotonicity of the spectral radius in $(A/F, \pi C)$.

Example For fixed $n \in \mathbb{N}, n > 1$ consider the vector space $A = \mathbb{C}^n$. This is a Banach algebra under componentwise addition, scalar multiplication and multiplication; with norm $||(z_k)|| = \sup_k |z_k|$. Let

$$C = \{(z_k) \in A : z_1 \in \mathbb{R}, z_1 \geq 0 \text{ and } z_1 \geq |z_k| \text{ for all } 1 < k \leq n\}.$$

It is easy to check that C is a normal algebra cone in A, so that the spectral radius in (A, C) is monotone. Now let

$$F = \{(z_k) \in A : z_1 \in \mathbb{C}, z_k = 0 \text{ for all } 1 < k \leq n\}.$$

Clearly, F is a closed ideal of A and we consider the OBA $(A/F, \pi C)$. Let $a = (1, 1, 0, \ldots, 0), b = (2, 0, 0, \ldots, 0) \in C$. Then $0 \leq a \leq b$ with respect to C, so that $F \leq a + F \leq b + F$ with respect to πC. Because $b \in F$, we have that $b + F = F$ and $r(b + F) = 0$. On the other hand, $r(a + F) \neq 0$ since, for instance, $1 \in \sigma(a + F)$. This shows that the spectral radius in $(A/F, \pi C)$ is not monotone.

We will prove Theorem 2.2, which provides natural conditions that guarantee that the spectral radius will be monotone in the quotient OBA, provided it is monotone in the original OBA. The following lemma will be required.

Lemma 2.1 *If I is a maximal inessential ideal in a Banach algebra A, then the Banach algebra A/I is semisimple.*

Proof We first note that since the closure $\bar{I}$ of I is also an inessential ideal, maximality of I implies that $I = \bar{I}$, so that A/I is indeed a Banach algebra under the standard norm $|||a + I||| = \inf_{b \in I} ||a - b||$. Now let $a + I \in \text{Rad}(A/I)$. Then $\sigma(a + I) = \{0\}$. We show that a is an inessential element. If $\sigma(a)$ is infinite, then by compactness of $\sigma(a)$, there is an accumulation point $\alpha \in \sigma(a)$ of $\sigma(a)$. Obviously α is not a Riesz point of $\sigma(a)$, and so it follows from [2, Theorem 5.7.4] that $\alpha \in \sigma(a + I)$. Thus $\sigma(a)$ is either finite or a sequence converging to 0, which means that a is an inessential element. Next we show $a \in I$. Let $J = \{a \in A : a + I \in \text{Rad(A/I)}\}$. Then, clearly, J is an inessential ideal of A and $I \subset J$. Maximality of I then means that $I = J$ and consequently, any inessential element a such that $a + I \in \text{Rad(A)}$ will be in I. And thus $a + I = I$, which implies that $\text{Rad}(A/I) = \{I\}$. Hence A/I is semisimple. □

Theorem 2.2 *Let (A, C) be an OBA and I a maximal inessential ideal of A. If πC is a proper algebra cone in A/I and if the spectral radius is monotone in (A, C), then it is monotone in $(A/I, \pi C)$.*

Proof By Lemma 2.1, the Banach algebra A/I is semisimple. It follows from [3, Proposition 2.1] that all Banach algebra norms on A/I are equivalent. Consider the real-valued map $||| \cdot |||_I$ defined on A/I by

$$|||a + I|||_I = \begin{cases} 0 & \text{if } a \in I \\ ||a|| & \text{if } a \notin I \end{cases}$$

It can easily be shown that $||| \cdot |||_I$ is a Banach algebra norm for A/I, which is equivelent to the standard norm $|||a + I||| = \inf_{b \in I} ||a - b||$.

Let $a, b \in A$ such that $0 \leq a \leq b$ with respect to C. Then $I \leq a + I \leq b + I$ with respect to πC. If $b \in I$, then $a \in I$ by Raubenheimer and Rode [10, Theorem

6.1] and so $r(a+I) = r(b+I) = 0$. If $b \notin I$, then $|||b+I|||_1 = ||b||$ and from the spectral radius formula we obtain that $r(b+I) = r(b)$, where $b+I$ is considered in the Banach algebra $(A/I, |||\cdot|||_1)$. Because the norms $|||\cdot|||_I$ and $|||\cdot|||$ are equivalent, the spectral radius formula yields that we still have $r(b+I) = r(b)$ with $b+I$ considered in the Banach algebra $(A/I, |||\cdot|||)$. Monotonicity of the spectral radius in (A, C) then implies that $r(a+I) \leq r(a) \leq r(b) = r(b+I)$. Hence the spectral radius is monotone in $(A/I, \pi C)$. □

It is well known that if H is a separable Hilbert space, the ideal $K(H)$ of compact operators on H is the unique proper closed ideal of the Banach algebra $B(H)$ of all bounded linear operators on H. This necessarily implies that $K(H)$ is a maximal inessential ideal in $B(H)$, and this fact leads to the following corollary of Theorem 2.2.

Corollary 2.3 *Let $K(H)$ be the ideal of compact operators on a separable Hilbert space H and $P(H)$ the set of all positive operators on H. Let $M(H)$ be any closed, commutative subalgebra of $B(H)$ containing the identity operator. Then $(M(H), M(H) \cap P(H))$ is an OBA and $\pi(M(H) \cap P(H))$ is a normal algebra cone in the quotient OBA $M(H)/(K(H) \cap M(H))$. In addition, the spectral radius is monotone in $(M(H)/(K(H) \cap M(H)), \pi(P(H) \cap M(H))$.*

Proof It is well known and easily verifiable that $P(H) \cap M(H)$ is a normal algebra cone in $M(H)$. Since $K(H) \cap M(H)$ is a maximal inessential ideal of $M(H)$, the result follows from Theorem 2.2. □

Corollary 2.3 can also be verified using properties of C^*-algebras.

Theorem 2.2 may be applied to obtain corollaries of [8, Theorem 4.3, 4.4, 4.5, 4.6, 4.7, 5.5, 5.6] by assuming monotonicity of the spectral radius in the OBA (A, C) rather than the OBA $(A/I, \pi C)$. In addition we can apply Theorem 2.2 to obtain the following corollary of [10, Theorem 4.6].

Corollary 2.4 *Let (A, C) be an OBA such that the spectral radius is monotone with respect to C and B be a semisimple Banach algebra such that $\mathbf{1} \in B \subset A$. Suppose that I is a maximal inessential ideal of both A and B. If $a, b \in B$ such that $0 \leq a \leq b$ with respect to $B \cap C$ then $r(a+I, B/I) \leq r(b+I, B/I)$ and $r(a+I, A/I) \leq r(b+I, A/I)$.*

3 Normal Algebra Cones

Let (A, C) be an OBA, where C is normal in A, and let B be a Banach algebra such that $\mathbf{1} \in B \subset A$. Under the norm of A, the algebra cone $B \cap C$ will be normal in B. Therefore if B is semisimple, regardless of the norm considered, $B \cap C$ will be normal in B by equivalence of all Banach algebra norms on a semisimple Banach algebra. The requirement that B be semisimple is not restrictive because most Banach algebras of interest are semisimple.

We consider quotient algebras. By combining [8, Example 3.2] and [10, Example 4.2] we see that if F is a closed ideal in A, normality of C in A does not guarantee normality of πC in A/F.

From the results of the previous section, we deduce the following proposition regarding normal algebra cones in quotients.

Proposition 3.1 *Let (A, C) be an OBA and I a maximal inessential ideal of A. If πC is a proper algebra cone in A/I and if C is normal in A, then πC is normal in A/I.*

Proof Let $|||\cdot|||$ be the standard norm and $|||\cdot|||_I$ the norm on A/I as defined in Theorem 2.2. Since these norms are equivalent, there are scalars α and β such that

$$|||a+I|||_I \leq \alpha|||a+I||| \leq \beta|||a+I|||_I. \tag{3.1}$$

Let $a, b \in A$ such that $0 \leq a \leq b$ with respect to C. Then $I \leq a+I \leq b+I$ with respect to πC. If $b \in I$, then $|||b+I||| = 0$ and because πC is proper in A/I, we have that $a \in I$ by Raubenheimer and Rode [10, Theorem 6.1]. Thus $|||a+I||| = 0$, and therefore $|||a+I||| = |||b+I||| = 0$. If $b \notin I$ then $|||b+I|||_I = ||b||$ and normality of C in A implies that $|||a+I|||_I \leq ||a|| \leq \mu||b|| = \mu|||b+I|||_I$ for some scalar $\mu > 0$. It follows from this and the inequality (3.1) that $|||a+I||| \leq |||a+I|||_I \leq \mu|||b+I|||_I \leq \gamma|||b+I|||$, where $\gamma = \alpha\mu$. Hence πC is normal in A/I. □

It turns out that the condition that the ideal in Proposition 3.1 is maximal and inessential can be dropped; it only needs to be closed. To prove this we will need the following lemma on ordered Banach spaces.

Lemma 3.2 *Let K be a closed, convex subset of a Banach space X ordered by a normal positive cone P. Suppose that K has the following two properties:*

1. *If $x, y \in X$ with $0 \leq x \leq y$ and if $y \in K$, then $x \in K$.*
2. *There is an $x' \in K$ such that $|||x+K||| = ||x-x'||$.*

If $x, y \in X$ with $0 \leq x \leq y$, then $|||x+K||| \leq |||y+K|||$.

Proof Let $x, y \in X$ with $0 \leq x \leq y$. By hypothesis there exist elements $x', y' \in K$ such that $||x+K|| = ||x-x'||$ and $||y+K|| = ||y-y'||$. Without loss of generality we assume that $0, x, x'$ are aligned such that 0 and x' can be connected by a line segment passing through x. Then

$$x' = x + ||x-x'||\left(\frac{x}{||x||}\right)$$

and similarly,

$$y' = y + ||y-y'||\left(\frac{y}{||y||}\right).$$

Now consider the element

$$x^* = x + ||y - y'|| \left(\frac{x}{||y||} \right).$$

Because $||x|| \leq ||y||$ by normality of P, from direct calculation, we get that $||x^*|| \leq ||y'||$. Since $y' \in K$, from the hypothesis, we obtain that $x^* \in K$. Let

$$S_x = \left\{ t \in \mathbb{R}^+ : x + t \left(\frac{x}{||x||} \right) \in K \right\}.$$

Clearly $||x - x'|| \in S_x$ and $||x - x'|| = \inf S_x$. In addition, because $x^* \in K$ and $\left|\left| \frac{x}{||y||} \right|\right| \leq \left|\left| \frac{x}{||x||} \right|\right|$ we have that $||b - b'|| \in S_x$. Hence $||x - x'|| \leq ||y - y'||$ and the result follows. □

Theorem 3.3 *Let A be an OBA with a normal algebra cone C and F a closed ideal in A. If πC is a proper algebra cone in A/F, then πC is normal in A/F.*

Proof If $F = \{0\}$ then $|||a + F||| = ||a||$ and the result follows by normality of C. Suppose that $F \neq \{0\}$. We first show that for any $a \in A$, there is a $b \in F$ such that $|||a + F||| = ||a - b||$. Since $|||a + F||| = \inf_{b \in F} ||a - b||$, for every $n \in \mathbb{N}$ there is a $b_n \in F$ such that $|||a + F||| \leq ||a - b_n|| < |||a + F||| + \frac{1}{n}$. We show that the sequence (b_n) converges in F. Suppose to the contrary that it does not converge in F. Then there is an $\epsilon > 0$ such that for infinitely many values of n, we have $||b_n - c|| > \epsilon$ for all $c \in F$. Let us fix n_0 such that $||b_{n_0} - c|| > \epsilon$ for all $c \in F$. If we consider the open ball $B(b_{n_0}, \epsilon)$, we see that $B(b_{n_0}, \epsilon) \cap F = \emptyset$. Now since $B(b_{n_0}, \epsilon)$ is bounded and F is an ideal, there is a c' in $F \setminus B(b_{n_0}, \epsilon)$. Then for t in the real interval $(0, 1)$, the line segment $tb_{n_0} + (1 - t)c'$ is not a subset of F. This is a contradiction of the fact that F is an ideal and $b_{n_0}, c' \in F$. Therefore (b_n) converges in F, say $b_n \to b$ as $n \to \infty$. Taking limits as $n \to \infty$ in the inequalities $|||a + F||| \leq ||a - b_n|| < |||a + F||| + \frac{1}{n}$, it follows from continuity of the norm that $|||a + F||| = ||a - b||$.

Now let $a, b \in A$ such that $0 \leq a \leq b$ with respect to C. Since F is an ideal, it is convex in A. Also since F is closed in A and πC is proper in A/F, it follows from [10, Theorem 6.1] and Lemma 3.2 that $|||a + F||| \leq |||b + F|||$. Hence πC is normal in A/F. □

We end by observing that the topological closure of an algebra cone C in a Banach algebra A is also an algebra cone. For this reason, the property of an algebra cone being closed may be taken for granted. In addition if B is a closed subalgebra of A containing **1**, then the algebra cone $B \cap C$ in B is obviously closed. By an elementary argument involving limits and continuity of the norm, it can also be shown that the canonical map $\pi : A \to A/F$ maps closed sets in A to closed set in A/F, so that the algebra cone πC will be closed in A/F.

Acknowledgments We would like to thank the anonymous reviewer for suggestions that have resulted in improvements to the substance and readability of the paper.

References

1. A.E. Alekhno, The irreducibility in ordered Banach algebras. Positivity **16**, 143–176 (2012)
2. B. Aupetit, *A Primer on Spectral Theory*. (Springer, New York, 1991)
3. B. Aupetit, Spectrum-preserving linear mappings between Banach algebras or Jordan-Banach algebras. J. London Math. Soc. **62**(2), 917–924 (2000)
4. D. Behrendt, H. Raubenheimer, On domination of inessential elements in ordered Banach algebras. Illinois J. Math. **51**(3), 927–936 (2007)
5. R. Benjamin, S. Mouton, Fredholm theory in ordered Banach algebras. Quaest. Math. **39**(5), 643–664 (2016)
6. G. Braatvedt, R. Brits, H. Raubenheimer, Gelfand-Hille type theorems in ordered Banach algebras. Positivity **13**, 39–50 (2009)
7. S. Mouton, Converegence properties of positive elements in Banach algebras. Math. Proc. R. Irish Acad. **102A**(2), 149–162 (2002)
8. S. Mouton, K. Muzundu, Commutatively ordered Banach algebras. Quaest. Math. **36**, 1–29 (2013)
9. S. Mouton, K. Muzundu, Domination by ergodic elements in ordered Banach algebras. Positivity **18**(1), 119–130 (2014)
10. H. Raubenheimer, S. Rode, Cones in Banach Algebras. Indag. Math. (N.S). **7**, 489–502 (1996)

Universally Complete Spaces of Continuous Functions

Jan Harm van der Walt

Dedicated to the memory of Coenraad Labuschagne.

Abstract We characterise Tychonoff spaces X so that $\mathrm{C}(X)$ is universally σ-complete and universally complete, respectively.

Keywords Vector lattices · Continuous functions · P-spaces

Mathematics Subject Classification (2010) Primary 46E05; Secondary 46A40, 54G10

1 Introduction

Recently, Mozo Carollo [2] showed, in the context of point-free topology, that the vector lattice $\mathrm{C}(X)$ of continuous, real valued functions on a Tychonoff (completely regular T_1) space X is universally complete if and only if X is an extremally disconnected P-space. This paper aims to make this result and its proof accessible to those members of the positivity community who, like the author, are less familiar with point-free topology. In so doing, and based on results due to Fremlin [7] and Veksler and Geĭler [15], we obtain a refinement of Mozo Carollo's result. In particular, we characterise those Tychonoff spaces X for which $\mathrm{C}(X)$ is laterally σ-complete. We also include some remarks on σ-order continuous duals of spaces $\mathrm{C}(X)$ which are universally σ-complete.

The paper is organised as follows. In Sect. 2 we introduce definitions and notation used throughout the paper, and recall some results from the literature. Section 3

J. H. van der Walt (✉)
Department of Mathematics and Applied Mathematics, University of Pretoria, Hatfield, Pretoria, South Africa
e-mail: janharm.vanderwalt@up.ac.za

E. Kikianty et al. (eds.), *Positivity and its Applications*, Trends in Mathematics,
https://doi.org/10.1007/978-3-030-70974-7_15

contains the main results of the paper, namely, characterisations of those Tychonoff spaces X for which $\mathrm{C}(X)$ is universally complete and universally σ-complete, respectively.

2 Preliminaries

Throughout this paper X denotes a Tychonoff space; that is, a completely regular T_1 space. $\mathrm{C}(X)$ stands for the lattice of all real-valued and continuous functions on X. For $u \in \mathrm{C}(X)$, $Z(u)$ denotes the zero set of u; that is, $Z(u) = u^{-1}[\{0\}]$. The co-zero set of u is $Z^c(u) = X \setminus Z(u)$. The collection of zero sets in X is denoted $\mathbf{Z}(X)$, while $\mathbf{Z}^c(X)$ consists of all co-zero sets in X. For $x \in X$ the collection of open neighbourhoods of x is denoted $\mathcal{N}_x$, and $\mathcal{N}_x^*$ denotes the set of clopen neighbhourhoods of x. A zero-neighbourhood of $x \in X$ is a set $V \in \mathbf{Z}(X)$ so that x belongs to the interior of V. The collection of all zero-neighbourhoods of $x \in X$ is denoted $\mathcal{N}_x^z$, and $\mathcal{N}_x^c = \mathcal{N}_x \cap \mathbf{Z}^c(X)$. Observe that $\mathbf{Z}^c(X)$ is a basis for the topology of X. Hence for each $x \in X$ and every $V \in \mathcal{N}_x$ there exists $U \in \mathcal{N}_x^c$ so that $U \subseteq V$. Furthermore, for every $V \in \mathcal{N}_x$ there exists $W \in \mathcal{N}_x^z$ so that $W \subseteq V$. The standard reference for all of this is [9].

We write $\mathbf{1}$ for the function which is constant one on X. More generally, for $A \subseteq X$, the indicator function of A is $\mathbf{1}_A$. The constant zero function is $\mathbf{0}$.

We recall, see for instance [9], that X is

(i) *basically disconnected* if the closure of every co-zero set is open;
(ii) *extremally disconnected* if the closure of every open set is open.

Every extremally disconnected space is basically disconnected, but not conversely [9, Problem 4N]. Since $\mathbf{Z}^c(X)$ is a basis for the topology on X, every basically disconnected space is zero-dimensional;[1] that is, it has a basis consisting of clopen sets. The converse is false. For instance, $\mathbb{Q}$ is zero-dimensional, the set of all open intervals with irrational endpoints forming a basis of clopen sets, but not basically disconnected, since $(0, 1)$ is a co-zero set whose closure is not open.

Each of the properties (i) and (ii) of X corresponds to order-theoretic properties of $\mathrm{C}(X)$, see for instance [13, Theorems 43.2, 43.3, 43.8 & 43.11]. In particular, X is

(i*) basically disconnected if and only if $\mathrm{C}(X)$ is Dedekind σ-complete, if and only if $\mathrm{C}(X)$ has the principle projection property;
(ii*) extremally disconnected if and only if $\mathrm{C}(X)$ is Dedekind complete, if and only if $\mathrm{C}(X)$ has the projection property.

[1]The term zero-dimensional should be understood in terms of small inductive dimension [6, Definition 1.1.1 & Proposition 1.2.1], as opposed to the Lebesgue covering dimension used in [9].

X is a P-space [8] if the intersection of countably many open sets in X is open. Equivalently, X is a P-space if $Z(u)$ is open (hence clopen) for every $u \in C(X)$. Clearly, every discrete space is a P-space, but the converse is false, see [9, Problem 4N]. In fact, there exists a P-space without any isolated points [9, Problem 13P]. Evidently, every P-space is basically disconnected (in particular, every $Z \in \mathbf{Z}(X)$ is open), but not conversely, see [9, Problem 4M].

The following basic lemma may well be known, but we have not found it in the literature. We include the simple proof for the sake of completeness.

Lemma 2.1 *Let X be zero-dimensional. Then the following statements are equivalent.*

(i) *X is a P-space.*
(ii) *The intersection of countably many clopen sets is clopen.*
(iii) *The union of countably many clopen sets is clopen.*

Proof By definition, (i) implies (ii) and (iii), and, (ii) and (iii) are equivalent. It therefore suffices to show that (ii) implies (i).

Assume that (ii) is true. For each $n \in \mathbb{N}$ let U_n be an open subset of X. Let $U = \bigcap\{U_n \ : \ n \in \mathbb{N}\}$. If $U = \emptyset$ we are done, so assume that $U \neq \emptyset$. Fix any $x \in U$. Since X is zero-dimensional, there exists for each $n \in \mathbb{N}$ a set $V_n \in \mathcal{N}_x^*$ so that $V_n \subseteq U_n$. Let $V = \bigcap_{n\in\mathbb{N}} V_n$. Then $x \in V \subseteq U$ and, by assumption, V is clopen, hence open. Therefore U is open so that X is a P-space. □

We recall, for later use, the following results of Fremlin [7] and Veksler and Geĭler [15], respectively; see also [1].

Theorem 2.2 *Let L be a Dedekind complete vector lattice. Then the following statements are equivalent.*

(i) *L is universally complete.*
(ii) *L is universally σ-complete and has a weak order unit.*

Theorem 2.3 *Let L be an Archimedean vector lattice. The following statements are true.*

(i) *If L is laterally complete then L has the projection property.*
(ii) *If L is laterally σ-complete then L has the principle projection property.*

3 Universally Complete C(X)

We begin this section with a characterisation of those X for which C(X) is universally σ-complete.

Theorem 3.1 *The following statements are equivalent.*

(i) $C(X)$ *is laterally* σ*-complete.*
(ii) $C(X)$ *is universally* σ*-complete.*
(iii) X *is a P-space.*

Proof Assume that $C(X)$ is laterally σ-complete. It follows from Theorem 2.3 (ii) that $C(X)$ has the principle projection property. Therefore $C(X)$ is Dedekind σ-complete, hence universally σ-complete. Conversely, if $C(X)$ is universally σ-complete then, by definition, it is laterally σ-complete. Hence (i) and (ii) are equivalent.

Assume that $C(X)$ is laterally σ-complete. Then $C(X)$ has the principle projection property so that X is basically disconnected, hence zero-dimensional. We show that X is a P-space.

By Lemma 2.1 it suffices to show that the intersection of countably many clopen subsets of X is clopen. Assume that $U_k \subseteq X$ is clopen for each $k \in \mathbb{N}$, and let $U = \bigcap\{U_k \ : \ k \in \mathbb{N}\}$. We claim that U is clopen.

Let $V_0 = X$, $V_1 = U_1$ and, for each natural number $n > 1$, let $V_n = U_1 \cap \ldots \cap U_n$. Then V_n is clopen for each $n \in \mathbb{N}$, $U = \bigcap\{V_n \ : \ n \in \mathbb{N}\}$ and $V_{n+1} \subseteq V_n$ for every $n \in \mathbb{N}$. For each $n \in \mathbb{N}$, let $W_n = V_{n-1} \setminus V_n$. Then each W_n is clopen and $W_n \cap W_m = \emptyset$ whenever $n \neq m$. Moreover, $\bigcup\{W_n \ : \ n \in \mathbb{N}\} = X \setminus U$. Indeed, the inclusion $\bigcup\{W_n \ : \ n \in \mathbb{N}\} \subseteq X \setminus U$ is immediate. For the reverse inclusion, consider some $x \in X \setminus U$. There exists $n \in \mathbb{N}$ so that $x \in X \setminus V_n$. Let $n_0 = \min\{n \in \mathbb{N} \ : \ x \in X \setminus V_n\}$. Then, since $V_0 = X$, $x \in V_{n-1} \setminus V_n = W_n$. Hence $x \in \bigcup\{W_n \ : \ n \in \mathbb{N}\}$.

Let $w_n = n\mathbf{1}_{W_n}$, $n \in \mathbb{N}$, and $F = \{w_n \ : \ n \in \mathbb{N}\}$. Then $F \subseteq C(X)^+$ and the w_n are mutually disjoint. Therefore, since $C(X)$ is universally σ-complete, $w = \sup F$ exists in $C(X)$.

Fix $x \in U$. There exists $V \in \mathcal{N}_x$ so that $w(y) < w(x) + 1$ for all $y \in V$. Fix a natural number $N_0 \geq w(x)+1$. Then, for all $n \geq N_0$ and $y \in W_n$, $w(y) \geq w_n(y) = n \geq N_0 \geq w(x) + 1$ so that $y \notin V$. Therefore $V \cap W_n = \emptyset$ for all $n \geq N_0$. Let $W = V_{N_0} \cap V$. Then $W \in \mathcal{N}_x$ and, since $W_n \cap V_{N_0} = \emptyset$ for all $n < N_0$, $W \cap W_n = \emptyset$ for all $n \in \mathbb{N}$. Therefore $W \subseteq X \setminus \bigcup\{W_n \ : \ n \in \mathbb{N}\} = U$. This shows that U is open, and, since each U_k is closed, U is also closed, hence clopen. By Lemma 2.1, X is a P-space. Hence (i) implies (iii)

Assume that X is a P-space. Consider a countable set F of mutually disjoint elements of $C(X)^+$. We observe that for each $x \in X$ there is at most one $u \in F$ so that $u(x) > 0$. Hence the function

$$w : X \ni x \to \sup\{u(x) \ : \ u \in F\} \in \mathbb{R}^+$$

is well defined. We claim that $w \in C(X)$ so that $w = \sup F$ in $C(X)$.

Fix $x \in X$. Assume that $w(x) > 0$. Then there exists $u \in F$ and $V \in \mathcal{N}_x$ so that $u(y) = w(y) > 0$ for all $y \in V$. Hence w is continuous at x. Suppose $w(x) = 0$. Then $u(x) = 0$ for all $u \in F$. Since X is a P-space there exists for each $u \in F$ some $V_u \in \mathcal{N}_x$ so that $u(y) = 0$ for every $y \in V_u$. The set $V = \cap\{V_u \ : \ u \in F\}$ is an open neighbourhood of x, and $w(y) = 0$ for all $y \in V$. Hence w is continuous at x. Thus w is continuous at every $x \in X$, hence on X. Therefore $\mathrm{C}(X)$ is laterally σ-complete. Hence (iii) implies (i). □

Mozo Carollo's characterisation of those X for which $\mathrm{C}(X)$ is universally complete now follows easily.

Corollary 3.2 *The following statements are equivalent.*

(i) $\mathrm{C}(X)$ *is laterally complete.*
(ii) $\mathrm{C}(X)$ *is universally complete.*
(iii) X *is an extremally disconnected P-space.*

Proof Assume that $\mathrm{C}(X)$ is laterally complete. By Theorem 2.3 (i), $\mathrm{C}(X)$ has the projection property and is therefore Dedekind complete, hence universally complete. Conversely, if $\mathrm{C}(X)$ is universally complete, then it is laterally complete. Therefore (i) and (ii) are equivalent.

Assume that $\mathrm{C}(X)$ is universally complete. Then, since $\mathrm{C}(X)$ is Dedekind complete, X is extremally disconnected, and by Theorem 3.1, X is a P-space.

Suppose that $\mathrm{C}(X)$ is an extremally disconnected P-space. Then $\mathrm{C}(X)$ is Dedekind complete and, by Theorem 3.1, laterally σ-complete. Since $\mathrm{C}(X)$ has a weak order unit, it is universally complete by Theorem 2.2. □

Remark 3.3 Isbell [11] showed that if X is an extremally disconnected P-space, and X has non-measurable cardinal, then X is discrete. It is consistent with ZFC that every cardinal is non-measurable.

Combining Theorem 3.1 with [4, Theorem 10.2] we have the following.

Corollary 3.4 *The following statements are equivalent.*

(i) $\mathrm{C}(X)$ *is laterally* σ*-complete.*
(ii) $\mathrm{C}(X)$ *is universally* σ*-complete.*
(iii) X *is a P-space.*
(iv) $\mathrm{C}(X)$ *is a von Neumann regular*[2] *ring.*
(v) $\mathrm{C}(X)$ *is* z*-regular.*[3]

Remark 3.5 Recall that a space X is called realcompact[4] if for every Tychonoff space Y containing X as a proper dense subspace, the map $\mathrm{C}(Y) \ni f \mapsto f|_X \in \mathrm{C}(X)$ is not onto; that is, X is not C-embedded in Y, see [5, page 214]

If X is a realcompact P-space, then $\mathrm{C}(X)^{\sim}$ has a peculiar structure. Indeed, due to a result of Fremlin [7, Proposition 1.15], every $\varphi \in \mathrm{C}(X)^{\sim}$ is a finite linear combination of linear lattice homomorphisms form $\mathrm{C}(X)$ into $\mathbb{R}$. Xiong [16] showed

[2]For every $u \in \mathrm{C}(X)$ there exists $v \in \mathrm{C}(X)$ so that $u = vu^2$.

[3]Every proper prime z-ideal in $\mathrm{C}(X)$ is a minimal prime z-ideal, see [3, 4] for details.

[4]Realcompact spaces were introduced by Hewitt [10] under the name "Q-spaces", and defined as follows: X is a Q-space if every free maximal ring ideal in $\mathrm{C}(X)$ is hyper-real. See for instance [9, Problem 8A no. 1] for the equivalence of our definition and Hewitt's.

that every such homomorphism is a positive scalar multiple of a point evaluation. Hence

$$\mathrm{C}(X)^{\sim} = \mathrm{span}\{\delta_x \,:\, x \in X\} = c_{00}(X).$$

However, each δ_x is σ-order continuous. Indeed, consider a decreasing sequence (u_n) in $\mathrm{C}(X)^+$ so that $\inf\{u_n(x) \,:\, n \in \mathbb{N}\} > 0$ for some $x \in X$. Then there exists a real number $\epsilon > 0$ so that for every $n \in \mathbb{N}$ there exists $V_n \in \mathcal{N}_x$ such that $u_n(y) > \epsilon$ for every $y \in V_n$. Since X is a P-space, $V = \bigcap\{V_n \,:\, n \in \mathbb{N}\}$ is open. Therefore there exists $v \in \mathrm{C}(X)$ so that $\mathbf{0} < v \leq \epsilon\mathbf{1}$ and $v(y) = 0$ for $y \in X \setminus V$. Since $u_n(y) > \epsilon$ for all $y \in V$ and $n \in \mathbb{N}$ it follows that $\mathbf{0} \leq v \leq u_n$ for all $n \in \mathbb{N}$; hence u_n does not decrease to $\mathbf{0}$ in $\mathrm{C}(X)$. This shows that $\delta_x \in \mathrm{C}(X)^{\sim}_{\mathrm{c}}$.

Combining all of the above, we see that

$$\mathrm{C}(X)^{\sim} = \mathrm{span}\{\delta_x \,:\, x \in X\} = c_{00}(X) = \mathrm{C}(X)^{\sim}_{\mathrm{c}}.$$

Remark 3.6 The condition that $\delta_x \in \mathrm{C}(X)^{\sim}_{\mathrm{c}}$ for all $x \in X$ does not imply that X is a P-space. In fact, this property characterises the so called almost-P-spaces introduced by Veksler [14], see also [12]. A space X is an almost-P-space if the nonempty intersection of countably many open sets has nonempty interior; equivalently, every $Z \in \mathbf{Z}(X)$ has nonempty interior. Thus every P-space is an almost-P-space, but not conversely, see [12].

De Pagter and Huijsmans [4] showed that $\mathrm{C}(X)$ has the σ-order continuity property if and only if X is an almost-P-space. Hence, if X is an almost-P-space, then $\delta_x \in \mathrm{C}(X)^{\sim}_{\mathrm{c}}$ for every $x \in X$. For the converse, suppose that X is not an almost-P-space. Then there exists $u \in \mathrm{C}(X)^+$ so that $Z(u)$ has empty interior. For each $n \in \mathbb{N}$, let $u_n = \mathbf{1} \wedge (nu)$. Then (u_n) is increasing and bounded above by $\mathbf{1}$. Let $v \in \mathrm{C}(X)$ be an upper bound for (u_n). If $x \in X \setminus Z(u)$ then $\sup\{u_n(x) \,:\, n \in \mathbb{N}\} = 1$ so that $v(x) \geq 1$. Since $Z(u)$ has empty interior and v is continuous, it follows that $v(x) \geq 1$ for all $x \in X$. Therefore $u_n \uparrow \mathbf{1}$ in $\mathrm{C}(X)$. But if $x \in Z(u)$, then $\delta_x(u_n) = u_n(x) = 0$ for every $n \in \mathbb{N}$ so that $\delta_x \notin \mathrm{C}(X)^{\sim}_{\mathrm{c}}$.

Remark 3.7 Shortly before this paper went to print, the author was informed that Theorem 3.1 and Corollary 3.2 had been proven by Buskes in [17] in the classical case considered in this paper. The method of proof used in [17] differs from that presented in this paper.

References

1. C.D. Aliprantis, O. Burkinshaw, *Locally Solid Riesz Spaces* (Academic Press, New York-London, 1978)
2. I. Mozo Carollo, On the universal completion of pointfree function spaces. J. Pure Appl. Algebra **225**(2), 106490 (2021)

3. B. de Pagter, C.B. Huijsmans, On z-ideals and d-ideals in Riesz spaces. I. Nederl. Akad. Wetensch. Indag. Math. **42**(2), 183–195 (1980)
4. B. de Pagter, C.B. Huijsmans, On z-ideals and d-ideals in Riesz spaces. II. Nederl. Akad. Wetensch. Indag. Math. **42**(4), 391–408 (1980)
5. R. Engelking, *General Topology*. Sigma Series in Pure Mathematics, vol. 6, 2nd ed. (Heldermann Verlag, Berlin, 1989). Translated from the Polish by the author
6. R. Engelking, *Theory of Dimensions Finite and Infinite*. Sigma Series in Pure Mathematics, vol. 10 (Heldermann Verlag, Lemgo, 1995)
7. D.H. Fremlin, Inextensible Riesz spaces. Math. Proc. Camb. Philos. Soc. **77**, 71–89 (1975)
8. L. Gillman, M. Henriksen, Concerning rings of continuous functions. Trans. Am. Math. Soc. **77**, 340–362 (1954)
9. L. Gillman, M. Jerison, *Rings of Continuous Functions*. The University Series in Higher Mathematics (D. Van Nostrand, Princeton-Toronto-London-New York, 1960)
10. E. Hewitt, Rings of real-valued continuous functions. I. Trans. Am. Math. Soc. **64**, 45–99 (1948)
11. J.R. Isbell, Zero-dimensional spaces. Tohoku Math. J. (2) **7**, 1–8 (1955)
12. R. Levy, Almost-P-spaces. Can. J. Math. **29**(2), 284–288 (1977)
13. W.A.J. Luxemburg, A.C. Zaanen, *Riesz Spaces. Vol. I* (North-Holland Publishing, Amsterdam-London; American Elsevier Publishing, New York, 1971)
14. A.I. Veksler, P'-points, P'-sets, P'-spaces. A new class of order-continuous measures and functionals. Dokl. Akad. Nauk SSSR **212**, 789–792 (1973)
15. A.I. Veksler, V.A. Geĭler, Order completeness and disjoint completeness of linear partially ordered spaces. Sibirsk. Mat. Ž **13**, 43–51 (1972)
16. H.Y. Xiong, A characterization of Riesz spaces which are Riesz isomorphic to $C(X)$ for some completely regular space X. Nederl. Akad. Wetensch. Indag. Math. **51**(1), 87–95 (1989)
17. G.J.H.M. Buskes, Disjoint sequences and completeness properties. Nederl. Akad. Wetensch. Indag. Math. **47**(1), 11–19 (1985)

Applications of Generalized B*-Algebras to Quantum Mechanics

Martin Weigt

Abstract In this paper, we discuss quantum mechanical systems of which the observables are represented by self-adjoint elements of a GB*-algebra. More specifically, we investigate the phenomenon of quantum entanglement within this framework. Motivated by this, we also give results on pure states of GB*-algebras, and provide an integral representation theorem for states of nuclear metrizable locally convex quasi $*$-algebras.

Keywords GB*-algebra · Quantum system · Separable state · Entangled state · Quantum entanglement · Linear nuclear space

1 Introduction

Generalized B*-algebras (GB*-algebras for short) are topological $*$-algebras which are generalizations of C*-algebras, and were first studied in 1967 by G. R Allan in [3]. GB*-algebras are, up to $*$-isomorphism, $*$-algebras of unbounded operators on Hilbert spaces in that the Gelfand-Naimark representation theorems for C*-algebras extend to GB*-algebras (see [18], Theorems 7.6 and 7.11) (we refer the reader to Definition 2.2 for the precise definition of a GB*-algebra). For every GB*-algebra $A[\tau]$, there exists a C*-algebra dense in A (see [7], Theorem 2, or [3], Theorem 2.6), so that one may think of a GB*-algebra as an enlargement of a C*-algebra with unbounded linear operators.

GB*-algebras have been investigated by various authors (see, for instance, [3, 7, 18, 19], and the literature in [23]). Reasons for their importance are that they occur among unbounded Hilbert algebras [29] which are essential for the development of Tomita-Takesaki theory of unbounded operator algebras developed in [30]. On the other hand, P. G. Dixon, in [18], extended the notion of GB*-algebra to a wider class

M. Weigt (✉)
Department of Mathematics, Nelson Mandela University, Port Elizabeth, South Africa
e-mail: Martin.Weigt@mandela.ac.za

E. Kikianty et al. (eds.), *Positivity and its Applications*, Trends in Mathematics,
https://doi.org/10.1007/978-3-030-70974-7_16

of topological $*$-algebras, which are not necessarily locally convex, such as various types of measurable operators affiliated with a von Neumann algebra [19].

Examples of GB*-algebras, besides C*-algebras, include pro-C*-algebras, i.e. complete locally convex $*$-algebras defined by directed families of C*-seminorms (see [21, Chapter 2] for further information). Another example of a GB*-algebra is the Arens algebra $L^{\omega}([0, 1]) = \bigcap_{p\geq 1} = L^{p}([0, 1])$, where the locally convex topology is defined by the family of L^p norms [21]. The algebra $L^{\omega}([0, 1])$ is a GB*-algebra which is not a pro-C*-algebra.

The theory of tensor products of C*-algebras is well developed (see, for instance, [13, 15, 32, 34] and [35]), and this provides motivation for a general investigation of tensor products of GB*-algebras. In this regard, the author, along with M. Fragoulopoulou and A. Inoue, initiated a general investigation of tensor products of GB*-algebras in [24].

Observables of a quantum mechanical system are regarded as unbounded self-adjoint linear operators on a Hilbert space, and the time evolution of the system can be modelled by a one-parameter group of automorphisms of a $*$-algebra of unbounded operators on a Hilbert space. Bearing in mind that GB*-algebras are significant for quantum mechanics, due to them being algebras of unbounded linear operators, we emphasize that there is a physical justification for using tensor products: Tensor products can be used to describe two quantum systems as one joint system (see [1]).

In the joint system of two or more quantum systems, one can have states which are entangled: In systems of particles, a state of a particle cannot be described independently of the states of the other particles. We say that the state is entangled. Quantum entanglement has led to paradoxes such as the EPR paradox, first described in 1935 by Albert Einstein, Boris Podolsky and Nathan Rosen. In entanglement, one state cannot be fully described without considering the other states. An entangled state can therefore be regarded as a state which cannot be expressed as a product of states of its constituents.

If one describes the states of a quantum system to be a Hilbert space, then a mathematical description of an entangled state would be as follows: Let H_1 and H_2 be Hilbert spaces describing the states of two quantum systems. Then the states of the joint system of the two systems would be the set of all unit vectors in $H := H_1 \widehat{\otimes} H_2$ (see [1]). We say that a state $\psi \in H$ is separable if it is an elementary tensor $\xi_1 \otimes \xi_2$ in the algebraic tensor product $H_1 \otimes H_2$ of H_1 and H_2. We say that $\psi \in H$ is entangled if it is not separable.

As explained in Sect. 3, we generally model the observables of a quantum mechanical system to be self-adjoint elements in a linearly nuclear GB*-algebra $A[\tau]$, i.e. a GB*-algebra which is nuclear as a locally convex space. The states will be taken to be positive linear functionals ϕ on A with $\phi(1) = 1$ (also called a state of A). Motivation from quantum physics for taking linearly nuclear GB*-algebras as the model of observables in a quantum system is also discussed in Sect. 3. An advantage of this approach is that one is not constrained to a specific Hilbert space in the model. The question is now how to think of quantum entanglement in this framework, i.e. in terms of states and self-adjoint elements in a GB*-algebra. In Sect. 4, we study quantum entanglement within the framework of GB*-algebras, culminating in Definition 4.1 of an entangled state of the joint system of two quantum systems. This definition is designed in such a way as to be in perfect agreement with the definition of an entangled state within the Hilbert space framework, as described above. One drawback with this definition is that it is dependent, in some way, on a specific representation of the GB*-algebra as a $*$-algebra of unbounded linear operators. For this reason, a second definition of an entangled state, within the GB*-algebra framework, is proposed in Definition 5.7 in Sect. 5. This definition is more in agreement with the notion of independence in probability theory and makes use of an integral representation theorem of states of a Fréchet GB*-algebra, which we also obtain in Sect. 5 directly from a result due to A. Inoue in [28]. An entangled state in the sense of Defintion 5.7 is also an entangled state in the sense of Definition 4.1, but the converse appears not to be true. Definition 5.7 of an entangled state immediately motivates the question as to the structure of states of a tensor product GB*-algebra, which were first studied in [24]. This motivated Sect. 6, which deals with the structure of pure states of a tensor product GB*-algebra, and the results in this section are extensions of the corresponding well known results in the C*-algebra framework, as can be found, for instance, in [42, Section IV.4].

There is one shortcoming in modelling a quantum system as self-adjoint elements of a locally convex $*$-algebra (which is taken to be a GB*-algebra above): A locally convex $*$-algebra is not closed under taking thermodynamical limits, and one requires a locally convex quasi $*$-algebra (Definition 2.9) to obtain closure under taking thermodynamical limits. This is described in some detail at the start of Sect. 8, where we also obtain an integral decomposition theorem of a state of a metrizable linearly nuclear locally convex quasi $*$-algebra into pure states (see Theorem 8.1). This result is an extension of a corresponding result by H. J. Borchers, J. Yngvason and G. C. Hegerfeldt (see [11] and [26]), and a brief exposition of the Borchers-Yngvason-Hegerfeldt result, including its importance to quantum mechanics, can be found in Sect. 7.

In Sect. 9, we give an example of a locally convex quasi $*$-algebra of operators, on the Hilbert space $L^2(X, \Sigma, \mu)$, over a C*-algebra, which satisfies the well known Heisenberg anti-commutation relation in quantum mechanics. The latter C*-algebra is the irrational rotation C*-algebra, and the von Neumann algebra generated by this C*-algebra is known to be of type II. For this reason, there can be no GB*-algebra over the irrational rotation C*-algebra which satisfies the Heisenberg anti-commutation relation (and this is also explained in Sect. 9).

Finally, Sect. 2 contains all the background information on GB*-algebras required to understand the main results of this paper.

2 Preliminaries

All vector spaces are assumed to be over the field $\mathbb{C}$ of complex numbers and all topological vector spaces are assumed to be Hausdorff. Moreover, all algebras are assumed to have an identity element denoted by 1.

A *topological algebra* is an algebra which is also a topological vector space such that the multiplication is separately continuous [21]. A *topological $*$-algebra* is a topological algebra endowed with a continuous involution. A topological $*$–algebra which is also a locally convex space is called a *locally convex $*$–algebra*. The symbol $A[\tau]$ will stand for a topological $*$-algebra A endowed with given topology τ.

Definition 2.1 ([3]) Let $A[\tau]$ be a topological $*$-algebra and $\mathcal{B}^*$ a collection of subsets B of A with the following properties:

(i) B is absolutely convex, closed and bounded,
(ii) $1 \in B$, $B^2 \subset B$ and $B^* = B$.

For every $B \in \mathcal{B}^*$, denote by $A[B]$ the linear span of B, which is a normed algebra under the Minkowski functional $\|\cdot\|_B$ of B. If $A[B]$ is complete for every $B \in \mathcal{B}^*$, then $A[\tau]$ is called *pseudo-complete*.

Every sequentially complete locally convex algebra is pseudo-complete [2, Proposition 2.6].

An element $x \in A$ is called *bounded* if for some nonzero complex number λ, the set $\{(\lambda x)^n : n = 1, 2, 3, \dots\}$ is bounded in A. We denote by A_0 the set of all bounded elements in A.

A topological $*$-algebra $A[\tau]$ is called *symmetric* if, for every $x \in A$, $(1+x^*x)^{-1}$ exists and belongs to A_0.

In [18], the collection $\mathcal{B}^*$ in the definition above is defined to be the same, except that $B \in \mathcal{B}^*$ is no longer assumed to be absolutely convex.

Definition 2.2 ([3]) A symmetric pseudo-complete locally convex $*$–algebra $A[\tau]$ such that the collection $\mathcal{B}^*$ has a greatest member denoted by B_0, is called a *GB^*–algebra* over B_0.

In [18], P. G. Dixon extended the notion of GB*-algebras to topological $*$-algebras which are not necessarily locally convex. In this definition, GB*-algebras are not assumed to be pseudo-complete nor locally convex. Every GB*-algebra in the sense of Definition 2.2 is a GB*-algebra in Dixon's sense. For a survey on GB*-algebras, see [23].

Proposition 2.3 ([3, Theorem 2.6], [7, Theorem 2]) *If $A[\tau]$ is a GB*-algebra, then the Banach $*$-algebra $A[B_0]$ is a C^*–algebra sequentially dense in A, and $(1 + x^*x)^{-1} \in A[B_0]$ for every $x \in A$. Furthermore, B_0 is the unit ball of $A[B_0]$.*

It follows easily that if $A[\tau]$ is a GB*-algebra, then the relative topology of τ on $A[B_0]$ is weaker than the $\|\cdot\|_{B_0}$-topology on $A[B_0]$.

A *pro-C^*-algebra* is a complete locally convex $*$-algebra $A[\tau]$, whose topology τ is defined by a directed family $(p_\alpha)_{\alpha\in\Lambda}$ of C^*-seminorms, i.e., $p_\alpha(x^*x) = p_\alpha(x)^2$ for all $x \in A$ and $\alpha \in \Lambda$. For every $\alpha \in \Lambda$, let $A_\alpha = A/N_\alpha$, where $N_\alpha = \{x \in A : p_\alpha(x) = 0\}$. Then A_α is a C*-algebra with respect to the C*-norm $\|x + N_\alpha\|_\alpha = p_\alpha(x)$ for all $\alpha \in \Lambda$ (it can be shown, for every $\alpha \in \Lambda$, that $A_\alpha = A/N_\alpha$ is complete with respect to the C*-norm $\|\cdot\|_\alpha$), and A is topologically $*$-isomorphic to the inverse limit of the C*-algebras A_α. We refer the reader to [21, Chapter 2] for further detail.

Furthermore, if $A[\tau]$ is a pro-C*-algebra, then

$$A_b := \{x \in A : \sup_\alpha p_\alpha(x) < \infty\}$$

is a C*-algebra dense in A, with respect to the C*-norm $\|x\|_b := \sup_\alpha p_\alpha(x), x \in A_b$ [21, Theorem 10.23]. Every pro-C^*-algebra is a GB^*-algebra with $A[B_0] = A_b$ (see [3, Example 3]).

Consider the Arens algebra $L^\omega([0, 1])$, defined to be $\cap_{p\geq 1} L_p([0, 1])$. Then $L^\omega([0, 1])$ is a GB*-algebra with respect to the L_p-norms $\|\cdot\|_p$, which is not a pro-C*-algebra [3].

For a $*$-algebra A, the set

$$A^+ = \{\sum_{i=1}^{n} x_i^* x_i : n \in \mathbb{N}, x_i \in A\}$$

induces a partial ordering $\leq$ on the set of all self-adjoint elements of A. That is, if x and y are self-adjoint elements of A, then $x - y \geq 0$ if and only if $x - y \in A^+$. For a self-adjoint element $x \in A$, we say that $x \in A$ is *positive*, written $x \geq 0$, if $x \in A^+$.

If $A[\tau]$ is a GB*-algebra, then a self-adjoint element $x \in A$ is positive if and only if there exists $y \in A$ such that $x = y^*y$ [18, Proposition 5.1].

Theorem 2.4 ([24, Theorem 4.3]) *Let $A_1[\tau_1]$ and $A_2[\tau_2]$ be GB*-algebras with jointly continuous multiplication. Let τ be a $*$-admissable topology (in the sense of [21, Chapter IV]) on $A_1 \otimes A_2$. The following statements are equivalent:*

(i) *$A_1 \widehat{\otimes}_\tau A_2$ is a GB*-algebra.*

(ii) *There is a C*-crossnorm $\|\cdot\|$ such that $A_0 = A_1[B_0^1] \widehat{\otimes}_{\|\cdot\|} A_2[B_0^2]$ is a C*-algebra contained in $A_1 \widehat{\otimes}_\tau A_2$ and such that τ is weaker than $\|\cdot\|$ on A_0.*

Definition 2.5 ([24, Definition 4.5]) Let $A_1[\tau_1]$ and $A_2[\tau_2]$ be GB*-algebras with jointly continuous multiplication. Let τ be a $*$-admissable topology on $A_1 \otimes A_2$. If the equivalent conditions of Theorem 2.4 are satisfied, the locally convex $*$-algebra $A_1[\tau_1]$ and $A_2[\tau_2]$ will be called a GB*-tensor product of $A_1[\tau_1]$ and $A_2[\tau_2]$.

A *positive linear functional* of a $*$-algebra A is a linear functional ϕ of A such that $\phi(x^*x) \geq 0$ for all $x \in A$.

Let $D = \bigoplus_{\phi\in F} A/N_\phi$, where F denotes the set of all positive linear functionals on A, and $N_\phi = \{x \in A : \phi(x^*x) = 0\}$ for all $\phi \in F$. For all $\phi \in F$, observe that A/N_ϕ is an inner product space under the inner product $\langle x + N_\phi, y + N_\phi\rangle = \phi(y^*x)$, $x, y \in A$. Using this, we note that D is an inner product space under the inner product $\langle(\xi_\phi)_{\phi\in F}, (\eta_\phi)_{\phi\in F}\rangle = \sum_{\phi\in F}\langle\xi_\phi, \eta_\phi\rangle$. Let H denote the Hilbert space completion of D. The *universal representation* π is defined by $\pi(x)((\xi_\phi)_{\phi\in F}) = (\pi_\phi(x)\xi_\phi)_{\phi\in F}$, for every $x \in A$, $(\xi_\phi)_{\phi\in F} \in D$, where $\pi_\phi(x)(y + N_\phi) = xy + N_\phi$, for all $x, y \in A, \phi \in F$. It is easily seen that $\pi(1)$ is the identity operator of D into itself. For every positive linear functional ϕ on A, we say that the representation π_ϕ, defined above, is the *GNS-representation* of A associated with ϕ.

Theorem 2.6 ([18, Theorem 7.6]) *If $A[\tau]$ is a GB*-algebra, then the universal representation π is faithful.*

It follows that every GB*-algebra can be represented faithfully as a $*$-algebra consisting of unbounded linear operators on the space D, and this generalizes the well known result that every C*-algebra can be faithfully represented as a norm closed $*$-subalgebra of bounded linear operators on some Hilbert space.

Let $\mathcal{L}^\dagger(D)$ denote the set

$$\{T : D \to D \text{ is a closable linear map} : D \subset D(T^*),\ T^*(D) \subset D\},$$

where $D(T^*)$ is the domain of the adjoint T^* of the densely defined operator T. Here, a closable linear map is understood to be a linear map which has a closed extension.

If D is a dense domain of a Hilbert space H, then $\mathcal{L}^\dagger(D)$ is a $*$-algebra of closable operators with involution given by $T^\dagger = T^*|_D$, and was introduced by G. Lassner in [36]. The algebra $\pi(A)$ is a $*$-subalgebra of $\mathcal{L}^\dagger(D)$.

A $*$-subalgebra of $\mathcal{L}^\dagger(D)$ containing the identity operator on D is called an *O*-algebra* on D [36].

Definition 2.7 ([27, Definition 2.1]) A $*$-representation of a $*$-algebra A with identity element 1 is a mapping of A into an O*-algebra with common dense domain $D(\pi)$ on a Hilbert space H, such that $\pi(1) = I$, where I is the identity operator on H, and such that the following conditions hold.

(i) $\pi(\alpha x + \beta y)\xi = \alpha\pi(x)\xi + \beta\pi(y)\xi$ for all $\xi \in D(\pi)$ and $\alpha, \beta \in \mathbb{C}$.
(ii) $\pi(x)D(\pi) \subseteq D(\pi)$ for all $x \in A$ and $\pi(x)\pi(y)\xi = \pi(xy)\xi$ for all $x, y \in A$ and $\xi \in D(\pi)$.
(iii) $\langle\pi(x)\xi, \eta\rangle = \langle\xi, \pi(x^*)\eta$ for all $x \in A$ and $\xi, \eta \in D(\pi)$, i.e. $\pi(x^*) \subseteq \pi(x)^*$ for all $x \in A$.

If $\pi : A \to \mathcal{L}^\dagger(D)$ is a $*$-representation of a $*$-algebra A, then we say that $\xi \in D$ is *strongly cyclic* for π if $\pi(A)\xi$ is dense in D [27, Definition 3.6].

A set of closed operators with dense domains on a Hilbert space H will be called a *$*$-algebra of closed operators* on H if it forms a $*$-algebra under the operations $x, y \mapsto \overline{x + y}$ for addition, $x, y \mapsto \overline{xy}$ for multiplication, and $x \mapsto x^*$ for involution [18, Definition 7.1], where $\overline{x}$ denotes the closure of the operator x.

A $*$-algebra A of closed operators on a Hilbert space H is called an *EC*-algebra* on H if $A \cap B(H)$ is a C*-algebra and $(1 + x^*x)^{-1} \in A$ for all $x, y \in A$. An

EC*-algebra A on H is called an *EW*-algebra* if $A \cap B(H)$ is a von Neumann algebra.

Theorem 2.8 ([18, Theorem 7.13]) *Every GB*-algebra $A[\tau]$ is algebraically $*$-isomorphic to an EC*-algebra on a Hilbert space with B_0 coinciding with the closed unit ball of $A \cap B(H)$.*

We recall that a nuclear C*-algebra is a C*-algebra A such that for every C*-algebra B, there is only one C*-norm on $A \otimes B$. A *nuclear GB*-algebra* is a GB*-algebra (in the sense of Definition 2.2) for which $A[B_0]$ is a nuclear C*-algebra [24]. This definition was motivated in [24] by the fact that a pro-C*-algebra is nuclear (as a pro-C*-algebra) if and only if A_b is a nuclear C*-algebra [10, Theorem 4.5]. Every commutative GB*-algebra is a nuclear GB*-algebra due to every commutative C*-algebra being a nuclear C*-algebra. For other examples of nuclear GB*-algebras, we refer the reader to [24].

The following concept of locally convex quasi $*$-algebra is required for Sect. 8, and a detailed exposition of locally convex quasi $*$-algebras can be found in [4].

Definition 2.9 Let A be a vector space and A_0 a $*$-algebra contained in A. We say that A is a quasi $*$-algebra over A_0 if

(i) the left multiplication ax and the right multiplication xa of an element $a \in A$ and $x \in A_0$, which extends the multiplication of A_0, is defined and bilinear;
(ii) $x_1(x_2a) = (x_1x_2)a$ and $x_1(ax_2) = (x_1a)x_2$ for all $x_1, x_2 \in A_0$ and $a \in A$;
(iii) A admits an involution $*$ which extends the involution on A_0 and $(ax)^* = x^*a^*$ and $(xa)^* = a^*x^*$ for each $x \in A_0$ and $a \in A$.

A quasi $*$-algebra is said to have an identity element if there exists $1 \in A_0$ such that $1a = a1 = a$ for all $a \in A$.

If $A[\tau]$ is a locally convex space and a quasi $*$-algebra over the $*$-algebra A_0, such that $a \mapsto ax$ and $a \mapsto xa$ are continuous on A for each $x \in A_0$, and A_0 is dense in A, then we say that $A[\tau]$ is a locally convex quasi $*$-algebra over A_0.

3 The Joint System of Quantum Systems

Let $A_1[\tau_1]$ be a GB*-algebra with jointly continuous multiplication. We say that $A_1[\tau_1]$ has the property UTP if the following is true: For any other GB*-algebra $A_2[\tau_2]$ having jointly continuous multiplication, there can only be one $*$-admissible topology τ on $A_1 \otimes A_2$ (algebraic tensor product) such that $A_1 \widehat{\otimes}_\tau A_2$ is a tensor product GB*-algebra (the topology τ is also assumed to make the multiplication on $A_1 \otimes A_2$ jointly continuous).

Proposition 3.1 *If $A_1[\tau_1]$ is a GB*-algebra having jointly continuous multiplication, and $A_1[B_0^1]$ has UTP, then $A[\tau_1]$ is a nuclear GB*-algebra.*

Proof Let A_2 be a C*-algebra. Since $A_1[B_0^1]$ has UTP, it is now immediate that

$$A_1[B_0^1] \mathop{\widehat{\otimes}}_{\max} A_2 = A_1[B_0^1] \mathop{\widehat{\otimes}}_{\min} A_2,$$

i.e. $A_1[B_0^1]$ is a nuclear C*-algebra, and hence $A_1[\tau_1]$ is a nuclear GB*-algebra, where max, min denote the maximal, respectively, minimal C*-cross norms. □

By the example in [44, Section 4], we see that not every nuclear GB*-algebra with jointly continuous multiplication has UTP.

In the algebraic approach to quantum physics, one uses a topological $*$-algebra to house the observables of the quantum system (see the introduction). Namely, they are the self-adjoint elements of the system. If one wants a GB*-algebra to be home to all observables, then what types of GB*-algebras would be adequate for such purposes? If we consider the joint system of two quantum systems, whereby the observables of the quantum systems are self-adjoint elements of GB*-algebras $A_1[\tau_1$ and $A_2[\tau_2]$ respectively, then one would want to model the joint system as the tensor product GB*-algebra $A_1 \mathop{\widehat{\otimes}}_{\tau} A_2$. If the tensorial GB*-topology τ is not unique, then there is more than one way to form the joint system, which is not good. One would therefore only want one topology τ such that $A_1 \mathop{\widehat{\otimes}}_{\tau} A_2$ is a GB*-algebra. For this purpose, we can either use a nuclear pro-C*-algebra (since they have UTP within the class of pro-C*-algebras, [10, Theorem 4.2]) or a linearly nuclear GB*-algebra (the reason is that if $A_1[\tau_1]$ is linearly nuclear, then $A_1 \otimes_\epsilon E[\tau_2] = A_1 \otimes_\pi E[\tau_2]$ for every locally convex space $E[\tau_2]$, by [39, Theorem, p. 115]). By a linearly nuclear algebra, we mean a (locally convex) algebra which is nuclear as a locally convex space. Linearly nuclear GB*-algebras are very useful if one wants to consider the joint system of *three or more* quantum systems: Let $A_1[\tau_1]$ and $A_2[\tau_2]$ be linearly nuclear GB*-algebras, then $A_1 \mathop{\widehat{\otimes}}_{\tau} A_2$ is also linearly nuclear [39, Proposition 5.4.1]. Therefore $A_1 \mathop{\widehat{\otimes}}_{\tau} A_2$ has UTP. If we consider general GB*-algebras with UTP, will $A_1 \mathop{\widehat{\otimes}}_{\tau} A_2$ have UTP if A_1 and A_2 have UTP, and $A_1 \mathop{\widehat{\otimes}}_{\tau} A_2$ is a GB*-algebra? If not, then using GB*-algebras with UTP to house all obervables will only be good for joining *two* quantum systems, not more. Linearly nuclear GB*-algebras are therefore very appealing. Any inverse limit of finite dimensional C*-algebras is a linearly nuclear pro-C*-algebra [10], which is not a C*-algebra unless the inverse system of finite dimensional C*-algebras consists only of one finite dimensional C*-algebra.

Another reason for considering linearly nuclear GB*-algebras as a home for the observables of a quantum system, is if one considers the role that rigged Hilbert spaces play in quantum mechanics, as discussed in [16]. Let Φ be an inner product space, and H the Hilbert space completion of Φ. If Φ is also equipped with a topology t which is stronger than the norm topology on Φ, then we say that $(\Phi \subseteq H \subseteq \Phi^\times)$ is a *rigged Hilbert space*, where $\Phi^\times$ is the space of t-continuous sesquilinear forms on Φ [4, Section 10.1]. We give a brief outline as to why rigged Hilbert spaces are important for quantum mechanics (see [16]).

1. For (operator) observables with discrete spectrum, a Hilbert space is sufficient for representing observables as an unbounded operator algebra on that Hilbert space: The eigenvectors corresponding to the eigenvalues (the values of the observables) are in that Hilbert space.
2. For (operator) observables with continuous spectrum, a Hilbert space will not be sufficient: For a rigged Hilbert space ($\Phi \subset H \subset \Phi^{\times}$), the eigenvectors generally belong to $\Phi^{\times}$. For example, the eigenfunctions for the finite square well potential (see [16, Section 3]) are generally not in the Hilbert space L^2, but in $\Phi^{\times}$. One therefore requires a rigged Hilbert space. In Dirac's Bra-ket formalism for quantum mechanics, the rigged Hilbert space is the underlying mathematical setting. Note that the usual notation of bras and kets are $\langle\cdot|$, respectively, $|\cdot\rangle$, which we shall also apply in what follows. The bras belong to Φ', where Φ' is the space of t-continuous linear functionals on Φ, and the kets belong to $\Phi^{\times}$ [16]. Observe that ($\Phi \subset H \subset \Phi'$) is rigged Hilbert space associated with ($\Phi \subset H \subset \Phi^{\times}$).
3. The following discussion is due to I. M. Gelfand and collaborators, and we use [20, Subsection 6.3.2] as a reference. Consider the rigged Hilbert space ($\Phi[t]$, H, $\Phi^{\times}$), where $\Phi[t]$ is a (linearly) nuclear space. We mention here that the notion of rigged Hilbert space is originally due to Gelfand and his collaborators, with $\Phi[t]$ being (linearly) nuclear as part of their definition of rigged Hilbert space. Let $A : \Phi \to \Phi$ be a linear operator. A linear functional $f : \Phi \to \mathbb{C}$ such that $f(A\phi) = \lambda f(\phi)$ for all $\phi \in \Phi$ is called a *generalized eigenvector* of A with respect to the eigenvalue λ of A. The *generalized eigenspace* of A with respect to the eigenvalue λ, denoted Φ'_{λ}, is the set of all generalized eigenvectors of A with respect to the eigenvalue λ. Let $\phi \in \Phi$ and $\lambda \in \mathbb{C}$. We define a map $\widetilde{\phi}_{\lambda} : \Phi'_{\lambda} \to \mathbb{C}$ by $f \mapsto f(\phi)$. The map $\phi \to \widetilde{\phi}_{\lambda}$ is called the *spectral decomposition* of $\phi \in \Phi$ with respect to A. If $\widetilde{\phi}_{\lambda} = 0$ implies $\phi = 0$, then we say that Φ'_{λ} is complete. The linear operator A above is said to be *self-adjoint* if its closure $\overline{A}$ is self-adjoint.

Theorem 3.2 (Gelfand et. al.) *Let A be a self-adjoint linear operator in a rigged Hilbert space ($\Phi[t]$, H, $\Phi^{\times}$), where $\Phi[t]$ is a (linearly) nuclear space. Then, for each (real) eigenvalue of A, the operator A admits a complete system of generalized eigenvectors.*

This theorem can be regarded as an extension, or continuation, of the spectral theorem for unbounded self-adjoint linear operators on a Hilbert space.

In the event where $H = L^2(\mathbb{R})$, this theorem takes on the following form [16]: If A is a self-adjoint unbounded linear operator on H (an operator observable), then to each element of the spectrum of A, there corresponds a left and a right eigenvector [16, p. 5]. If discrete eigenvalues of A are denoted by a_n, and continuous eigenvalues by a, then, for the corresponding right eigenvectors (i.e. the kets), we have $A|a_n\rangle = a_n|a_n\rangle$ and $A|a\rangle = a|a\rangle$. For the corresponding left eigenvectors (i.e. the bras), we have $\langle a_n|A = a_n\langle a_n|$ and $\langle a|A = a\langle a|$.

In the Dirac formalism of quantum mechanics, any wave function ϕ has the Dirac basis expansion as determined by the operator A above [16, p. 6]:

$$\phi = \sum_{n}^{\infty} |a_n\rangle\langle a_n, \phi\rangle + \int |a\rangle\langle a, \phi\rangle da.$$

Furthermore,

$$A = \sum_{n}^{\infty} a_n |a_n\rangle\langle a_n, \phi\rangle + \int a|a\rangle\langle a, \phi\rangle da.$$

4. Now we note that $\Phi = \mathcal{S}(\mathbb{R} - \{a, b\})$, as defined in [16], is a linearly nuclear space (this follows immediately from [39, Theorem 6.2.2 and Proposition 5.1.1]). By (4) above, we note that the physics of the system described by a particle in finite square well leads naturally to a linearly nuclear space Φ on which the operator observables act. The author in [16] remarked that it is not very easy to choose Φ linearly nuclear (see [16, p. 26]). However, by [16, p. 26], if V is infinitely many times differentiable except on a closed subset of $\mathbb{R}$ with zero Lebesgue measure, then one can choose Φ in such a way as to be linearly nuclear.

   For the nucleus of an atom, the Coulomb potential is of the form $V(r) = \frac{k}{r}$, and V is infinitely many times differentiable except at $\{0\}$, which is a closed subset of $\mathbb{R}$ with zero Lebesgue measure. Therefore, for a proton trapped inside the nucleus of an atom by the Coulomb potential, we can choose Φ to be a linearly nuclear space.

All of the above motivates the following result, which also informs us that linearly nuclear GB*-algebras form an appropriate mathematical setting for housing the observables of a quantum system (which are seen as self-adjoint elements of the linearly nuclear GB*-algebra).

Proposition 3.3 *Let $A[\tau]$ be a Fréchet GB*-algebra. Then there exists a faithful $*$-representation $\pi : A \to \mathcal{L}^\dagger(D)$ for some domain D. The domain $D[t]$ is part of a rigged Hilbert space $(D, H, D^\times)$, where H is the Hilbert space completion of D, and $D^\times$ is the space of t-continuous sesquilinear forms on D, and t is a topology on D which is stronger than the norm topology on D. If, in addition, $A[\tau]$ is linearly nuclear, then $D[t]$ is a linearly nuclear space.*

Proof Choose the universal representation π of A (Proposition 2.6): For any positive linear functional ϕ of A, let

$$N_\phi = \{x \in A : \phi(x^*x) = 0\}.$$

Let $D_\phi = A/N_\phi$ and $D = \oplus_\phi D_\phi$. Then $\pi(a) = \oplus_\phi \pi_\phi(a)$ for all $a \in A$.

Since $A[\tau]$ is a Fréchet algebra, we know that $A[\tau]$ has jointly continuous multiplication and that all positive linear functionals of A are continuous [17, Theorem 11.1]. Thus N_ϕ is a closed vector subspace of A (that N_ϕ is a vector space follows from the Cauchy-Schwartz inequality for positive linear functionals). Thus $D_\phi = A/N_\phi$ can be equipped with the quotient topology t making it a Hausdorff locally convex space. Observe that t is stronger than the norm topology on D_ϕ defined by its inner product: This follows from the fact that the norm topology on D_ϕ, induced by its inner product, makes the quotient map of A into D_ϕ continuous. Here, we use continuity of ϕ and the involution, as well as joint continuity of multiplication of A.

It follows that $(D_\phi, H_\phi, D_\phi^\times)$ is a rigged Hilbert space, where H_ϕ is the norm completion of D_ϕ. It follows that $(D, H, D^\times)$ is a rigged Hilbert space. We therefore also have the associated rigged Hilbert space (D, H, D').

If, in addition, $A[\tau]$ is linearly nuclear, then D_ϕ is linearly nuclear (see [39, p. 88, 5.1.6, Proposition]). Hence $D = \oplus_\phi D_\phi$ is linearly nuclear [39, p. 90, 5.2.1, Proposition, and Proposition 5.1.1]. □

4 Quantum Entanglement

Let $A_1[\tau_1]$ and $A_2[\tau_2]$ be GB*-algebras and assume that $A_1 \widehat{\otimes}_\tau A_2$ is a tensor product GB*-algebra, and let ϕ_1, ϕ_2 be positive linear functionals on A_1 and A_2 respectively. By [27, Theorem 3.18], there exists $*$-representation $\pi_1 : A_1 \to \mathcal{L}^\dagger(D_1)$ of A_1 and $\xi_1 \in D_1$ such that $\phi_1(x) = \langle \pi_1(x)\xi_1, \xi_1 \rangle$ for all $x \in A_1$. Likewise, there exists $*$-representation $\pi_2 : A_2 \to \mathcal{L}^\dagger(D_2)$ of A_2 and $\xi_2 \in D_2$ such that $\phi_2(y) = \langle \pi_2(y)\xi_2, \xi_2 \rangle$ for all $y \in A_2$. It is easily verified that

$$(\phi_1 \otimes \phi_2)(x \otimes y) = \langle \big((\pi_1 \otimes \pi_2)(x \otimes y)\big)(\xi_1 \otimes \xi_2), \xi_1 \otimes \xi_2 \rangle$$

for all $x \in A_1$ and $y \in A_2$.

Let ϕ be a positive linear functional on $A := A_1 \widehat{\otimes}_\tau A_2$. Assume that there exists a $*$-representation $\pi : A \to \mathcal{L}^\dagger(D)$ and $\xi \in D$ such that

$$\phi(x \otimes y) = \langle \pi(x \otimes y)\xi, \xi \rangle$$

for all $x \in A_1$ and $y \in A_2$, where for two given Hilbert spaces H_1 and H_2 and their respective Hilbert space tensor product $H_1 \widehat{\otimes} H_2$, D is a dense subspace of $H_1 \widehat{\otimes} H_2$. Therefore

$$\phi\Big(\sum_{i=1}^{n} x_i \otimes y_i\Big) = \sum_{i=1}^{n} \langle \pi(x_i \otimes y_i)\xi, \xi \rangle \tag{1}$$

for all $x_i \in A_1$, $y_i \in A_2$ and $n \in \mathbb{N}$.

If $\phi = \phi_1 \otimes \phi_2$, then it follows easily that

$$\phi(\sum_{i=1}^{n} x_i \otimes y_i) = \sum_{i=1}^{n} \langle ((\pi_1 \otimes \pi_2)(x_i \otimes y_i))(\xi_1 \otimes \xi_2), \xi_1 \otimes \xi_2 \rangle$$

for all $x_i \in A_1$ and $y_i \in A_2$, where $\xi_1 \in H_1$ and $\xi_2 \in H_2$.

In quantum mechanics, the state space of the joint system of two quantum systems with state spaces H_1 and H_2, is the Hilbert space tensor product $H_1 \widehat{\otimes} H_2$ of H_1 and H_2. A separable state in H is regarded as an elementary tensor of two vectors in H_1 and H_2 respectively (of norm one) on which the (unbounded operator) observables act. A state in H is said to be entangled if it is not separable (see also Definition 5.5 in Sect. 5). The state vectors in a Hilbert space are identified with positive linear functionals on the $*$-algebra of observables. In light of this, one is led to the following definition of quantum entanglement in the setting of linearly nuclear GB*-algebras, which is consistent with the definition of an entangled state in terms of Hilbert spaces given above.

Definition 4.1 Let $A_1[\tau_1]$ and $A_2[\tau_2]$ be linearly nuclear GB*-algebras such that $A := A_1 \widehat{\otimes}_\tau A_2$ is a tensor product GB*-algebra (so $\epsilon = \tau = \pi$). We say that a state ϕ on A (i.e. a positive linear functional ϕ of A with $\phi(1) = 1$) is separable if it can be expressed in the form (1) above, with ξ an elementary tensor. We say that the state ϕ is entangled if ϕ can be expressed in the form (1) above, and ξ can never be chosen to be an elementary tensor.

This definition makes sense: If ϕ on A does not have the form (1), then ϕ cannot be faithfully represented as an unbounded operator algebra such that one has Eq. (1) above. So ϕ cannot be identified with an elementary tensor. So ϕ represents a state which is physically entangled in the quantum-mechanical sense.

Remark 4.2 If ϕ is a state on A with $\phi = \phi_1 \otimes \phi_2$, where ϕ_1 and ϕ_2 are states on A_1 and A_2 respectively, then it follows from a computation above that ϕ is a separable state of A. If ϕ is a separable state, it appears, however, that it need not be of the form $\phi = \phi_1 \otimes \phi_2$, where ϕ_1 and ϕ_2 are states on A_1 and A_2 respectively.

Example 4.3 If $A_1[\tau_1]$ and $A_2[\tau_2]$ are linearly nuclear pro-C*-algebras (hence GB*-algebras), then $A := A_1 \widehat{\otimes}_\tau A_2$ is clearly a pro-C*-algebra, and hence a GB*-algebra. In fact, $\epsilon = \tau = \pi$, implying that there is only one topology τ for which $A_1 \widehat{\otimes}_\tau A_2$ is a pro-C*-algebra.

In the above, we represent GB*-algebras as unbounded operator algebras on a Hilbert space. What if we represent a GB*-algebra $A[\tau]$ as a rigged Hilbert space $(D, H, D^\times)$ instead? Then states on A (i.e. positive linear functionals ϕ on A with $\phi(1) = 1$) do not correspond to *all* physical states of the joint quantum system A. The remaining states are in $D^\times$, the space of all continuous sesquilinear forms on D with respect to a topology on D stronger than the norm topology (see [16]). Recall that elements of D correspond to positive linear functionals on A: If $\xi \in D$, then $\phi(x) = \langle a\xi, \xi\rangle$ for all $a \in A$.

Now if h is a bounded sesquilinear form on D, then $h \in D^{\times}$. Then h can be extended to a bounded sesquilinear form on H, and by Riesz's theorem, there exists a bounded linear operator $S : H \to H$ such that $h(\xi, \eta) = \langle S\xi, \eta\rangle$ for all $\xi, \eta \in D$. Therefore, other physical states of the system are to be found in $A[B_0]$ (not only amongst the positive linear functionals of A). Note, however, that not every $h \in D^{\times}$ is bounded.

Let $A_1[\tau_1]$ and $A_2[\tau_2]$ be linearly nuclear GB*-algebras such that $A := A_1\widehat{\otimes}_{\tau} A_2$ is a tensor product GB*-algebra (so $\epsilon = \tau = \pi$). Let $x_1 \in A_1[B_0^1]$ and $y_1 \in A_2[B_0^2]$, and let $x = x_1 \otimes y_1$. Now represent A_1 and A_2 faithfully as $*$-algebras of unbounded linear operators on D_1 and D_2 respectively (see Theorem 2.6). By the proof of (i) implies (ii) of Theorem 4.3 in [24], one has that $x \in (A_1\widehat{\otimes}_{\tau} A_2)[B_0]$. Therefore $h(\xi, \eta) = \langle (x_1 \otimes y_1)\xi, \eta\rangle$ is a bounded sesquilinear form on $D_1 \otimes D_2$. It is easily verified that

$$h\Big(\sum_{i=1}^{n} \xi_{1,i} \otimes \xi_{2,i}, \sum_{j=1}^{m} \eta_{1,j} \otimes \eta_{2,j}\Big) = (h_1 \otimes h_2)\Big(\sum_{i=1}^{n} \xi_{1,i} \otimes \xi_{2,i}, \sum_{j=1}^{m} \eta_{1,j} \otimes \eta_{2,j}\Big),$$

where h_1 and h_2 are those bounded sesquilinear forms on D_1 and D_2, defined by x_1 and y_1 respectively. It therefore makes sense to define a physical state $a \in (A_1\widehat{\otimes}_{\tau} A_2)[B_0]$ to be *separable* if $a = x_1 \otimes y_1$, where all x_1, y_1 are in $A_1[B_0^1]$, respectively $A_2[B_0^2]$. We say that the physical state $a \in (A_1\widehat{\otimes}_{\tau} A_2)[B_0]$ is *entangled* if it is not of this form.

We now give an example of an entangled state of a tensor product GB*-algebra. For this, we require knowledge of unbounded $*$-representations of topological $*$-algebras induced by unbounded C*-seminorms, as defined in [9].

Let p be an unbounded C*-seminorm of a $*$-algebra A, i.e. p is an unbounded C*-seminorm on a $*$-subalgebra $D(p)$ of A. Let

$$I_p = \{x \in D(p) : ax \in D(p) \text{ for all } a \in A\}.$$

Suppose that I_p is not a subset of $\mathrm{Ker}(p)$. We use this to construct an unbounded $*$-representation of A, as described in [9].

Observe that $D(p)/\mathrm{Ker}(p)$ is a C*-normed algebra in the obvious manner. Denote by A_p the completion of $D(p)/\mathrm{Ker}(p)$. Then A_p is a C*-algebra. Let Π_p be a faithful nondegenerate $*$-representation of A_p on the Hilbert space H_{Π_p}. Let $\pi_p^0(x) = \Pi_p(x + \mathrm{Ker}(p))$ for all $x \in D(p)$. Then π_p^0 is a bounded $*$-representation of $D(p)$ on H_{Π_p}. Let $a \in A$ and $\sum_{\text{finite}} \Pi_p(x_k + \mathrm{Ker}(p))\xi_k \in D(\pi_p)$, where

$$D(\pi_p) = \mathrm{span}\{\Pi_p(x + \mathrm{Ker}(p))\xi : x \in I_p, \xi \in H_{\Pi_p}\} \subseteq H_{\Pi_p}.$$

Then the linear map

$$\pi_p(a)\Big(\sum_k \Pi_p(x_k + \mathrm{Ker}(p))\xi_k\Big) = \sum_k \Pi_p(ax_k + \mathrm{Ker}(p))\xi_k$$

is a well defined linear operator on $D(\pi_p)$, where the sums are taken to be finite sums. Let H_{π_p} denote the norm closure of $D(\pi_p)$ in H_{Π_p}. Since I_p is not contained in Ker(p), we get that π_p is non-trivial. If $H_{\pi_p} = H_{\Pi_p}$, then π_p is called a *well-behaved* $*$*-representation* of A [9].

Example 4.3 Let $A_1[\tau_1]$ and $A_2[\tau_2]$ be GB*-algebras such that $A := A_1 \widehat{\otimes}_\tau A_2$ is a tensor product GB*-algebra. Assume that A has a well-behaved $*$-representation $\pi_r : A \to \mathcal{L}^\dagger(D_r)$ induced by the unbounded C*-seminorm $r := \|\cdot\|_{B_0}$ on $A[B_0] = (A_1 \widehat{\otimes}_\tau A_2)[B_0]$. Suppose that $\pi_r = \pi_1 \otimes \pi_2$, where $\pi_1 : A_1 \to \mathcal{L}^\dagger(D_1)$ and $\pi_2 : A_2 \to \mathcal{L}^\dagger(D_2)$ are well-behaved $*$-representations of A_1 and A_2 respectively, defined by the unbounded C*-seminorms $\|\cdot\|_{B_0^1}$ and $\|\cdot\|_{B_0^2}$ on $A_1[B_0^1]$ and $A_2[B_0^2]$ respectively. Furthermore, we let Π_r denote the bounded $*$-representation $\Pi_r : A[B_0] \to B(H_r)$ on $A[B_0]$ (on which the unbounded C*-seminorm r is defined) defining π_r (see above). We also assume that $\Pi_r = \Pi_1 \otimes \Pi_2$, where Π_1 and Π_2 are the bounded $*$-representations on $A_1[B_0^1]$ and $A_2[B_0^2]$, defined on the Hilbert spaces H_1 and H_2 respectively, defining π_1 and π_2 respectively. Note that H_r, H_1, H_2 are the Hilbert space completions of D_r, D_1, D_2 respectively.

Let I_r, I_p and I_q be the left ideals in the definition of π_r, π_1 and π_2 respectively, where $p := \|\cdot\|_{B_0^1}$ and $q := \|\cdot\|_{B_0^2}$. Furthermore, $I_r = I_p \otimes I_q$ (see the proof of [22, Proposition 3.1]). Then

$$\pi_r(x)\xi = \pi_r(x)\Big(\sum_{k=1}^n \Pi_r(x_k)\xi_k\Big),$$

where $\xi = \sum_{k=1}^n \Pi_r(x_k)\xi_k$, and where $\xi_k \in H_r$, $x_k \in I_r$ for all $1 \leq k \leq n$.

Observe that $D_r = D_1 \otimes D_2$ and $H_r = H_1 \widehat{\otimes} H_2$. Let

$$\xi_k = \sum_{i=1}^m \xi_{k,1}^{(i)} \otimes \xi_{k,2}^{(i)} \in H_1 \otimes H_2,$$

and let

$$x_k = \sum_{j=1}^l x_{k,p}^{(j)} \otimes x_{k,q}^{(j)} \in I_p \otimes I_q = I_r.$$

Let

$$\xi = \sum_{k=1}^{n} \Pi_r(x_k)\xi_k$$
$$= \sum_{k=1}^{n} [\Pi_r(\sum_{j=1}^{l} x_{k,p}^{(j)} \otimes x_{k,q}^{(j)})](\sum_{i=1}^{m} \xi_{k,1}^{(i)} \otimes \xi_{k,2}^{(i)}).$$

Therefore

$$\pi_r(x)\xi = \pi_r(x)\big(\sum_{k=1}^{n} \Pi_r(x_k)\xi_k)\big)$$
$$= (\pi_1 \otimes \pi_2)(x)\big(\sum_{k=1}^{n}[\Pi_r(\sum_{j=1}^{l} x_{k,p}^{(j)} \otimes x_{k,q}^{(j)})](\sum_{i=1}^{m} \xi_{k,1}^{(i)} \otimes \xi_{k,2}^{(i)})\big)$$
$$= (\pi_1 \otimes \pi_2)(x)\big([\sum_{k}\sum_{j} \Pi_1(x_{k,p}^{(j)}) \otimes \Pi_2(x_{k,q}^{(j)})](\sum_{i=1}^{l} \xi_{k,1}^{(i)} \otimes \xi_{k,2}^{(i)})\big)$$
$$= (\pi_1 \otimes \pi_2)(x)[\sum_{k,j,i}[\Pi_1(x_{k,p}^{(j)})](\xi_{k,1}^{(i)}) \otimes [\Pi_2(x_{k,q}^{(j)})](\xi_{k,2}^{(i)})].$$

Let

$$\eta = \sum_{k,j,i}[\Pi_1(x_{k,p}^{(j)})](\xi_{k,1}^{(i)}) \otimes [\Pi_2(x_{k,q}^{(j)})](\xi_{k,2}^{(i)}),$$

where $\eta \in D_r$ is a unit strongly cyclic vector for π. Suppose that η is so chosen that it cannot be expressed as an elementary tensor.

Let $\phi(x) = \langle \pi_r(x)\eta, \eta\rangle$ for all $x \in A$. We show that ϕ is an entangled state. To do this, consider a $*$-representation $\pi_0 : A \to \mathcal{L}^\dagger(D)$, where D is a dense subspace of the Hilbert space tensor product $K_1\widehat{\otimes}K_2$ (where K_1 and K_2 are Hilbert spaces), such that $\phi(x) = \langle \pi_0(x)\eta_0, \eta_0\rangle$ for all $x \in A$, where $\eta_0 \in D$ is a unit strongly cyclic vector for π (we may assume, without loss of generality, that η_0 is strongly cyclic for π_0 by replacing D with $D_0 = \pi_0(A)\eta_0$, and by replacing $K_1\widehat{\otimes}K_2$ with the norm closure of D_0 in $K_1\widehat{\otimes}K_2$. It is clear that π_0 is invariant under D_0). We show that $\eta_0 \in D$ is not an elementary tensor.

Observe that

$$\phi(x) = \langle \pi_0(x)\eta_0, \eta_0\rangle = \langle \pi_r(x)\eta, \eta\rangle$$

for all $x \in A$. Therefore π_r and π_0 are unitarily equivalent [27, Proposition 3.12], i.e. there exists a unitary operator mapping from $H_1\widehat{\otimes}H_2$ onto $K_1\widehat{\otimes}K_2$, such that

$U(D_r) = D$, $U\eta = \eta_0$ and $U^*\pi_0(x)U\xi = \pi_r(x)\xi$ for all $\xi \in D_r$. Recall here that $D = D_1 \otimes D_2$.

Therefore, since η can never be expressed as an elementary tensor, it follows that η_0 can also not be expressed as an elementary tensor. It follows that ϕ is an entangled state.

Let $A_1[\tau_1]$ and $A_2[\tau_2]$ be GB*-algebras which both have well-behaved $*$-representations π_1 and π_2 respectively, defined by the unbounded C*-norms $\|\cdot\|_{A_1[B_0^1]}$ and $\|\cdot\|_{A_2[B_0^2]}$ respectively. Suppose that $A_1\widehat{\otimes}_\tau A_2$ is a tensor product GB*-algebra such that

$$(A_1\widehat{\otimes}_\tau A_2)[B_0] = A_1[B_0^1] \widehat{\otimes}_{\max} A_2[B_0^2].$$

By [22, Proposition 4.1] and its proof, one has that $A_1\widehat{\otimes}_\tau A_2$ admits a well-behaved $*$-representation induced by $\|\cdot\|_{B_0}$. By Example 4.3, this gives rise to an entangled state on $A_1\widehat{\otimes}_\tau A_2$. Note that not all GB*-algebras have well-behaved $*$-representations.

5 Integral Representations of Positive Linear Functionals of GB*-Algebras and an Alternative Definition of Quantum Entanglement

For this section, we refer the reader to [29] for the definition of an unbounded Hilbert algebra.

Theorem 5.1 ([28, Proposition 3.1]) *Let $A[\tau]$ be a commutative Fréchet GB*-algebra and ϕ a positive linear functional on A. Then there exists a locally compact space X, a positive Radon measure μ on X with support X, and a $*$-representation $x \mapsto \hat{x}_\phi$ of A onto an unbounded Hilbert algebra generated by μ such that*

$$\phi(x) = \int_X \hat{x}_\phi(\xi)\, d\mu(\xi)$$

for all $x \in A$.

Remark 5.2

(1) Theorem 5.1 is a generalization of the Riesz representation theorem, which gives an integral representation of positive linear functionals of unital commutative C*-algebras: If ϕ is a positive linear functional of a unital commutative C*-algebra $A \cong C(X)$, then there exists a positive Radon measure μ on X such that $\phi(f) = \int_X f\, d\mu$. In the commutative C*-algebra case, we note that

the space X is independent of ϕ, whereas this is not the case for a general commutative GB*-algebra (as revealed by Theorem 5.1).

(2) The above result, due to A. Inoue, is given as Proposition 3.1 in [28], but without the assumption that A be a Fréchet algebra. The proof of [28, Proposition 3.1], however, shows that one needs joint continuity of multiplication and that all positive linear functionals of A are continuous. A Fréchet GB*-algebras has both of these properties, and this is why the assumption of A being a Fréchet algebra is in the hypothesis of Theorem 5.1.

Let A be a (unital) commutative C*-algebra, and let ϕ be a *state* on $A \cong C(X)$ (so X is compact, due to A being unital). Then there exists a positive Radon measure μ on X such that $\phi(f) = \int_X f \, d\mu$, where $\mu(X) = 1$, for all $f \in C(X)$. By [33, Proposition 4.30], it follows immediately that $\phi(f) = \int_{\mathbb{R}} t \, d\mu_f(t)$, where μ_f is the distribution function of f, for all *self-adjoint (i.e. real-valued)* $f \in C(X)$ (see also [33, Example 4.31]). Here one considers the C*-subalgebra of $C(X)$ generated by a given self-adjoint (i.e. real-valued) $f \in C(X)$. Therefore, if A is a commutative C*-algebra, and $x \in A$ is self-adjoint, one has that $\phi(x) = \int_{\mathbb{R}} t \, d\mu_x(t)$ (see also [43, Remark 2.3.2]).

Let $A[\tau]$ be a Fréchet GB*-algebra, and let ϕ be *state* on A. Let $x \in A$ be self-adjoint, and let B be the unital closed commutative $*$-subalgebra of A generated by x. Then $\psi := \phi|_B$ is a state on B, and B is a Fréchet GB*-algebra (by [3, Proposition 2.9], every unital closed $*$-subalgebra of a GB*-algebra is a GB*-algebra). By Theorem 5.1, there is a locally compact space X, a Radon measure μ on X, and a $*$-representation $a \mapsto \hat{a}_\psi$ of B onto an unbounded Hilbert algebra generated by μ such that

$$\psi(x) = \int_X \hat{x}_\psi(\xi) \, d\mu(\xi).$$

Therefore

$$\begin{aligned}\psi(x) &= \int_{\mathbb{R}} t \, d\mu_{\hat{x}_\psi}(t) \\ &= \int_{\mathbb{R}} t \, d\mu_x(t),\end{aligned}$$

say. Therefore $\phi(x) = \int_{\mathbb{R}} t \, d\mu_x(t)$. One therefore has the following result, which is an extension of Inoue's Theorem 5.1 for states to non-commutative GB*-algebras.

Corollary 5.3 *Let $A[\tau]$ be a Fréchet GB*-algebra, $x \in A$ be self-adjoint, and let ϕ be a state on A. Then there exists a measure $d\mu_x$ on $\mathbb{R}$ such that*

$$\phi(x) = \int_{\mathbb{R}} t \, d\mu_x(t).$$

The following concept is a non-commutative analogue of the notion of independence in measure and probability theory. For the measure theoretic notion of independence, see [33, Definition 3.18]. A *noncommutative probability space* (A, ϕ) is an algebra A with identity element 1 together with a linear functional ϕ such that $\phi(1) = 1$ [43, Definition 2.2.1].

Definition 5.4 ([43, Definition 2.4.1]) Let (A, ϕ) be a non-commutative probability space. A family of subalgebras A_α of A is independent if the algebras commute with each other, $\phi(a_1 \cdots a_n) = \phi(a_1) \cdots \phi(a_n)$ for all $a_k \in A_{\alpha_k}$, and $k \neq l$ implies $\alpha_k \neq \alpha_l$.

If (A, ϕ) and (B, ψ) are non-commutative probability spaces, then it is easily verified that the non-commutative probability space $(A \otimes B, \phi \otimes \psi)$ contains $A \otimes 1_B$ and $1_A \otimes B$ as independent subalgebras of $A \otimes B$ (see [43, p. 13]).

Let $A_1[\tau_1]$ and $A_2[\tau_2]$ be GB*-algebras (with jointly continuous multiplication). Assume that $A := A_1 \widehat{\otimes}_\tau A_2$ is a tensor product GB*-algebra with respect to some $*$-admissible topology τ. Let ϕ be a state on A. If ϕ is a state on A such that $\phi = \phi_1 \otimes \phi_2$ for some states ϕ_i on A_i, $i = 1, 2$, then, from the above, $A_1 \otimes 1_2$ and $1_1 \otimes A_2$ are independent subalgebras of $A_1 \otimes A_2$ with respect to the state ϕ on A.

Conversely, assume that $A_1 \otimes 1_2$ and $1_1 \otimes A_2$ are independent pairs of subalgebras in $A_1 \otimes A_2$ with respect to some state ϕ on A. For the state ϕ on A, we show that $\phi = \phi_1 \otimes \phi_2$ for some states ϕ_i on A_i, $i = 1, 2$. We first observe that $A_1 \otimes 1_2$ and $1_1 \otimes A_2$ already commute (regardless of whether they are independent). Note that

$$\begin{aligned}\phi(x \otimes y) &= \phi((x \otimes 1_2)(1_1 \otimes y)) \\ &= \phi(x \otimes 1_2) \cdot \phi(1_1 \otimes y)\end{aligned}$$

for all $x \in A_1$ and $y \in A_2$ (due to $A_1 \otimes 1_2$ and $1_1 \otimes A_2$ being independent). Let $\phi_1(x) = \phi(x \otimes 1_2)$ for all $x \in A_1$. Let $\phi_2(y) = \phi(1_1 \otimes y)$ for all $y \in A_2$. It is easily seen that $\phi_1(1_1) = 1_1$ and $\phi_2(1_2) = 1_2$. Therefore ϕ_1 and ϕ_2 are states on A_1 and A_2 respectively (it is easily checked that ϕ_1 and ϕ_2 are positive linear functionals on A_1 and A_2). It follows that $\phi(x \otimes y) = \phi_1(x)\phi_2(y)$ for all $x \in A_1$ and $y \in A_2$. Therefore $\phi(\sum_{i=1}^n x_i \otimes y_i) = \sum_{i=1}^n \phi_1(x_i)\phi_2(y_i)$ for all $x_i \in A_1$, $y_i \in A_2$ and $n \in \mathbb{N}$. So $\phi = \phi_1 \otimes \phi_2$.

Corollary 5.5 *Let $A_1[\tau_1]$ and $A_2[\tau_2]$ be GB*-algebras with jointly continuous multiplication. Assume that $A := A_1 \widehat{\otimes}_\tau A_2$ is a tensor product GB*-algebra with respect to some $*$-admissible topology τ. Let ϕ be a state on A. The following statements are equivalent.*

(i) *$\phi = \phi_1 \otimes \phi_2$, where ϕ_1 and ϕ_2 are states on A_1 and A_2 respectively.*
(ii) *$A_1 \otimes 1_2$ and $1_1 \otimes A_2$ are independent pairs of subalgebras in $A_1 \otimes A_2$ with respect to the state ϕ on A.*

Remark 5.6 Let $A_1[\tau_1]$ and $A_2[\tau_2]$ be Fréchet GB*-algebras. Assume that $A := A_1 \widehat{\otimes}_\tau A_2$ is a tensor product Fréchet GB*-algebra with respect to some $*$-admissible topology τ. Let ϕ be a state on A. Then, by Corollary 5.3, for all self-adjoint $x \in A_1$ and self-adjoint $y \in A_2$, we have that

$$\phi(x \otimes y) = \int_{\mathbb{R}} t \, d\mu_{x\otimes y}(t).$$

Let ϕ_1 and ϕ_2 be states on A_1 and A_2 respectively. Then

$$\phi_1(x) = \int_{\mathbb{R}} \lambda \, d\mu_x(\lambda),$$

and

$$\phi_2(y) = \int_{\mathbb{R}} s \, d\mu_y(s).$$

Observe that $\phi = \phi_1 \otimes \phi_2$ if and only if

$$\begin{aligned}\int_{\mathbb{R}} t \, d\mu_{x\otimes y}(t) &= \left(\int_{\mathbb{R}} \lambda \, d\mu_x(\lambda)\right)\left(\int_{\mathbb{R}} s \, d\mu_y(s)\right) \\ &= \int\int_{\mathbb{R}^2} \lambda s d\mu_x(\lambda) d\mu_y(s).\end{aligned}$$

One may therefore "identify" $d\mu_{x\otimes y}$ with $d\mu_x d\mu_y$ for all self-adjoint elements $x \in A_1$ and self-adjoint elements $y \in A_2$.

Using Corollary 5.5 and the remark above, we are lead to an alternative definition of quantum entanglement.

Definition 5.7 Let $A_1[\tau_1]$ and $A_2[\tau_2]$ be GB*-algebras with jointly continuous multiplication. Assume that $A := A_1 \widehat{\otimes}_\tau A_2$ is a tensor product GB*-algebra with respect to some $*$-admissible topology τ. We say that a state ϕ on A is separable if $\phi = \phi_1 \otimes \phi_2$ for some states ϕ_i on A_i, $i = 1, 2$. An entangled state on A is a state on A which is not separable.

If a state on A (as defined in Definition 5.7) is separable in the sense of Definition 5.7, then, as we have seen above, $A_1 \otimes 1_2$ and $1_1 \otimes A_2$ are independent pairs of subalgebras in $A_1 \otimes A_2$ with respect to the state ϕ. In particular, the obervables $x \otimes 1_2$ (identified with x) and $1_1 \otimes y$ (identified with y) can be thought of as independent observables with respect to ϕ. The alternative Definition 5.7 is therefore also a good definition as it incorporates independence, and therefore independent events in quantum mechanics, which plays a crucial role in the study of quantum entanglement.

Definition 5.7 for *separable* states clearly implies *separable* in the sense of Definition 4.1. Does separable in the sense of Definition 4.1 imply separable in the sense of Definition 5.7?

Remark 5.8 By Theorem 6.4, Corollary 6.8 and Proposition 6.12 below, it follows that if both A_1 and A_2 are non-commutative Fréchet GB*-algebras, where both A_1 and A_2 have sufficiently many irreducible $*$-representations to separate its points, then the GB*-tensor product A, from above, will admit at least one entangled pure state, in the sense of Definition 5.7.

6 Pure States of GB*-Algebras

Motivated by Definition 5.7, we ask the question as to what states on the tensor product GB*-algebra $A_1 \widehat{\otimes}_{\tau} A_2$ are separable, and which ones are not, i.e. entangled. In this section, we answer this question for pure states of $A_1 \widehat{\otimes}_{\tau} A_2$, thereby obtaining further results on tensor product GB*-algebras in addition to results obtained in [24].

A state ϕ on a GB*-algebra $A[\tau]$ is said to be a *pure state* if for any positive linear functional ψ on A with $\psi \leq \phi$, there exists $0 \leq \lambda \leq 1$ such that $\psi = \lambda\phi$.

If ϕ is a state on A, then ϕ is a pure state if and only if ϕ is an extreme point of the state space of A (same proof as in the C*-case, namely [14, Theorem 32.7]).

A positive linear functional ϕ on a GB*-algebra $A[\tau]$ is said to be *representable* [27] if for all $x \in A$, there exists $M_x > 0$ such that $\phi(y^*x^*xy) \leq M_x\phi(y^*y)$ for all $y \in A$. We first show that every pure state of a GB*-algebra is representable. To do this, we first show that every irreducible representation of a GB*-algebra is a bounded $*$-representation.

A $*$-representation is said to be closed if $D(\pi)$ is complete with respect to the graph topology, i.e. the topology defined by the family of seminorms $\xi \in D(\pi) \to \|\pi(x)\xi\|$, $x \in A$.

Let π be a $*$-representation of A on H. Then the commutant of $\pi(A)$, denoted by $\pi(A)'$, is the set of all bounded linear operators z on H such that $\langle z\pi(x)\xi, \eta\rangle = \langle z\xi, \pi(x^*)\eta\rangle$ for all $x \in A$ and $\xi, \eta \in D(\pi)$ [27, Definition 3.4].

A closed $*$-representation π of a GB*-algebra A is said to be irreducible if $\pi(A)' = \mathbb{C}I$ [27, Definition 3.9]. This is equivalent to: If D is a π invariant subspace of $D(\pi)$, then $D = \{0\}$ or $\overline{D} = H$ [27, Proposition 3.10].

If ϕ is a state on A, then its GNS representation π_ϕ is a closed $*$-representation of A (see the proof of [27, Theorem 3.18]). Now ϕ is a pure state of A if and only if π_ϕ is an irreducible $*$-representation of A [27, Theorem 3.18].

By the same proof as that of [8, Theorem 1], one has the following result (our notion of irreducible representation is more general than that in [8], and therefore one must check that the proof of the theorem in [8] holds for our more general notion of irreducibility).

Theorem 6.1 *If $A[\tau]$ is a GB*-algebra and π is an irreducible $*$-representation of A, then π is a bounded $*$-representation of A.*

Proof Let $\rho(x) = \overline{\pi(x)}$ for all $x \in A[B_0]$. Then, for all $x \in A[B_0]$, we get that $\rho(x) \in B(H)$, i.e. $D(\rho(x)) = H$ for all $x \in A[B_0]$, where H is the underlying Hilbert space of π [27, Lemma 3.1]. We show that ρ is an irreducible $*$-representation of $A[B_0]$ into $B(H)$.

Let H_1 be a norm-closed subspace of H such that $\rho(A[B_0])H_1 \subseteq H_1$. Let

$$D_1 = \{\xi \in D(\pi) : \rho(x)\xi \in H_1 \text{ for all } x \in A[B_0]\}.$$

Clearly, $\pi(A[B_0])D_1 \subseteq D_1$. We show that $\pi(A)D_1 \subseteq D_1$.

Let $\xi \in D_1$ and $x \in A$. Then $\pi(x)\xi \in D_1$ if for all $y \in A[B_0]$, we have $\rho(y)\pi(x)\xi = \pi(yx)\xi \in H_1$. We recall that there is a stronger locally convex GB*-topology τ_1 on A with same underlying C*-algebra $A[B_0]$ (see [18, Section 6]). Therefore, by Proposition 2.3, there exists a sequence (x_n) in $A[B_0]$ such that $x_n \to x$ with respect to τ_1. Let U be a τ_1-neighbourhood of $0 \in A$, and let $B = \{(x-x_n)^* : n \in \mathbb{N}\}$. Since $x_n \to x$ with respect to τ_1 and involution is τ_1-continuous, we get that B is τ_1-bounded. Therefore, by [18, Lemmas 6.2 and 6.3], it follows that there is a τ_1-neighbourhood V of $0 \in A$ such that $BV \subseteq U$. Let $y \in A[B_0]$. Since $x_n \to x$ with respect to τ_1, there exists $n \in \mathbb{N}$ such that $y^*y(x - x_n) \in V$ for all $n \geq N$. It follows that, for all $n \geq N$, $(x - x_n)^*y^*y(x - x_n) \in U$. Hence $(x - x_n)^*y^*y(x - x_n) \to 0$ with respect to τ_1 for all $y \in A[B_0]$. By, [8, Theorem B], we get that

$$\|\pi(yx)\xi - \pi(yx_n)\xi\|^2 = \langle \pi\big((x - x_n)^*y^*y(x - x_n)\big)\xi, \xi\rangle \to 0.$$

Here, we bear in mind that $A[\tau_1]$ can be faithfully represented as an EC*-algebra B. Via this faithful $*$- representation, we transfer the topology τ_1 so as to obtain a topological $*$-isomorphism. It is to the EC*-algebra B that we apply [8, Theorem B].

Now $\pi(yx_n)\xi \in H_1$: First, observe that $\pi(x_n)\xi \in H_1$ because $x_n \in A[B_0]$ for all $n \in \mathbb{N}$ and $\xi \in D_1$. But $\rho(y)H_1 \subseteq H_1$, because $y \in A[B_0]$. So $\pi(yx_n)\xi = \rho(y)\pi(x_n)\xi \in H_1$.

Since H_1 is norm-closed in H, we get that $\pi(yx)\xi \in H_1$, and therefore $\pi(A)D_1 \subseteq D_1$. Therefore, by hypothesis, $D_1 = \{0\}$ or $\overline{D_1} = H$.

If $D_1 = \{0\}$, then $H_1 = \{0\}$. If $\overline{D_1} = H$, then $H_1 = H$: If $\xi \in D_1$, then $\rho(x)\xi \in H_1$ for all $x \in A[B_0]$. Therefore $\xi = I\xi = \rho(1)\xi \in H_1$, due to $1 \in A[B_0]$. Thus $D_1 \subseteq H_1$. Therefore $H_1 \subseteq H = \overline{D_1} \subseteq H_1$, and hence $H_1 = H$.

Therefore $H_1 = \{0\}$ or $H_1 = H$, implying that ρ is topologically irreducible. By the Kadison transitivity theorem, it follows that ρ is algebraically irreducible. By definition of the $*$-representation ρ on $A[B_0]$, it follows that $\rho(x)D(\pi) \subseteq D(\pi)$ for all $x \in A[B_0]$, and hence $D(\pi) = H$. Since $I \in \mathcal{L}^\dagger(D(\pi))$ is closed (because $D(\pi) = H$), it follows from [4, Proposition 2.1.13] that π is bounded. In the last statement, we know that $I \in \mathcal{L}^\dagger(D(\pi))$ due to the fact that $\pi(1) = I$. □

Corollary 6.2 *Every pure state of a GB*-algebra is representable.*

Proof If ϕ is a pure state on a GB*-algebra $A[\tau]$, then the GNS representation π_ϕ is an irreducible $*$-representation (see above). By Theorem 6.1, it follows that π_ϕ is a bounded $*$-representation. Therefore, for every $x, y \in A$,

$$\begin{aligned}\phi(y^*x^*xy) &= \langle \pi_\phi(y^*x^*xy)\xi, \xi\rangle \\ &= \langle \pi_\phi(x^*x)\pi_\phi(y)\xi, \pi_\phi(y)\xi\rangle \\ &\leq \|\pi_\phi(x^*x)\| \cdot \|\pi_\phi(y)\xi\|^2 \\ &= \|\pi_\phi(x^*x)\| \cdot \langle \pi_\phi(y^*y)\xi, \xi\rangle \\ &= \|\pi_\phi(x^*x)\| \cdot \phi(y^*y) \\ &= M_x\phi(y^*y),\end{aligned}$$

where $M_x = \|\pi_\phi(x^*x)\|$. □

Theorem 6.3 *Let $A[\tau]$ be a Fréchet GB*-algebra. Then a state ϕ on A is a pure state if and only if $\psi := \phi|_{A[B_0]}$ is a pure state on $A[B_0]$.*

Proof Since $A[\tau]$ is a Fréchet locally convex $*$-algebra, it follows that all bounded $*$-representations of A and all positive linear functionals of A are continuous [17, Theorem 11.1].

Assume that ψ is a pure state on $A[B_0]$, and assume that $\lambda\phi_1 + (1-\lambda)\phi_2 = \phi$, where $\lambda \in (0, 1)$, and where ϕ_1 and ϕ_2 are states on A. Let $\psi_1 = \phi_1|_{A[B_0]}$ and $\psi_2 = \phi_2|_{A[B_0]}$. Therefore $\lambda\psi_1 + (1-\lambda)\psi_2 = \psi$. Since ψ is a pure state of $A[B_0]$, it follows that $\psi_1 = \psi_2 = \psi$. Since ϕ_1, ϕ_2 and ϕ are all τ-continuous, and $A[B_0]$ is dense in A (Proposition 2.3), we get that $\phi_1 = \phi_2 = \phi$. Therefore, ϕ is a pure state of A.

Conversely, let ϕ be a pure state of A. Observe that $\phi(x) = \langle \pi_\phi(x)\xi_\phi, \xi_\phi\rangle$ for all $x \in A$. Letting $\pi_0 = \pi_\phi|_{A[B_0]}$, we get that $\psi(x) = \langle \pi_0(x)\xi_\phi, \xi_\phi\rangle$ for all $x \in A[B_0]$. Since ϕ is a pure state of A, we deduce from Corollary 6.2 that ϕ is representable, and therefore π_ϕ is a bounded $*$-representation on H, say (see the proof of Corollary 6.2). Therefore π_0 is a bounded $*$-representation of $A[B_0]$ on H.

Let $z \in B(H)$ such that $\pi_0(y)z = z\pi_0(y)$ for all $y \in A[B_0]$. We show that $z \in \mathbb{C}I$. Let $x \in A$. Then there is a sequence (x_n) in $A[B_0]$ such that $x_n \to x$ with respect to τ, by Proposition 2.3. Since π_ϕ is $\tau - \|\cdot\|$ continuous, it follows that $\pi_0(x_n) = \pi_\phi(x_n) \to \pi_\phi(x)$ with respect to $\|\cdot\|$. Since all $x_n \in A[B_0]$, it follows that $\pi_0(x_n)z = z\pi_0(x_n)$ for all $n \in \mathbb{N}$. By uniqueness of limits, $\pi_\phi(x)z = z\pi_\phi(x)$, i.e. $x \in \pi(A)'$. Since ϕ is a pure state, we get that π_ϕ is an irreducible $*$-representation of A, and hence $z \in \mathbb{C}I$. Hence π_0 is an irreducible $*$-representation of $A[B_0]$. Therefore ψ is a pure state of $A[B_0]$. □

Theorem 6.4, below, is a partial extension of [42, Theorem IV.4.14] (and this has been extended to the setting of a pro-C*-algebra in [10]).

Theorem 6.4 *Let $A_1[\tau_1]$ and $A_2[\tau_2]$ be Fréchet GB*-algebras, with at least one of A_1 or A_2 commutative. Let $A := A_1 \widehat{\otimes}_\tau A_2$ be a tensor product Fréchet GB*-algebra under some $*$-admissible topology τ such that $A[B_0] = A_1[B_0^1] \widehat{\otimes}_{\|\cdot\|_{min}} A_2[B_0^2]$. If ϕ is a pure state on A, then there exist pure states ϕ_i on A_i, $i = 1, 2$, such that $\phi = \phi_1 \otimes \phi_2$.*

Proof Let ϕ be a pure state on A. Since A is a Fréchet GB*-algebra, it follows from Theorem 6.3 that $\psi := \phi|_{A[B_0]}$ is a pure state on $A[B_0]$. Since at least one of the C*-algebras $A_1[B_0^1]$ and $A_2[B_0^2]$ are commutative (which follows directly from the hypothesis), we get from [42, Theorem IV.4.14] that $\psi = \psi_1 \otimes \psi_2$, where ψ_i are pure states on $A_i[B_0^i]$, $i = 1, 2$. Therefore $\psi(x \otimes 1_2) = \psi_1(x)\psi_2(1) = \psi_1(x)$ for all $x \in A_1[B_0^1]$. Since ψ is continuous (due to A being a Fréchet locally convex $*$-algebra) and the tensor map $\otimes$ is continuous (because the topology τ is $*$-admissible), it follows that ψ_1 is τ_1-continuous on $A_1[B_0^1]$. Similarly, ψ_2 is τ_2-continuous on $A_2[B_0^2]$. Therefore ψ_i extends by continuity to a (τ-continuous) state ϕ_i on A_i, $i = 1, 2$.

It follows that $\phi = \phi_1 \otimes \phi_2$: Let $\sum_{i=1}^m a_i \otimes b_i \in A_1 \otimes A_2$. Since $A_i[B_0^i]$ are sequentially dense in A_i, $i = 1, 2$, it follows that there exist, for each $1 \leq i \leq m$, sequences $(a_{i,n})$ in $A_1[B_0^1]$ such that $a_{i,n} \to a_i$. Similarly, there exist, for each $1 \leq i \leq m$, sequences $(b_{i,n})$ in $A_2[B_0^2]$ such that $b_{i,n} \to b_i$. By continuity of ϕ and the tensor map $\otimes$, and the fact that $A[B_0] = A_1[B_0^1] \widehat{\otimes}_{\|\cdot\|_{\min}} A_2[B_0^2]$, it follows that

$$\begin{aligned}
\phi(\sum_{i=1}^m a_i \otimes b_i) &= \phi(\sum_{i=1}^m \lim_{n\to\infty} a_{i,n} \otimes \lim_{n\to\infty} b_{i,n}) \\
&= \phi(\lim_{n\to\infty} \sum_{i=1}^m a_{i,n} \otimes b_{i,n}) \\
&= \lim_{n\to\infty} \sum_{i=1}^m \phi(a_{i,n} \otimes b_{i,n}) \\
&= \lim_{n\to\infty} \sum_{i=1}^m \psi(a_{i,n} \otimes b_{i,n}) \\
&= \lim_{n\to\infty} \sum_{i=1}^m (\psi_1 \otimes \psi_2)(a_{i,n} \otimes b_{i,n}) \\
&= \lim_{n\to\infty} \sum_{i=1}^m \psi_1(a_{i,n})\psi_2(b_{i,n})
\end{aligned}$$

$$= \lim_{n\to\infty} \sum_{i=1}^{m} \phi_1(a_{i,n})\phi_2(b_{i,n})$$

$$= \sum_{i=1}^{m} \phi_1(a_i)\phi_2(b_i)$$

$$= \sum_{i=1}^{m} (\phi_1 \otimes \phi_2)(a_i \otimes b_i)$$

$$= (\phi_1 \otimes \phi_2)(\sum_{i=1}^{m} a_i \otimes b_i).$$

Since $A_1[\tau_1]$ are $A_2[\tau_2]$ are Fréchet GB*-algebras, it follows from Theorem 6.3 that ϕ_i is a pure state on A_i, $i = 1, 2$. □

In what follows, a converse of Theorem 6.4 will be given. For this, we require the following lemma.

Lemma 6.5 *Let $A_1[\tau_1]$ and $A_2[\tau_2]$ be GB*-algebras. Assume that $A := A_1 \widehat{\otimes}_\tau A_2$ is a tensor product Fréchet GB*-algebra under some $*$-admissible topology τ such that $A[B_0] = A_1[B_0^1] \widehat{\otimes}_{\|\cdot\|_{min}} A_2[B_0^2]$. If π_1 and π_2 are irreducible $*$-representations of A_1 and A_2 respectively, then $\pi := \pi_1 \otimes \pi_2$ is an irreducible $*$-representation of A.*

Proof Let $\pi_0 = \pi|_{A[B_0]}$, $\pi_{1,b} = \pi_1|_{A_1[B_0^1]}$ and $\pi_{2,b} = \pi_2|_{A_2[B_0^2]}$. Then $\pi_0 = \pi_{1,b} \otimes \pi_{2,b}$. Since π_1 and π_2 are irreducible $*$-representations of A_1 and A_2 respectively, it follows from Theorem 6.1 that π_1 and π_2 are bounded $*$-representations of A_1 and A_2 respectively, say $\pi_1 : A_1 \to B(H_1)$ and $\pi_2 : A_2 \to B(H_2)$, where H_1 and H_2 are Hilbert spaces. Therefore π is a bounded $*$-representation of A on $H := H_1 \widehat{\otimes}_{\|\cdot\|} H_2$. By the proof of Theorem 6.1, or [8, Theorem 1], it follows that $\pi_{1,b}$ and $\pi_{2,b}$ are irreducible $*$-representations of $A_1[B_0^1]$ and $A_2[B_0^2]$ respectively. By [42, Theorem IV.4.13], it follows that $\pi_0 := \pi_{1,b} \otimes \pi_{2,b}$ is an irreducible $*$-representation of $A[B_0] = A_1[B_0^1] \widehat{\otimes}_{\|\cdot\|_{\min}} A_2[B_0^2]$.

Let $\psi(x) = \langle \pi_0(x)\xi, \xi \rangle$ for all $x \in A[B_0]$, where $\xi \in H$. Since π_0 is an irreducible $*$-representation of $A[B_0]$, it follows that ψ is a pure state on $A[B_0]$. Let $\phi(x) = \langle \pi(x)\xi, \xi \rangle$ for all $x \in A$. Clearly, $\psi = \phi|_{A[B_0]}$. Since A is a Fréchet GB*-algebra, and ψ is a pure state on $A[B_0]$, we get from Theorem 6.3 that ϕ is a pure state on A. Hence π is an irreducible $*$-representation of A. □

Corollary 6.6 *Let $A_1[\tau_1]$ and $A_2[\tau_2]$ be Fréchet GB*-algebras that each have at least one irreducible $*$-representation. Assume that $A := A_1 \widehat{\otimes}_\tau A_2$ is a tensor product Fréchet GB*-algebra under some $*$-admissible topology τ such that*

$A[B_0] = A_1[B_0^1] \underset{\|\cdot\|_{min}}{\widehat{\otimes}} A_2[B_0^2]$. *Assume that for every pure state* ϕ *on* A, *there exist pure states* ϕ_i *on* A_i $(i = 1, 2)$ *such that* $\phi = \phi_1 \otimes \phi_2$. *Then at least one of* A_1 *and* A_2 *is commutative and has only multiplicative linear functionals as irreducible* $*$*-representations.*

Proof Suppose to the contrary that both A_i, $i = 1, 2$, each have at least one irreducible $*$-representation which are not multiplicative linear functionals, say $\pi_i : A_i \to B(H_i)$. We recall here that every irreducible $*$-representation is bounded (see Theorem 6.1). Then $\dim(H_i) \geq 2$ for $i = 1, 2$. Recall that every vector is a cyclic vector for an irreducible $*$-representation of a GB*-algebra (by [27, Proposition 3.10]). By Lemma 6.5, it now follows from the same proof as (ii) ⇒ (i) in [42, Theorem IV.4.14] that one obtains a contradiction. Hence the result follows. □

Remark 6.7 If $A_1[\tau_1]$ and $A_2[\tau_2]$ are Fréchet GB*-algebras such that $A := A_1 \underset{\tau}{\widehat{\otimes}} A_2$ is a tensor product GB*-algebra, then it is not necessarily true that every pure state ϕ of A is such that $\phi = \phi_1 \otimes \phi_2$, where ϕ_i is a pure state on A_i $(i = 1, 2)$. For example, if $A_1[\tau_1]$ and $A_2[\tau_2]$ are both $L^\omega([0, 1])$ (see Sect. 2), then A_1 and A_2 have no multiplicative linear functionals.

Corollary 6.8 *Let* $A_1[\tau_1]$ *and* $A_2[\tau_2]$ *be Fréchet GB*-algebras that each have at least one irreducible* $*$*-representation. Assume that* $A := A_1 \underset{\tau}{\widehat{\otimes}} A_2$ *is a tensor product Fréchet GB*-algebra under some* $*$*-admissible topology* τ *such that* $A[B_0] = A_1[B_0^1] \underset{\|\cdot\|_{min}}{\widehat{\otimes}} A_2[B_0^2]$. *Assume that for every pure state* ϕ *on* A, *there exist pure states* ϕ_i *on* A_i $(i = 1, 2)$ *such that* $\phi = \phi_1 \otimes \phi_2$. *If, for both* A_1 *and* A_2, *the irreducible* $*$*-representations separate their points, then at least one of* A_1 *and* A_2 *is commutative.*

Proof By Corollary 6.6, A_1, say, only has multiplicative linear functionals as irreducible $*$-representations. Therefore, since the irreducible $*$-representations of A_1 separate its points, it follows that if $\phi(x) = 0$ for all multiplicative linear functionals ϕ on A_1, then $x = 0$. This implies that A_1 is commutative. □

Remark 6.9

(1) In Corollaries 6.6 and 6.8, we consider only *Fréchet* GB*-algebras in order to avoid any hidden subtle pathologies. For instance, by knowing that A_1 and A_2 are Fréchet GB*-algebras, we know that all multiplicative linear functionals are continuous: Since A_1 and A_2 are GB*-algebras, they are symmetric in the sense of Definition 2.1, and are therefore classically symmetric. By a classically symmetric algebra A with identity element 1 and involution $*$, we mean that $(1 + x^*x)^{-1} \in A$ (but not necessarily in A_0) for all $x \in A$. Every multiplicative linear functional of a Fréchet classically symmetric topological algebra, with continuous involution, is continuous ([38, Theorem 3]).

It is, however, still an open problem if every multiplicative linear functional of a Fréchet locally m-convex algebra is continuous.

(2) If the tensor product GB*-algebra A in Theorem 6.4 and Corollary 6.8 has at least one pure state ϕ, and there exist pure states ϕ_i on A_i ($i = 1, 2$) such that $\phi = \phi_1 \otimes \phi_2$, then A_1 and A_2 already have pure states, and hence irreducible $*$-representations.

Every C*-algebra and, more generally, every pro-C*-algebra has a separating family of irreducible $*$-representations. We now give examples of GB*-algebras satisfying the hypothesis of Corollaries 6.6 and 6.8.

Example 6.10

(1) Let $A[\tau]$ be a GB*-algebra having at least one minimal projection (i.e. atomic projection), and such that $A[B_0]$ is a W*-algebra. By Theorem 2.8, $A \cong B$, where B is an EW*-algebra having at least one minimal projection (in the underlying von Neumann algebra of B). By the proof of (b) $\Rightarrow$ (a) of [8, Proposition 3], it follows that B, and therefore A, admits an irreducible $*$-representation in the sense of [8], and therefore in the sense of the definition above (this representation is the GNS-representation with respect to some positive linear functional of A, which is closed by the proof of [27, Theorem 3.18]).
(2) By (1) above, if $A[\tau]$ is a GB*-algebra with $A[B_0]$ a W*-algebra having an atomic projection lattice, then $A[\tau]$ has a separating family of irreducible $*$-representations of A (see [8, p. 107]).

We considered above the question as to the conditions under which a pure state ϕ on a tensor product GB*-algebra $A_1 \widehat{\otimes}_\tau A_2$ is of the form $\phi_1 \otimes \phi_2$, where ϕ_1 and ϕ_2 are states on $A_1[\tau_1]$ and $A_2[\tau_2]$ respectively (i.e ϕ is separable). The conditions entailed having that ϕ_1 and ϕ_2 are pure states of $A_1[\tau_1$ and $A_2[\tau_2]$ respectively. Proposition 6.12, below, shows that this is by no means a severe restriction: It is demonstrated that if ϕ is of the form $\phi_1 \otimes \phi_2$, where it is only assumed that ϕ_1 and ϕ_2 are states on $A_1[\tau_1]$ and $A_2[\tau_2]$ respectively, then ϕ_1 and ϕ_2 must necessarily be pure states on $A_1[\tau_1]$ and $A_2[\tau_2]$ respectively. For this, we require the following lemma.

Lemma 6.11 *Let $A_1[\tau_1]$ and $A_2[\tau_2]$ be GB*-algebras with jointly continuous multiplication, and suppose that $A := A_1 \widehat{\otimes}_\tau A_2$ is a tensor product GB*-algebra under some $*$-admissible topology τ. If $x \in (A_1 \widehat{\otimes}_\tau A_2)^+$, then there exists a net (x_α) in $(A_1 \otimes A_2)^+$ such that $x_\alpha \to a$.*

Proof Since A is a GB*-algebra, there exists $y \in A$ such that $x = y^* y$. There exists a net (y_α) in $A_1 \otimes A_2$ such that $y_\alpha \to y$. Since the involution on A is continuous, and since multiplication on A is jointly continuous, it follows that $y_\alpha^* y_\alpha \to y^* y = x$. Let $x_\alpha = y_\alpha^* y_\alpha$ for all α. Then $x_\alpha \in (A_1 \otimes A_2)^+$ for all α and $x_\alpha \to x$. □

Proposition 6.12 *Let $A_1[\tau_1]$ and $A_2[\tau_2]$ be Fréchet GB*-algebras, and suppose that $A := A_1 \widehat{\otimes}_\tau A_2$ is a tensor product GB*-algebra under some $*$-admissible topology τ. Let ϕ be a pure state of A such that $\phi = \phi_1 \otimes \phi_2$, where ϕ_i are states on A_i, $i = 1, 2$. Then ϕ_1 and ϕ_2 are pure states on A_1 and A_2 respectively.*

Proof Since A_1 and A_2 are Fréchet GB*-algebras, all positive linear functionals of A_1 and A_2 are continuous. Assume that $0 \le \psi_1 \le \phi_1$, where ψ_1 is a positive linear functional on A_1. That is, $\psi_1(x) \le \phi_1(x)$ for all $x \in A_1^+$. Therefore $\psi_1(x)\phi_2(y^*y) \le \phi_1(x)\phi_2(y^*y)$ for all $x \in A_1^+$ and $y \in A_2$. Therefore

$$(\psi_1 \otimes \phi_2)(x \otimes y^*y) \le (\phi_1 \otimes \phi_2)(x \otimes y^*y)$$

for all $x \in A_1^+$ and $y \in A_2$. Since $x = w^*w$ for some $w \in A$ (because $x \in A_1^+$), it follows that

$$(\psi_1 \otimes \phi_2)(z) \le (\phi_1 \otimes \phi_2)(z)$$

for all $z \in (A_1 \otimes A_2)^+$. Since ϕ is continuous and $\otimes$ is continuous, we get from Lemma 6.11 that $(\psi_1 \otimes \phi_2)(a) \le (\phi_1 \otimes \phi_2)(a) = \phi(a)$ for all $a \in A$. Note here that in order to speak about $(\psi_1 \otimes \phi_2)(a)$ for $a \in A$, one requires that both ψ_1 and ϕ_2 are continuous, which follows from the start of this proof. Since ϕ is a pure state on A, it follows that $\psi_1 \otimes \phi_2 = \lambda\phi$ for some scalar $0 \le \lambda \le 1$. Therefore

$$\begin{aligned}
\psi_1(x) &= \psi_1(x)\phi_2(1) \\
&= (\psi_1 \otimes \phi_2)(x \otimes 1) \\
&= \lambda\phi(x \otimes 1) \\
&= \lambda\phi_1(x)\phi_2(1) \\
&= \lambda\phi_1(x)
\end{aligned}$$

for all $x \in A_1$. Therefore ϕ_1 is a pure state of A_1. Similarly, ϕ_2 is a pure state of A_2. □

7 Extremal Decomposition of States of Linearly Nuclear $*$-Algebras

The following result is due to H. J. Borchers and J. Yngvason.

Theorem 7.1 ([11, Theorem 3.10 (ii)]) *Let $A[\tau]$ be a linearly nuclear and separable locally convex $*$-algebra with identity. Suppose that A is either barrelled or has jointly continuous multiplication. Then, for every continuous state ϕ on A, there is a set Z, a weakly measurable map $\zeta \mapsto \phi_\zeta$ to the extremal (i.e. pure) states of A, and a measure μ on Z with $\mu(Z) = 1$ such that $\phi = \int_Z \phi_\zeta d\mu(\zeta)$.*

The proof of this result in [11] is complicated, owing to the fact that the authors could not apply the following theorem of Choquet due to the fact that the positive cone is not generally compact.

Theorem 7.2 ([45, p. 364]) *Let X be a metric space, K a compact subset of X, and E the extremal points in K. Then E is a G_δ-set and, for every $x \in K$, there exists a Baire measure μ_x defined on the Borel sets of X such that $\mu_x(X \setminus E) = 0$, $\mu_x(E) = 1$ and $x = \int_E y d\mu_x(y)$.*

The above theorem of Choquet is similar to the Krein-Milman theorem, which says that every compact convex subset of a real locally convex space is the closed convex hull of its extreme points.

In [26], G. C. Hegerfeldt gave a much simpler proof of the Borchers-Yngavason Theorem 7.1 by overcoming the compactness obstacle above: He considered a $*$-subalgebra of A which is countably generated, thereby making it possible to apply Choquet's theorem above (a proof of the Choquet theorem can be found in [42, p. 230–238]). In this process, he proved the following result, which is a modification of the Bochers-Yngvason result above (with a simpler proof).

Theorem 7.3 ([26, Theorem]) *Let $A[\tau]$ be a linearly nuclear $*$-algebra with identity. Then, for every continuous state ϕ on A with $x \mapsto \phi(x^*x)$ continuous, there is a set Z, a weakly measurable map $\zeta \mapsto \phi_\zeta$ to the extremal (i.e. pure) states of A, and a measure μ on Z with $\mu(Z) = 1$ such that $\phi = \int_Z \phi_\zeta d\mu(\zeta)$.*

Theorem 7.4 ([42, p. 238]) *Let A be a separable C^*-algebra. Then, for every state ϕ on A, we have that*

$$\phi(x) = \int_{P(A)} \omega(x)\, du(\omega)$$

for all $x \in A$, where $P(A)$ denotes the pure state space of A.

Now let $A[\tau]$ be a GB*-algebra. If ϕ is a state on A, then ϕ is a pure state on A if and only if

$$t\phi_1 + (1-t)\phi_2 = \phi \text{ (where } 0 \leq t \leq 1) \Rightarrow \phi = \phi_1 = \phi_2,$$

where ϕ_1, ϕ_2 are states on A. That is, ϕ is pure if and only if it is an extremal point of the state space of A. Therefore, if ϕ is not a pure state of A, then there exist states ϕ_1 and ϕ_2 of A, with $\phi_1 \neq \phi_2$, such that $t\phi_1 + (1-t)\phi_2 = \phi$, where $0 \leq t \leq 1$. This is what one calls a mixed state in the quantum mechanical sense. The interpretation is that the probability is t that one is in state ϕ_1, and the probability is $1-t$ that one is in state ϕ_2. A pure state of A is therefore pure in the quantum mechanical sense.

Now suppose that the GB*-algebra $A[\tau]$ is also linearly nuclear, and is either barrelled or has jointly continuous multiplication. Then, for every continuous state ϕ on A, there is a measure space Z, a weakly measurable map $\zeta \mapsto \phi_\zeta$ to the extremal states of A, and a measure μ on Z with $\mu(Z) = 1$ such that $\phi = \int_Z \phi_\zeta d\mu(\zeta)$ (by

Theorem 7.1). That is, ϕ is a weak $*$-limit of Riemann sums of pure states. There is a certain probability $P(\zeta)$ that one is in the pure state ϕ_ζ and $\int_\zeta P(\zeta)d\mu(\zeta) = 1$. This also demonstrates that linear nuclear GB*-algebras seem to be good for housing the observables (as self-adjoint elements) of a quantum mechanical system.

8 An Extremal Decomposition Theorem for Locally Convex Quasi $*$-Algebras

We already know that we can model the observables of a quantum mechanical system as being self-adjoint elements of a locally convex $*$-algebra. However, the shortcoming of this model is that locally convex $*$-algebras are not closed under taking thermodynamical limits. This is one of the reasons which eventually led to the notion of a quasi $*$-algebra in the early eighties. We shall expand on this, and we refer to [12, Introduction] and [6] (see also [5]). Consider a quantum system having an infinite number of particles (the system has such a large number of particles that one thinks of the system as having infinitely many particles). We break the system up into "local" regions V, where there are finitely (small) number of particles. Every local region V has its own observables as self-adjoint elements in a C*-algebra A_V, and dynamics governed by the Hamiltonian H_V. We assume that for any two local regions V_1 and V_2, there is a local region V_3 such that $V_1 \subset V_2$ and $V_2 \subset V_3$. One also has that $A_{V_1} \subset A_{V_3}$. Then $A_0 := \cup_V A_V$ is also a C*-algebra. In quantum statistical mechanics, one considers equilibrium states $\phi_{V,\alpha}$ of A_V, which are states on A_V. We would like to consider equilibrium states of A_0, and it seems natural to construct these states as $\phi_\lambda(x) = \lim_{V\to\infty} \phi_{V,\lambda}(x)$ for all $x \in A_V$ and for all local regions V. The parameter λ could represent temperature and density, for instance.

One can consider the Gibbs equilibrium states defined as

$$\phi_{V,\beta}(x) = \frac{\mathrm{Tr}(e^{-\beta H_V} x)}{\mathrm{Tr}(e^{-\beta H_V})}$$

for all $x \in A_V$, where β denotes the inverse temperature. Therefore, to construct the state ϕ_β defined above, one has to take the limit $\lim_{V\to\infty} H_V$, i.e. take the limit of the local dynamics of the system. One can check that one has the identity

$$\mathrm{Tr}(e^{-\beta H_V}(e^{itH_V} x e^{-itH_V})y) = \mathrm{Tr}(e^{-\beta H_V} y(e^{i(t+i\beta)H_V} x e^{-i(t+i\beta)H_V})).$$

Therefore, by definition of the Gibbs state $\phi_{V,\beta}$, and identifying $\alpha_t(x)$ as $\lim_{V\to\infty} e^{itH_V} x e^{-itH_V}$ (a group of $*$-automorphisms of A_0), one would expect the thermodynamic limit $\phi_\beta = \lim_{V\to\infty} \phi_{V,\beta}$ to satisfy the KMS condition $\phi_\beta(\alpha_t(x)y) = \phi_\beta(y\alpha_{t+i\beta}(x))$ for all $x, y \in A_0$ and $t \in \mathbb{R}$. Now the (norm) limit $\lim_{V\to\infty} e^{itH_V} x e^{-itH_V}$ does not always exist, so that C*-algebras are inadequate for accommodating thermodynamical limits. We can get this limit to exist in some locally convex topology τ on the C*-algebra A_0 which is weaker than the norm topology on A_0. One then takes the completion $\widetilde{A_0}[\tau]$ of A_0 with respect to the topology τ, which contains this (dynamic) thermodynamical limit. Since the multiplication on A_0 is only separately continuous in general, instead of being jointly continuous, one has that $\widetilde{A_0}[\tau]$ is only a locally convex quasi $*$-algebra. In [6] and [5], the topology τ is considered to be one of the physical topologies, making $\widetilde{A_0}[\tau]$ a locally convex quasi C*-normed algebra. This can be directly applied to the BCS model (as explained in [5] and [6]), which can therefore contain all thermodynamical limits of the system.

For these reasons, and the previous section, one is led to wanting to have an extremal decomposition theorem of positive linear functionals on locally convex quasi $*$-algebras instead of only locally convex $*$-algebras (see also the last two paragraphs of Sect. 7).

There exist notions of positive element and positive linear functional in the locally convex quasi $*$-algebra setting. Namely, an element x in a locally convex quasi $*$-algebra $A[\tau]$ over a $*$-algebra A_0 is said to be positive if it is a τ-limit of elements of the form $\Sigma_{k=1}^{n} x_k^* x_k$, where $n \in \mathbb{N}$ and all $x_k \in A_0$ (see [25, p. 1184]). A linear functional ϕ on a locally convex quasi $*$-algebra over a $*$-algebra A_0 is said to be positive if $\phi(x) \geq 0$ for all positive elements $x \in A$ (see [25, Definition 3.3]).

Theorem 8.1 *Let $A[\tau]$ be a linearly nuclear metrizable locally convex quasi $*$-algebra over a (locally convex with respect to $\tau|_{A_0}$) $*$-algebra A_0. Let T be a continuous positive linear functional on A with $x \mapsto T(x^*x)$ continuous on A_0. Then there exists a measure space Z, a weakly measurable map $\zeta \mapsto \widetilde{T}_\zeta$ of Z to the extremal states of A, and a positive measure ρ on Z with $\rho(Z) = 1$ such that $T(x) = \int_Z \widetilde{T}_\zeta(x)\, d\rho(\zeta)$ for all $x \in A$.*

Proof Since $A[\tau]$ is linearly nuclear, so is $A_0[\tau]$ (since all subspaces of a linearly nuclear space are linearly nuclear, by [39, Proposition 5.1.1]). Therefore, by Theorem 7.3, there exists a measure space Z, a weakly measurable map $\zeta \mapsto T_\zeta$ of Z to the extremal (i.e. pure) states of A_0, and a positive measure ρ on Z with $\rho(Z) = 1$ such that $S(x) := T|_{A_0}(x) = \int_Z T_\zeta(x)\, d\rho(\zeta)$ for all $x \in A_0$.

All T_ζ are τ-continuous (in Hegerfeldt's proof given in [26], all T_ζ are indirectly shown to be τ-continuous). Hence, all T_ζ extend uniquely to τ-continuous linear functionals $\widetilde{T}_\zeta$ on A. Let $f(\zeta) = \widetilde{T}_\zeta$ for all $\zeta \in Z$. Then $f(\zeta)(x) = \widetilde{T}_\zeta(x)$ for all $\zeta \in Z$ and for all $x \in A$. Let $g_x(\zeta) = f(\zeta)(x)$ for all $\zeta \in Z$ and for all $x \in A$ (so $g_x(\zeta) = \widetilde{T}_\zeta(x)$).

For all $y \in A$, we show that g_y is a measurable function on Z, i.e. $\zeta \mapsto \widetilde{T}_\zeta$ is weakly measurable. Let $y \in A$. Then there exists a sequence (y_n) in A_0 such that $y_n \to y$ with respect to the topology τ. Observe that $\zeta \mapsto T_\zeta(y_n)$ is a measurable function for all $n \in \mathbb{N}$. Let $g_n(\zeta) = T_\zeta(y_n)$ for all $\zeta \in Z$ and $n \in \mathbb{N}$. Then g_n is a measurable function for all $n \in \mathbb{N}$. Now

$$g_n(\zeta) = T_\zeta(y_n) = \widetilde{T}_\zeta(y_n) \to \widetilde{T}_\zeta(y) = g_y(\zeta).$$

Therefore $g_n \to g_y$ pointwise, and hence g_y is a measurable function for all $y \in A$, i.e. $\zeta \to \widetilde{T}_\zeta(y)$ is a measurable function for all $y \in A$.

We show that $\widetilde{T}_\zeta$ is a positive linear functional on A for all $\zeta \in Z$. Let $0 \leq x \in A$. By definition (see [25, p. 1184]), there exists a net (a_n) in A_0^+ such that $a_n \to x$ with respect to τ. Hence $0 \leq T_\zeta(a_n) = \widetilde{T}_\zeta(a_n) \to \widetilde{T}_\zeta(x)$ for all $\zeta \in Z$. Therefore $\widetilde{T}_\zeta(x) \geq 0$ for all $\zeta \in Z$.

By the short argument as in the beginning of the proof of Theorem 6.3, one has that all $\widetilde{T}_\zeta$ are extremal (i.e. pure) states of A (we recall that all T_ζ are pure states on A_0).

Now T is the unique continuous extension of $S := T|_{A_0}$. We use this to show that $x \in A \mapsto \int_Z \widetilde{T}_\zeta(x)\, d\rho(\zeta)$ is the unique continuous extension of $x \in A_0 \mapsto \int_Z T_\zeta(x)\, d\rho(\zeta)$. We therefore only have to show that $x \in A \mapsto \int_Z \widetilde{T}_\zeta(x)\, d\rho(\zeta)$ is continuous. Let (x_n) be a sequence in A with $x_n \to x \in A$ with respect to the topology τ.

Observe that

$$\int_Z \widetilde{T}_\zeta(x)\, d\rho(\zeta) = \int_Z f(\zeta)(x)\, d\rho(\zeta) = \int_Z g_x(\zeta)\, d\rho(\zeta).$$

From this, it follows that

$$\int_Z \widetilde{T}_\zeta(x_n)\, d\rho(\zeta) = \int_Z g_{x_n}(\zeta)\, d\rho(\zeta)$$

for all $n \in \mathbb{N}$. Let $f_n = g_{x_n}$ for all $n \in \mathbb{N}$. Then

$$\int_Z \widetilde{T}_\zeta(x_n)\, d\rho(\zeta) = \int_Z f_n(\zeta)\, d\rho(\zeta)$$

for all $n \in \mathbb{N}$. Observe that $|f_n(\zeta)| = |g_{x_n}(\zeta)| = |f(\zeta)(x_n)| = |\widetilde{T}_\zeta(x_n)|$ for all $n \in \mathbb{N}$ and for all $\zeta \in Z$. Since $x_n \to x$ with respect to the topology τ, and $\widetilde{T}_\zeta$ is τ-continuous for every $\zeta \in Z$, it follows that $\widetilde{T}_\zeta(x_n) \to \widetilde{T}_\zeta(x)$ for all $\zeta \in Z$. Hence, for all $\zeta \in Z$, $\widetilde{T}_\zeta(x_n)$ is a bounded sequence of complex numbers, say $|f_n(\zeta)| = |\widetilde{T}_\zeta(x_n)| \leq |h(\zeta)|$ for all $\zeta \in Z$ and for all $n \in \mathbb{N}$. The function h can be chosen to be measurable.

Also, $f_n \to g_x$ pointwise: $f_n(\zeta) = \widetilde{T}_\zeta(x_n) \to \widetilde{T}_\zeta(x) = g_x(\zeta)$ for all $\zeta \in Z$. By the Lebesgue dominated convergence theorem,

$$\int_Z f_n(\zeta)\, d\rho(\zeta) \to \int_Z g_x(\zeta)\, d\rho(\zeta)$$

as $n \to \infty$. Hence

$$\begin{aligned}\int_Z \widetilde{T}_\zeta(x_n)\, d\rho(\zeta) &= \int_Z g_{x_n}(\zeta)\, d\rho(\zeta) \\ &= \int_Z f_n(\zeta)\, d\rho(\zeta) \\ &\to \int_Z g_x(\zeta)\, d\rho(\zeta) \\ &= \int_Z f(\zeta)\,(x) d\rho(\zeta) \\ &= \int_Z \widetilde{T}_\zeta(x)\, d\rho(\zeta).\end{aligned}$$

Therefore $x \in A \mapsto \int_Z \widetilde{T}_\zeta(x)\, d\rho(\zeta)$ is continuous, and is therefore the unique continuous extension of $x \in A_0 \mapsto \int_Z T_\zeta(x)\, d\rho(\zeta)$, which is S. Hence $T(x) = \int_Z \widetilde{T}_\zeta(x)\, d\rho(\zeta)$. □

9 An Example

Consider a quantum system consisting of a single particle with one degree of freedom. Let p and q be the momentum and position operators respectively on the Schwartz space $\mathcal{S}(\mathbb{R})$, contained in $H = L^2(\mathbb{R})$, corresponding to the particle. Let $u_t = e^{itp}$ and $v_s = e^{isq}$ for all $t, s \in \mathbb{R}$. Then $(u_t)_{t\in\mathbb{R}}$ and $(v_s)_{s\in\mathbb{R}}$ are groups of unitary operators on H. Observe that $u_t v_s = e^{-\frac{h}{2\pi}st} v_s u_t$ for all $t, s \in \mathbb{R}$ (the Weyl commutation relation, which is equivalent to the Heisenberg uncertainty principle). Let $\lambda = -\frac{h}{4\pi^2}$, and let $u = u_1 = e^{ip}$ and $v = v_1 = e^{iq}$. Then $uv = e^{2\pi i\lambda} vu$. Let A_λ denote the irrational rotation C^*-algebra, i.e. the C*-algebra generated by u and v. Observe that $u_t = u^t$ and $v_t = v^t$ for all $t, s \in \mathbb{R}$. Therefore, by the functional calculus, $u_s, v_t \in A_\lambda$ for all $t, s \in \mathbb{R}$. See, for example, [40, Introduction] as a reference for the definition and some properties of the irrational rotation C*-algebra.

1. Form the completion $\widetilde{A}_\lambda[\tau]$ of A_λ with respect to a locally convex topology τ weaker than the norm $\|\cdot\|_\lambda$ on A_λ. Then $\widetilde{A}_\lambda[\tau]$ is a locally convex quasi $*$-algebra if the multiplication on A is not jointly continuous with respect to the topology τ [23, Section 4].

Question Are $p, q \in \widetilde{A}_\lambda[\tau]$?

From semigroup theory,

$$p\xi = \lim_{t\to 0} -i\frac{u_t - 1}{t}\xi$$

for all $\xi \in D(p)$. Observe that $q\xi$, for $\xi \in D(q)$, is defined similarly. Let τ_1 be the locally convex topology on A_λ defined by the family of seminorms $p_\xi(x) = \|x\xi\|_\lambda + \|x^*\xi\|_\lambda$ for all $\xi \in D(p) \cap D(q)$. Observe that $D(p) \cap D(q)$ is dense in H (both $D(p)$ and $D(q)$ contain the set of all Hermite polynomials, which is a dense subspace of H). Therefore $A_\lambda[\tau_1]$ is a Hausdorff space. Assume that $\tau = \tau_1$. Observe that $\left(- i\frac{u_t - 1}{t}\right)$ is a τ_1-Cauchy net in A_λ (due, in part, to the fact that the operators in the limit are normal), and is therefore a τ-Cauchy net. Therefore $p \in \widetilde{A}_\lambda[\tau]$. Similarly, $q \in \widetilde{A}_\lambda[\tau]$. We now consider the case the where τ is one of the physical topologies, as described in Section 7 of [6].

Let $\pi_\alpha : A_\lambda \to \mathcal{L}^\dagger(D_\alpha)$ be a separating family of $\tau_1 - \tau_{u,\alpha}$ continuous $*$-representations of A_λ, where $\tau_{u,\alpha}$ denotes the uniform topology on $\mathcal{L}^\dagger(D_\alpha)$, as defined in [36]. The weakest topology τ_{phys} on A_λ making all π_α continuous, where $\mathcal{L}^\dagger(D_\alpha)$ is equipped with the topology $\tau_{u,\alpha}$ for all α, is called the *physical topology* of A_λ. Observe that τ_{phys} is weaker than the topology τ_1 on A, so that $p, q \in \widetilde{A}[\tau_{\text{phys}}]$.

2. Form the completion $\widetilde{A}_\lambda[\tau]$ of A_λ with respect to a Hausdorff locally convex topology τ weaker than the norm $\|\cdot\|_\lambda$ on A_λ, making the multiplication on A_λ jointly continuous and making the involution on A_λ continuous. Then $\widetilde{A}_\lambda[\tau]$ is a GB*-algebra over the τ-closure of the unit ball of A_λ [23, Corollary 3.2]. Assume that τ is weaker than the topology τ_1 on A_λ, defined as in (1) above. By the same reasoning as in (1) above, it follows that $p, q \in \widetilde{A}_\lambda[\tau]$.

   The condition that τ is weaker than τ_1 on A_λ might be a strong requirement. In an attempt to circumvent this difficulty, we note that the Heisenberg uncertainty principle $pq - qp = -i\frac{h}{2\pi}i$ is equivalent to the Weyl form $u_t v_s = e^{-\frac{h}{2\pi}st} v_s u_t$ for all $t, s \in \mathbb{R}$. The Heisenberg uncertainty principle is therefore equivalent to a statement about bounded linear operators inside the C*-algebra A_λ, and this might still be sufficient as far as quantum physics is concerned. The only problem is that p and q need not be in $\widetilde{A}_\lambda[\tau]$. Assume for now that this is not going to be a problem.

   Recall that A_λ contains all u_s, v_t. Then form the completion $\widetilde{A}_\lambda[\tau]$ of A_λ with respect to a locally convex topology τ weaker than the norm $\|\cdot\|_\lambda$ on A_λ. Then it is no longer necessary to instill the requirement that τ is weaker than τ_1 on A_λ.
3. In (1) and (2), one can replace A_λ with the von Neumann algebra generated by A_λ. However, the position and momentum operators, q and p respectively, are not in any GB*-algebra $A[\tau]$ with $A[B_0]$ equal to the von Neumann algebra generated by A_λ: Assume, to the contrary, that there is a GB*-algebra $A[\tau]$ such that $A[B_0]$ is the von Neumann algebra generated by A_λ, and such that p and q are in A. Then A can be identified with an EW*-algebra of operators on a common dense domain D, over a von Neumann algebra $*$-isomorphic to $A[B_0]$ (see Theorem 2.8). Since A is a GB*-algebra, $(1 + p^*p)^{-1}$ and $(1 + q^*q)^{-1}$ are in $A[B_0]$, by Proposition 2.3. Therefore p and q are affiliated with $A[B_0]$ (See [19, Proposition 2.4]). By [31], $A[B_0]$ is a type II_1 factor. By [37, Theorem

4.3.4 or Corollary 4.3.5], p and q are not affiliated with any type II_1 factor. This is a contradiction. The same reasoning applies to a single particle in a quantum system with n degrees of freedom: For each $1 \leq j \leq n$, the particle has corresponding momentum and position operators p_j and q_j on the Schwartz space $\mathcal{L}(\mathbb{R}^n)$, contained in the Hilbert space $L^2(\mathbb{R}^n)$. Now replace A_λ with the C*-algebra, or von Neumann algebra, generated by the corresponding operators $u_{j,t} = e^{itp_j}$ and $v_{j,s} = e^{isq_j}$, for all $1 \leq j \leq n$ and $t, s \in \mathbb{R}$.
4. In (1)–(3) above, the same arguments will work for a finite (small) number of particles, instead only a single particle.
5. For a quantum system with an infinite number of particles, i.e. with a large number of particles so that one thinks of the system as having an infinite number of particles, one requires a type III von Neumann algebra in order to incorporate quantum statistical effects.

Remark 9.1 Let A be the O*-algebra on the Schwartz space $D := \mathcal{L}(\mathbb{R}^n)$ generated by the momentum and position operators p_j and q_j, $1 \leq j \leq n$. Then there exists $c \in A$ of which the inverse is a compact operator, hence completely continuous (see [41, Section 3, p. 120]) By [41, Theorem 2.2 and Lemma 2.3], every strongly positive linear functional ϕ on A is a trace functional, i.e. $\phi(x) = \mathrm{Tr}(tx)$ for all $x \in A$, where t is a positive density operator. By a *strongly positive linear functional* [41, p. 114–115], we mean a linear functional ϕ on A such that $\phi(x) \geq 0$ for all positive operators x in A, i.e. for all $x \in A$ such that $\langle x\xi, \xi\rangle \geq 0$ for all $\xi \in D$.

Now any O*-algebra B on D containing p_j and q_j, $1 \leq j \leq n$, contains A, and therefore $c \in A \subseteq B$. Therefore, by [41, Theorem 2.2 and Lemma 2.3], every strongly positive linear functional on B is a trace functional.

Let $A[\tau]$ be the GB*-algebra as in Example (2) and (3) above. If A is also an O*-algebra on $\mathcal{L}(\mathbb{R}^n)$, then every strongly positive linear functional ϕ on A is a trace functional, i.e. $\phi(x) = \mathrm{Tr}(tx)$ for all $x \in A$, where t is a positive density operator.

Acknowledgments

1. This work is wholly supported by the National Research Foundation of South Africa (NRF).
2. The author expresses his gratitude to Prof Nadia Boudi of the Mohammed V University in Rabat, Morocco, for bringing references [16] and [20] to his attention.
3. The author expresses his sincere gratitude to the referee for a very careful reading of the manuscript and for his/her numerous suggestions which greatly improved the manuscript.

References

1. D. Aerts, I. Daubechies, Physical justification for using the tensor product to describe two quantum systems as one joint system. Helv. Phys. Acta. **51**, 661–675 (1978)
2. G.R. Allan, A spectral theory for locally convex algebras. Proc. Am. Math. Soc. **15**, 399–421 (1965)
3. G.R. Allan, On a class of locally convex algebras. Proc. Lond. Math. Soc. **17**, 91–114 (1967)
4. J.-P Antoine, A. Inoue, C. Trapani, *Partial *-Algebras and Their Operator Realizations* (Kluwer Academic Press, Dordecht, 2003)

5. F. Bagarello, C. Trapani, The Heisenberg dynamcs of spin systems: a quasi $*$-algebras approach. J. Math. Phys. **37**, 4219–4234 (1996)
6. F. Bagarello, M. Fragoulopoulou, A. Inoue, C. Trapani, Locally convex quasi C*-normed algebras. J. Math. Anal. Appl. **366**, 593–606 (2010)
7. S.J. Bhatt, A note on generalized B*-algebras. J. Indian Math. Soc. **43**, 253–257 (1979)
8. S.J. Bhatt, Irreducible representations of a class of unbounded operator algebras. Jpn. J. Math. **11**, 103–108 (1985)
9. S.J. Bhatt, M. Fragoulopoulou, A. Inoue, Spectral well-behaved $*$-representations. Banach Center Publ. **67**, 1–9 (2005)
10. S.J. Bhatt, D.J. Karia, Complete positivity, tensor products and C*-nuclearity for inverse limits of C*-algebras. Proc. Indian. Acad. Sci (Math Sci) **10**, 149–167 (1991)
11. H.J. Borchers, J. Yngvason, On the algebra of field operators. The weak commutant and integral decomposition of states. Commun. Math. Phys. **42**, 231–252 (1975)
12. O. Bratteli, D.W. Robsinson, *Operator Algebras and Quantum Statistical Mechanics 1* (Springer, Berlin, 1979)
13. M.-D. Choi, E.G. Effros, Nuclear C*-algebras and the approximation property. Am. J. Math. **100**, 61–79 (1978)
14. J.B. Conway, *A Course in Operator Theory*. Graduate Studies in Mathematics, vol. 21 (American Mathematical Society, Providence, 2000)
15. M. Dadarlat, S. Eilers, On the classification of nuclear C*-algebras. Proc. Lond. Math. Soc. **85**, 168–210 (2002)
16. R. de la Madrid, The role of Rigged Hilbert space in Quantum mechanics (2005). arXiv:quant-ph/0502053v1
17. H.G. Dales, Automatic continuity: a survey. Bull. Lond. Math. Soc. **10**, 129–183 (1978)
18. P.G. Dixon, Generalized B*-algebras. Proc. Lond. Math. Soc. **21**, 693–715 (1970)
19. P.G. Dixon, Unbounded operator algebras. Proc. Lond. Math. Soc. **23**, 53–69 (1971)
20. Z. Ennadifi, *An Introduction to the Mathematical Formalism of Quantum Mechanics*, Master of Science Thesis, 2018, Mohammed V University, Rabat, Morocco
21. M. Fragoulopoulou, *Topological Algebras with Involution*. (North-Holland, Amsterdam, 2005)
22. M. Fragoulopoulou, A. Inoue, Unbounded $*$-representations of tensor product locally convex $*$-algebras induced by unbounded C*-seminorms. Stud. Math. **183**, 259–270 (2007)
23. M. Fragoulopoulou, A. Inoue, K.-D. Kursten, Old and new results on GB*-algebras. Banach Center Publ. **91**, 169–178 (2010)
24. M. Fragoulopoulou, A. Inoue, M. Weigt, Tensor products of generalized B*-algebras. J. Math. Anal. Appl. **420**, 1787–1802 (2014)
25. M. Fragoulopoulou, C. Trapani, S. Triolo, Locally convex quasi $*$-algebras with sufficiently many $*$-representations. J. Math. Anal. Appl. **338**, 1180–1193 (2012)
26. G.C. Hegerfeldt, Extremal decomposition of Wightman functions and of states on nuclear $*$-algebras by Choquet theory. Commun. Math. Phys. **45**, 133–135 (1975)
27. A. Inoue, Representations of GB*-algebras I. Fukuoka Univ. Sci. Rep. **5**, 63–78 (1975)
28. A. Inoue, Positive linear functionals on topological $*$-algebras. J. Math. Anal. Appl. **68**, 17–24 (1979)
29. A. Inoue, Unbounded Hilbert algebras as locally convex $*$-algebras. Math. Rep. Coll. Gen. Educ. Kyushu Univ. **X**, 113–129 (1976)
30. A. Inoue, *Tomita-Takesaki Theory in Algebras of Unbounded Operators*. Lecture Notes in Math., vol. 1699 (Springer, Berlin, 1998)
31. Y. Kawahigashi, One parameter automorphism groups of the injective II_1 factor arising from the irrational rotation C*-algebra. Am. J. Math. **112**, 499–523 (1990)
32. E. Kirchberg, C*-nuclearity implies CPAP.. Math. Nachr. **76**, 203–212 (1977)
33. E. Kopp, *From Measures to Itô Integrals*. AIMS Library Series (Cambridge University Press, Cambridge, 2011)
34. C. Lance, On nuclear C*-algebras. J. Funct. Anal. **12**, 157–176 (1973)
35. C. Lance, Tensor products and nuclear C*-algebras. Proc. Sympos. Pure Math. **38**, 379–399 (1982)

36. G. Lassner, Topological algebras of operators. Rep. Math. Phys. **3**, 279–293 (1972)
37. Z. Liu, *Von Neumann Algebras, Affilated Operators and Representations of the Heisenberg Relation*, Doctoral dissertation, University of New Hampshire, May 2010
38. S.B. Ng, S. Warner, Continuity of positive and multiplicative functionals. Duke Math. J. **39**, 281–284 (1972)
39. A. Pietsch, *Nuclear Locally Convex Spaces* (Springer, Berlin, 1972)
40. M.A. Rieffel, C^*-algebras associated with irrational rotations. Pac. J. Math. **93**, 415–429 (1981)
41. K. Schmudgen, On trace representation of linear functionals on unbounded operator algebras. Commun. Math. Phys. **63**, 113–130 (1978)
42. M. Takesaki, *Theory of Operator Algebras I* (Springer, Berlin, 1979)
43. D.V. Voiculescu, K.J. Dykema, A. Nica, *Free Random Variables*. CRM Monograph Series (American Mathematical Society, Providence, 1992)
44. M. Weigt, On nuclear generalized B^*-algebras, in *Proceedings of the International Conference on Topological Algebras and Their Applicarions-ICTAA2018*. Math. Stud. (Tartu), vol. 7 (Est Math. Soc, Tartu, 2018), pp. 37–164
45. K. Yosida, *Functional Analysis*, 5th edn. (Springer, Berlin, 1978)

List of Participants

1. **Azouzi, Youssef** (University of Tunis El Manar, Tunisia)
2. **Banasiak, Jacek** (University of Pretoria, South Africa)
3. **Bartosiewicz, Zbigniew** (Bialystok University of Technology, Poland)
4. **Benjamin, Ronalda** (Stellenbosch University, South Africa)
5. **Blanco, Ariel** (Queen's University Belfast, United Kingdom)
6. **Blecher, David** (University of Houston, USA)
7. **Budde, Christian** (Bergische Universität Wuppertal, Germany)
8. **Chitanga, Painos** (University of Pretoria, South Africa)
9. **Claassens, Floris** (University of Kent, United Kingdom)
10. **Conradie, Jurie** (University of Cape Town, South Africa)
11. **Cornet, Bernard** (University of Kansas and University of Paris 1, USA)
12. **de Clercq, Adriaan** (University of Pretoria, South Africa)
13. **de Jager, Pierre** (North West University, South Africa)
14. **de Jeu, Marcel** (Leiden University/University of Pretoria, The Netherlands/South Africa)
15. **Deng, Yang** (Leiden University, The Netherlands)
16. **Drnovsek, Roman** (University of Ljubljana, Slovenia)
17. **Duru, Hülya** (Istanbul University, Turkey)
18. **Fouche, Willem** (University of South Africa, South Africa)
19. **Ger, Roman** (Institute of Mathematics, Silesian University of Katowice, Poland)
20. **Glück, Jochen** (Formerly Ulm University, Germany)
21. **Grobler, Koos** (North-West University and University of Johannesburg, South Africa)
22. **Heymann, Retha** (Stellenbosch University, South Africa)
23. **Jardón Sánchez, Héctor** (University of Oviedo, Spain)
24. **Jiang, Xingni** (Sichuan University, China)
25. **Kalauch, Anke** (TU Dresden, Germany)
26. **Kandić, Marko** (University of Ljubljana, Slovenia)

E. Kikianty et al. (eds.), *Positivity and its Applications*, Trends in Mathematics,
https://doi.org/10.1007/978-3-030-70974-7

27. **Kaplin, Michael** (Institute of Mathematics, Physics and Mechanics, Slovenia)
28. **Karn, Anil Kumar** (National Institute of Science Education and Research, Bhubaneswar, India)
29. **Kassay, Gabor** (Babes-Bolyai University Cluj, Romania)
30. **Kgosana, Tebogo** (University of Pretoria, South Africa)
31. **Khan, M. Ali** (Johns Hopkins University, USA)
32. **Kikianty, Eder** (University of Pretoria, South Africa)
33. **King, Malcolm** (University of Pretoria, South Africa)
34. **Kramar Fijavž, Marjeta** (University of Ljubljana, Slovenia)
35. **Kuo, Wen-Chi** (University of the Witwatersrand, South Africa)
36. **Labuschagne, Louis** (North-West University, South Africa)
37. **Lemmens, Bas** (University of Kent, United Kingdom)
38. **Letsoalo, Sabelo** (University of Pretoria, South Africa)
39. **Leung, Denny** (National University of Singapore, Singapore)
40. **Mabula, Mokhwetha** (University of Pretoria, South Africa)
41. **Maepa, Malesela Salthiel "Charles"** (University of Pretoria, South Africa)
42. **Majozi, Nduduzo** (University of Pretoria, South Africa)
43. **Mbambo, Sello** (University of Pretoria, South Africa)
44. **Messerschmidt, Miek** (University of Pretoria, South Africa)
45. **Mouton, Sonja** (Stellenbosch University, South Africa)
46. **Muzundu, Kelvin** (University of Zambia, Zambia)
47. **Naudé, Alma** (North-West University, South Africa)
48. **Ndumba, Brian** (University of Pretoria, South Africa)
49. **Pinchuck, Andrew** (Rhodes University, South Africa)
50. **Polyrakis, Ioannis** (National Technical University of Athens, Greece)
51. **Ranjbari, Asghar** (University of Tabriz, Iran)
52. **Riihentaus, Juhani** (University of Eastern Finland, Finland)
53. **Roberts, Stephan** (University of Mississippi, USA)
54. **Roelands, Mark** (University of Kent, United Kingdom)
55. **Schouten-Straatman, Mayke** (Leiden University, the Netherlands)
56. **Schwanke, Christopher** (Lyon College/North-West University, USA/South Africa)
57. **Sifi, Mohamed** (Université Tunis El Manar, Tunisia)
58. **Sinnamon, Gord** (University of Western Ontario, Canada)
59. **Snyman, Dillon** (Nelson Mandela University, South Africa)
60. **Snyman, Machiel** (University of Pretoria, South africa)
61. **Steyn, Claud** (North-West university, South Africa)
62. **Tahir, Hassan** (Ocean University of China, China)
63. **Tantrawan, Made** (National University of Singapore, Singapore)
64. **Taylor, Mitchell** (UC Berkeley, USA)
65. **ter Horst, Sanne** (North West University, South Africa)
66. **Troitsky, Vladimir** (University of Alberta, Canada)
67. **van Amstel, Walt** (University of Pretoria, South Africa)
68. **van der Walt, Jan Harm** (University of Pretoria, South Africa)
69. **van Gaans, Onno** (Leiden University, Netherlands)

70. **van Zyl, Gusti** (University of Pretoria, South Africa)
71. **Walsh, Cormac** (Inria and Ecole Poytechnique, France)
72. **Watson, Bruce** (University of the Witwatersrand, South Africa)
73. **Weigt, Martin** (Nelson Mandela University, South Africa)
74. **Wortel, Marten** (University of Pretoria, South Africa)
75. **Xanthos, Foivos** (Ryerson University, Canada)
76. **Zabeti, Omid** (University of Sistan and Baluchestan, Iran)
77. **Zaharopol, Radu** (No Affiliation, USA)